Praise for Louisa Gilder's

The Age of Entanglement

"A delightfully unconventional history.... Especially enjoyable are the portraits of the less famous physicists.... Gilder has done her homework."   —*Nature*

"[Gilder] displays an ability to capture a personality in a few words."
—*The Washington Post*

"A witty, charming, and accurate account of the history of that bugaboo of physics—quantum entanglement.... There are many books out there on the history or foundations of quantum mechanics. Some are more technical, others more historical, but none take the unique approach that Gilder has—to focus on the quantum weirdness of entanglement itself as her book's unifying theme and to present it in an inviting and accessible way.... I was enthralled and found the book delightful."   —Jonathan Dowling, *Science*

"A welcome addition to the genre.... [Gilder's] book really shines.... [She] proves that the neglected last fifty years of quantum mechanics is ... full of brilliant, quirky personalities and mind-bending discoveries.... She is a very compelling writer, and she clearly understands what makes science exciting and science history interesting."
—ScientificBlogging.com

"Astonishing.... The courage and even audacity of a nonscientist to investigate the evolution of ideas about the most esoteric aspects of quantum physics are truly remarkable.... Gilder is a phenomenal writer."   —*The Post & Courier* (Charleston)

"This is not a book about quantum mechanics. This is a book about how science is done. It is the clearest and most intriguing history of the manner in which the scientific method continues to advance knowledge (in spite of—or even because of—the people involved in the investigations). An amazing story."
—*Sacramento News & Review*

Louisa Gilder

The Age of Entanglement

Louisa Gilder was born in Tyringham, Massachusetts, and graduated from Dartmouth College in 2000. This is her first book.

www.ageofentanglement.com

The Age of Entanglement

WHEN QUANTUM PHYSICS WAS REBORN

Louisa Gilder

Vintage Books
A Division of Random House, Inc.
New York

The Library of Congress has cataloged the Knopf edition as follows:
Gilder, Louisa.
The age of entanglement: when quantum physics was reborn / Louisa Gilder.
p. cm.
Includes bibliographical references and index.
1. Quantum theory.   I. Title.
QC174.12.G528 2008
530.12—dc22
2008011796

Vintage ISBN: 978-1-4000-9526-1

Book design by Robert C. Olsson

www.vintagebooks.com

For my father

If people do not believe that mathematics is simple, it is only because they do not realize how complicated life is.

—John von Neumann

CONTENTS

LIST OF ILLUSTRATIONS

A NOTE TO THE READER

Werner Heisenberg, the pioneer who first laid down the laws of the fundamental behavior of matter and light, was an old man when he sat down to write about his life. The book he wrote is not an autobiography of the man but an autobiography of his intellect, entirely a series of reconstructed conversations. His two most famous papers are solo affairs—one introducing quantum mechanics (the laws of the fundamental behavior of matter and light) and the other on the uncertainty principle (which declares that at any given time, the more specific a particle's position, the more vague its speed and direction, and vice versa). But the roots of each solitary paper reach deep into months of heated and careful conversation with most of the great names of quantum physics. "Science rests on experiments," wrote Heisenberg, but "science is rooted in conversations."

Nothing could be further from the impression physics textbooks give to students. There, physics seems to be a perfect sculpture sitting in a vacuum-sealed case, as if brains, only tenuously connected to bodies, had given birth to insights fully formed. These Athena-like theories and Zeus-like theorists seem shiny, glassy, smooth—sometimes, if the light is right, you can see through them into the mysteries and beauties of the physical universe; but there is hardly a trace of humanity, or any sense of questions still to be answered.

Physics, in actuality, is a never-ending search made by human beings. Gods and angels do not come bearing perfectly formed theories to disembodied prophets who instantly write textbooks. The schoolbook simplifications obscure the crooked, strange, and fascinating paths that stretch out from each idea, not only back into the past but also onward into the future. While we aspire to universality and perfection, we are lying if we write as if we have achieved it.

Conversations are essential to science. But the off-the-cuff nature of conversation poses a difficulty. It is rare, even in these digital times, to have a complete transcript of every word spoken between two people on a given day, even if that conversation someday leads to a new understanding of the world. The result is that history books rarely have much of the to-and-fro of human interaction. Heisenberg's statement suggests that something is therefore lost.

When I first started poring through the memoirs and biographies of the quantum physicists of the twentieth century, I felt as if I were watching a movie—the cast of characters was so vivid and the plot twists so unexpected. While the strength of science is its ability to slough off the contingencies of history and reach toward pure knowledge, this knowledge is built, one puzzle piece at a time, by people living their lives in specific times and places with specific passions. Science unfolds in some directions rather than in others because of circumstances. Characters (not disembodied brains) and plot twists (not the relentless forward march of truth) almost guarantee that this is true.

As Tom Wolfe wrote at the beginning of *The Electric Kool-Aid Acid Test:* "I have tried not only to tell what the Pranksters did but to re-create the mental atmosphere or subjective reality of it. I don't think their adventure can be understood without that." Wolfe was recounting a very different kind of mental history, but his point, I find, is even more true about the portentous history of science and intellect that unfolded as the age of entanglement.

This is a book of conversations, a book about how the give-and-take between physicists repeatedly changed the direction in which quantum physics developed, just as conversations, subtly or dramatically, change the world we live in and experience every day. All the conversations in this book occurred in some form, on the date specified in the text, and I have fully documented the substance of every one. (The endnotes detailing the source of each quote speak for themselves.) Most are composed of direct quotes (or close paraphrases) from the trove of letters, papers, and memoirs that these physicists left behind. When occasional connective tissue (e.g., "Nice to see you," or "I agree") was necessary, I tried to keep it both innocuous and also sensitive to the character, beliefs, and history of the people involved. A glance at the notes will separate quote from filler.

Here is a sample from the text, from a conversation that took place in the summer of 1923 on a streetcar in Copenhagen between two of the founders of the quantum theory, Albert Einstein and Niels Bohr, and its first great teacher, Arnold Sommerfeld.

"It's good to see you doing so well," says Einstein.

Bohr shakes his head, smiling: "My life from the scientific point of view passes off in periods of over-happiness and despair . . . as I know that both of you understand . . . of feeling vigorous and over-worked, of starting papers and not getting them published"—his face is earnest—"because all the time I am gradually changing my views about this terrible riddle which the quantum theory is."

"I know," says Sommerfeld, "I know."

Einstein's eyes almost close; he is nodding. "That is a wall before which I am stopped. The difficulties *are* terrible." His eyes open. "The theory of relativity was only a sort of respite which I gave myself during my struggles with the quanta."

We know that the conversation (of which this interchange represents a tiny piece) happened, because Bohr mentioned it in an interview late in his life with his son and one of his closest colleagues. The content of the conversation is easy to gather from a look at what the three men were working on and writing friends about around the same time. Here Bohr, in the interview, describes that day in 1923:

> Sommerfeld was not impractical, not quite impractical; but Einstein was not more practical than I and, when he came to Copenhagen, I naturally fetched him from the railway station. . . .
>
> We took the streetcar from the station and talked so animatedly about things that we went much too far past our destination. So we got off and went back. Thereafter we again went too far, I can't remember how many stops, but we rode back and forth in the streetcar because Einstein was really interested at that time; we don't know whether his interest was more or less skeptical—but in any case we went back and forth many times in the streetcar and what people thought of us, that is something else.

Here is the first quote on which this particular short section of the conversation is based. It comes from a letter Bohr wrote to a British colleague in August of 1918:

> I know that you understand . . . how my life from the scientific point of view passes off in periods of over-happiness and despair, of feeling vigorous and overworked, of starting papers and not getting them published, because all the time I am gradually

changing my views about this terrible riddle which the quantum theory is.

How can a passage written five years earlier be relevant? Some things had changed for Bohr in the intervening years, but what he touches on in the letter had remained the same—the excitement, dejection, and overwork (during this whole period he was building his institute of physics in Copenhagen); the long, arduous papers only partially published; and, most of all, his struggle to understand the quantum theory, which until Heisenberg's breakthrough in 1925 stood on shifting sand.

Here is the second quote, from a journey by train taken a year before. The astronomer of the Paris Observatory rode with Einstein from Belgium to Paris and asked him about the quantum problem. "That is a wall before which one is stopped," Einstein replied. "The difficulties are terrible; for me, the theory of relativity was only a sort of respite which I gave myself during their examination." His opinions on the subject were the same at the time of our scene in the summer of 1923; by the following summer, an unexpected letter from India would help him chip a crack in this quantum wall.

As for the filler: Bohr was the kind of person whose happiness was infectious—he would indeed have been looking well when he picked up Einstein to show him his newly completed institute, no matter how overworked and secretly despairing he might actually be. And Sommerfeld, always intellectually engaged with Bohr during those early years of the quantum theory, would have known intimately what Bohr meant by "this terrible riddle."

I believe the risks of telling the story in this way are outweighed by the reward: a sense of how, through minds meeting minds, the quantum theory unfolded. Please check the notes (found on page 351) if ever it seems that someone "couldn't have said that!," and the glossary on page 337 for any unfamiliar physical terms. I am hopeful of earning your trust, and of honoring Heisenberg's sense of how science is really done.

—L. G., October 2007

POSTSCRIPT

In the Note to the Reader above, I had picked an example that would bring to the fore the speculative, collage-like nature of the conversations in the book. So it is not such a surprise that this example I used was the one that proved to conflict partially with further research.

Sommerfeld, it turns out, was not in the streetcar with Bohr that day in 1923, and in fact, neither was Einstein.

Bohr told the story of his streetcar ride with Einstein in the course of a rambling conversation during a long Scandinavian summer day at his Zealand beach house on July 12, 1961. Bohr was almost seventy-six (he would die the next year), and his son Aage and his most devoted assistant, Leon Rosenfeld, were interviewing him. Bohr guessed that Einstein had come after giving a speech in the Swedish town of Göteburg (in 1923, when the town celebrated its three hundredth anniversary).

But the Bohr archive, comparing notes with the Einstein archive, now believes that Einstein visited Copenhagen only once—in 1920, after a speaking trip to Kristiania (as Oslo was called until 1924, in honor of a long-ago Norwegian and Danish king). Immediately afterward Einstein wrote to his father figure, the great Dutch physicist Hendrik Lorentz:

> The trip to Kristiania was really beautiful; the most beautiful were the hours I spent with Bohr in Copenhagen. He is a highly gifted and excellent man.
> It is a good omen for physics that prominent physicists are mostly also splendid people.

Einstein's and Bohr's biographer, Abraham Pais, wrote in *Niels Bohr's Times*, "I have not found any comment by Bohr on this visit." It is rather nice to realize now that Bohr's streetcar story is in fact describing those "beautiful" hours.

Sommerfeld's name came into Bohr's telling of this tale not because he was there (as I had believed), but because talking of those events brought him to Bohr's mind. Bohr, Sommerfeld, and Einstein in those days were in close communication—Bohr and Sommerfeld as they elaborated their atomic model, Einstein and Sommerfeld as they discussed general relativity during its genesis. The twenty years between 1908 and 1928 were Sommerfeld's heyday as a mathematical physicist (during which he supervised the Ph.D. theses of an incredible total of four future Nobel laureates).

What, then, in 1920, would have engrossed Einstein and Bohr so much that they repeatedly missed their stop? Well, the focus of their discussions and disagreements in 1920 was the same as in 1923: the light-quantum. Again, let me emphasize that looking at the notes will show exactly when (and to whom) each man made the statement that I have woven into the conversation.

All of which is to say that the mission of this book is to complement the existing histories and textbooks with a reminder of the wonderful unsettledness of conversation, interchange, and experiment, and the moments in which these occasionally flash into inspired clarity. This book is a celebration of the glorious, human mess that incubates discovery, as well as of its flawed and heroic searchers.

—L. G., March 25, 2009

The Age of Entanglement

John Stewart Bell

Entanglement

ANY TIME TWO ENTITIES INTERACT, they entangle. It doesn't matter if they are photons (bits of light), atoms (bits of matter), or bigger things made of atoms like dust motes, microscopes, cats, or people. The entanglement persists no matter how far these entities separate, as long as they don't subsequently interact with anything else—an almost impossibly tall order for a cat or a person, which is why we don't notice the effect.

But the motions of subatomic particles are dominated by entanglement. It starts when they interact; in doing so, they lose their separate existence. No matter how far they move apart, if one is tweaked, measured, observed, the other seems to instantly respond, even if the whole world now lies between them. And no one knows how.

Strange as it seems, this kind of correlation is happening all the time—and we know it happens because of the work of John Bell. Raised in the chaos of Ireland during the Second World War, he spent his working years in peaceful Switzerland and died just after his sixty-second birthday, the year (unbeknownst to him) that he was nominated for the Nobel Prize. He called the work for which he is now most famous his hobby: probing into the logical foundations of quantum mechanics. His second paper on this subject, in 1964, briefly and beautifully demonstrates the utter lack of any commonsense mechanism to correlate these entangled particles. Bell had extended and deepened a hitherto sneered-at paper of Einstein's on the subject, written in 1935 with two little-known colleagues (Boris Podolsky and Nathan Rosen). Forty years after its rehabilitation by John Bell, the paper is, by a massive margin, the most cited of *all* Einstein's roster of glittering, earthshaking work,* and the

---

*A paper is considered famous if it has been cited in more than a hundred subsequent papers; Einstein's monumental papers on special relativity (1905) and the quantum theory (1917) have each been cited more than seven hundred times; his 1905 Ph.D. dissertation on the size of an atom, more than fifteen hundred times. By contrast, twenty-five hundred

most cited paper of the dominant physics journal of the second half of the twentieth century, *Physical Review.*

Entanglement's presence (especially over tiny distances, for example within a hydrogen molecule) began to be manifest in the early twentieth century, in the springtime of the quantum theory. But it was Bell, with his simple algebra and deep thinking, who laid open the central paradox.

The mysteries embedded in quantum mechanics provoked four major reactions from its founders: orthodoxy, heresy, agnosticism, and simple misunderstanding. Three of the theory's founders (Bohr, Heisenberg, and Wolfgang Pauli) gave it its orthodox exegesis, which came to be known as the Copenhagen interpretation. Three more founders (including Einstein) were heretics, believing that something was rotten in the quantum theory they had played such a role in developing. Finally, pragmatic people said, The time is not yet ripe for understanding these things, and confused people dismissed the mysteries with simplistic explanations.

This riot of different reactions had a huge impact on the future of quantum mechanics, because the theory needed interpretation the way a fish needs water. This fact alone was a drastic break with the past history of science. A classical (i.e., pre-quantum) equation, after its terms were defined, essentially explained itself. With the quantum revolution, the equations fell silent. Only an interpretation allowed them to speak about the natural world.

Take this analogy. A Bhutanese artist, flown to the Metropolitan Museum of Art and introduced to Western painting for the first time, would have no problem understanding the essentials of the gory story represented by any of the several paintings of Judith, sword in one elegant hand, the head of Holofernes swinging from the other. Before 1900, a painting could be relied on to speak about what the painter intended. Standing in the Guggenheim before a series of razor-edged swaths of browns that give an impression of motion, however, our Bhutanese artist will be pardoned for glancing quickly over to the little title card (in a now universal art-gallery ritual) to find out that this is actually a "Sad Young Man on a Train."

More scandalous than any Jewish maiden carrying a severed head, the companion painting to the sad young man—Marcel Duchamp's famous "Nude Descending a Staircase," which rocked the New York art world in

---

papers in physics journals have cited Bell's 1964 paper on entanglement; the same as for the 1935 Einstein-Podolsky-Rosen paper that inspired him.

1913—graces the cover of one of Heisenberg's books; quantum mechanics represented a perfectly contemporaneous and analogous break with the past. Just as much as the paintings of Duchamp and his successors, quantum mechanics needed that little title card to connect with a reality outside its beautiful mathematics, and in the 1920s and '30s physicists argued over who would get to script it.

Here are the protagonists.

1. THE COPENHAGEN INTERPRETATION

Niels Bohr, a lifelong friend and intellectual adversary of Einstein's, who founded the Institute for Theoretical Physics in Copenhagen, tried to make sense of the mysteries with a concept he called complementarity. For Bohr, complementarity was an almost religious belief that the paradoxes of the quantum world must be accepted as fundamental, not to be "solved" or trivialized by attempts to find out "what's really going on down there." Bohr used the word in an unusual way: the "complementarity" of waves and particles, for example (or of position and momentum), meant that when one existed fully, its complement did not exist at all.

In order to take this view, Bohr emphasized, there has to be a big "classical" world, devoid of complementarity—the world of circling planets and falling apples that Isaac Newton had explained so well—which serves as a platform from which to stare into the quantum abyss. In fact, instead of thinking of classical-sized things, like apples and cats, as being made of quantum things, like atoms, Bohr put the dependence the other way. In his famous Como lecture of 1927, he emphasized that waves and particles are "abstractions, their properties being definable and observable only through their interaction with other systems"—and these other systems must be "classical," like a measuring apparatus.

Rather than urging physicists to find a way to move beyond such "abstractions" to a more accurate description, Bohr further insisted that "these abstractions are indispensable to describe experience in connection with our ordinary space-time view." That is, quantum things must be talked about in a classical language ill-suited to describe them, and the existence of any property we can recognize in a quantum object must always depend upon finding another system that will interact with it in a "classical" way. Classical systems are paradoxically necessary to describe the quantum systems of which they are made.

His enthusiastic supporter *Werner Heisenberg* and best critic, *Wolfgang Pauli,* would go so far as to say that the quantum world is in a certain way created or transformed by our observation of it, since the atom seems to have no properties before measurement.

"Those who are not shocked when they first come across quantum theory," said Bohr in conversation with Heisenberg and Pauli, "cannot possibly have understood it."

2. "SOMETHING IS ROTTEN"*

Starting in 1909, only nine years after quantum theory's tentative debut, *Albert Einstein* began to worry that it implied a world composed of non-separable pieces that were "not . . . mutually independent." When he tried to treat the individual particles as individuals, they seemed to exert "a mutual influence . . . of a quite mysterious nature" on each other, or even seemed to affect each other in what he ridiculed as "spooky action-at-a-distance," or "a sort of telepathic coupling." To him it was clear that this meant a fatal flaw in the theory.

Erwin Schrödinger showed that, on its face, the quantum theory (and in particular its foundational equation that bears his name) leads to a bizarre paradox. If we do not firmly declare with Bohr that something big, like a cat, does *not* follow the laws of quantum mechanics (though it is indubitably constructed of particles that do), we can prove the cat to be alive and dead simultaneously. Schrödinger yearned to reject the Copenhagen dualism and believe in a single world described by his equation, but could never find a way to do it.

Louis de Broglie, a young Frenchman, came up with a version of the quantum theory in which the Schrödinger equation describes a long-range force that moves faster than the speed of light, spookily guiding the particles that make up our world.

This kind of interpretation is an example of what is commonly called "hidden variables." The concept to remember and link with this opaque designation—in the case of de Broglie's version of it—is "a quantum theory without observers": a quantum theory in which the reality of particles does not depend on whether they are observed.

3. THE TIME IS NOT RIPE

Paul Dirac (always known in public life by his first initials, P.A.M.), whose equation describing electrons[†] was one of the most astonishingly powerful results of the quantum theory, felt that it was too soon to be wasting time worrying about entanglement. It would make sense someday.

4. DISMISSIVE INCOMPREHENSION

Max Born, like Bohr a lifelong friend of Einstein's and contributor to the Copenhagen interpretation, could never understand why the others

---

*The words of both Hamlet and John Bell.
[†]Electricity-carrying subatomic particles that are a crucial component of all matter.

thought the meaning of the theory was such an important and difficult issue.

After the 1930s it seemed clear that the analyses of Einstein, Schrödinger, and de Broglie were dead ends, and, in fact, most of the great and lasting triumphs of the quantum theory did come from one of the other schools of thought.

. But no one following Bohr, Heisenberg, Pauli, Dirac, or Born dared grasp, measure, or even name the deepest of all the puzzles, entanglement. Then along came John Bell. An admirer of Einstein, Schrödinger, and de Broglie, he followed their minority views to their natural conclusions and brought unexpected clarity where before there had been fog. And what the fog had been hiding was vividly and wonderfully strange.

Bohr used to say, "Truth and clarity are complementary"—meaning that the more truthful you try to be, the more unclear will be your statements, and vice versa. This was certainly true of Bohr himself. But Bell wasn't buying it. As he once told one of Bohr's most famous postwar disciples, John Wheeler, "I'd rather be clear and wrong than foggy and right."

Bohr's books and papers—full of careful prohibitions about what cannot be contemplated and obscure statements about "complementarity," "indivisibility," and "irrationality"—have become holy writ to be interpreted and reinterpreted by each new generation of physicists. From the point of view of the history of entanglement, they are not worth one clear sentence from Einstein, Schrödinger, de Broglie, or John Bell, who each said, in a way that opened up a new world: "Hey, look at this."

1

The Socks

1978 and 1981

Reinhold Bertlmann

IN 1978, when John Bell first met Reinhold Bertlmann, at the weekly tea party at the Organisation Européenne pour la Recherche Nucléaire, near Geneva, he could not know that the thin young Austrian, smiling at him through a short black beard, was wearing mismatched socks. And Bertlmann did not notice the characteristically logical extension of Bell's vegetarianism—plastic shoes.

Deep under the ground beneath these two pairs of maverick feet, ever-increasing magnetic fields were accelerating protons (pieces of the tiny center of the atom) around and around a doughnut-shaped track a quarter of a kilometer in diameter. Studying these particles was part of the daily work of CERN, as the organization was called (a tangled history left the acronym no longer correlated with the name). In the early 1950s, at the age of twenty-five, Bell had acted as consultant to the team that designed this subterranean accelerator, christened in scientific pseudo-Greek "the Proton Synchrotron." In 1960, the Irish physicist returned to Switzerland to live, with his Scottish wife, Mary, also a physicist and a designer of accelerators. CERN's charmless, colorless campus of box-shaped buildings with protons flying through their foundations became

Bell's intellectual home for the rest of his life, in the green pastureland between Geneva and the mountains.

At such a huge and impersonal place, Bell believed, newcomers should be welcomed. He had never seen Bertlmann before, and so he walked up to him and said, his brogue still clear despite almost two decades in Geneva: "I'm John Bell."

This was a familiar name to Bertlmann—familiar, in fact, to almost anyone who studied the high-speed crashes and collisions taking place under Bell's and Bertlmann's feet (in other words, the disciplines known as particle physics and quantum field theory). Bell had spent the last quarter of a century conducting piercing investigations into these flying, decaying, and shattering particles. Like Sherlock Holmes, he focused on details others ignored and was wont to make startlingly clear and unexpected assessments. "He did not like to take commonly held views for granted but tended to ask, 'How do you know?,' " said his professor Sir Rudolf Peierls, a great physicist of the previous generation. "John always stood out through his ability to penetrate to the bottom of any argument," an early co-worker remembered, "and to find the flaws in it by very simple reasoning." His papers—numbering over one hundred by 1978—were an inventory of such questions answered, and flaws or treasures discovered as a result.

Bertlmann already knew this, and that Bell was a theorist with an almost quaint sense of responsibility who shied away from grand speculations and rooted himself in what was directly related to experiments at CERN. Yet it was this same responsibility that would not let him ignore what he called a "rottenness" or a "dirtiness" in the foundations of quantum mechanics, the theory with which they all worked. Probing the weak points of these foundations—the places in the plumbing where the theory was, as he put it, "unprofessional"—occupied Bell's free time. Had those in the lab known of this hobby, almost none of them would have approved. But on a sabbatical in California in 1964, six thousand miles from his responsibilities at CERN, Bell had made a fascinating discovery down there in the plumbing of the theory.

Revealed in that extraordinary paper of 1964, Bell's theorem showed that the world of quantum mechanics—the base upon which the world we see is built—is composed of entities that are either, in the jargon of physics, not *locally causal*, not *fully separable*, or even not *real unless observed*.

If the entities of the quantum world are not locally causal, then an action like measuring a particle can have instantaneous "spooky" effects across the universe. As for separability: "Without such an assumption of the mutually independent existence (the 'being-thus') of spatially distant things . . . ," Einstein insisted, "physical thought in the sense familiar to us would not be possible. Nor does one see how physical laws could be formulated and tested without such a clean separation." The most extreme version of nonseparability is the idea that the quantum entities do not become solid until they are observed, like the proverbial tree that makes no sound when it falls unless a listener is around. Einstein found the implications ludicrous: "Do you really believe the moon is not there if nobody looks?"

Up to that point, the idea of science rested on separability, as Einstein had said. It could be summarized as humankind's long intellectual journey away from magic (not locally causal) and from anthropocentrism (not real unless observed). Perversely, and to the consternation of Bell himself, his theorem brought physics to the point where it seemingly had to choose between these absurdities.

Whatever the ramifications, it would become obvious by the beginning of this century that Bell's paper had caused a sea change in physics. But in 1978 the paper, published fourteen years before in an obscure journal, was still mostly unknown.

Bertlmann looked with interest at his new acquaintance, who was smiling affably with eyes almost shut behind big metal-rimmed glasses. Bell had red hair that came down over his ears—not flaming red, but what was known in his native country as "ginger"—and a short beard. His shirt was brighter than his hair, and he wore no tie.

In his painstaking Viennese-inflected English, Bertlmann introduced himself: "I'm Reinhold Bertlmann, a new fellow from Austria."

Bell's smile broadened. "Oh? And what are you working on?"

It turned out that they were both engaged with the same calculations dealing with quarks, the tiniest bits of matter. They found they had come up with the same results, Bell by one method on his desktop calculator, Bertlmann by the computer program he had written.

So began a happy and fruitful collaboration. And one day, Bell happened to notice Bertlmann's socks.

Three years later, in an austere room high up in one of the majestic stone buildings of the University of Vienna, Bertlmann was curled over the screen of one of the physics department's computers, deep in the world of quarks, thinking not in words but in equations. His computer—at fifteen feet by six feet by six feet one of the department's smaller ones—almost filled the room. Despite the early spring chill, the air-conditioning ran, fighting the heat produced by the sweatings and whirrings of the behemoth. Occasionally Bertlmann fed it a new punch card perforated with a line of code. He had been at his work for hours as the sunlight moved silently around the room.

He didn't look up at the sound of someone's practiced fingers poking the buttons that unlocked the door, nor when it swung open. Gerhard Ecker, from across the hall, was coming straight at him, a sheaf of papers in hand. He was the university's man in charge of receiving preprints—papers that have yet to be published, which authors send to scientists whose work is related to their own.

Ecker was laughing. "Bertlmann!" he shouted, even though he was not four feet away.

Bertlmann looked up, bemused, as Ecker thrust a preprint into his hands: "You're famous now!"

The title, as Bertlmann surveyed it, read:

Bertlmann's Socks and the Nature of Reality
J. S. Bell
CERN, Geneve, Suisse

The article was slated for publication in a French physics periodical, *Journal de Physique*, later in 1981. Its title was almost as incomprehensible to Bertlmann as it would be for a casual reader.

"But what's this about? What possibly—"

Ecker said, "Read it, read it."

He read.

The philosopher in the street, who has not suffered a course in quantum mechanics, is quite unimpressed by Einstein-Podolsky-Rosen correlations. He can point to many examples of similar correlations in everyday life. The case of Bertlmann's socks is often cited.

My socks? What is he talking about? And EPR correlations? It's a big joke, John Bell is playing a big published joke on me.

"EPR"—short for the paper's authors, Albert Einstein, Boris Podolsky, and Nathan Rosen—was, like Bell's 1964 theorem, which it inspired thirty years later, something of an embarrassment for physics. To the question posed by their title, "Can Quantum-Mechanical Description of

Physical Reality Be Considered Complete?," Einstein and his lesser-known cohorts answered no. They brought to the attention of physicists the existence of a mystery in the quantum theory. Two particles that had once interacted could, no matter how far apart, remain "entangled"—the word Schrödinger coined in that same year—1935—to describe this mystery. A rigorous application of the laws of quantum mechanics seemed to force the conclusion that measuring one particle affected the state of the second one: acting on it at a great distance by those "spooky" means. Einstein, Podolsky, and Rosen therefore felt that quantum mechanics would be superseded by some future theory that would make sense of the case of the correlated particles.

Physicists around the world had barely looked up from their calculations. Years went by, and it became more and more obvious that despite some odd details, ignored like the eccentricities of a general who is winning a war, quantum mechanics was the most accurate theory in the history of science. But John Bell was a man who noticed details, and he noticed that the EPR paper had not been satisfactorily dealt with.

Bertlmann felt like laughing in confusion. He looked at Ecker, who was grinning: "Read on, read on."

Dr. Bertlmann likes to wear two socks of different colors. Which color he will have on a given foot on a given day is quite unpredictable. But when you see (Fig. 1) that the first sock is pink . . .

Les chaussettes
de M. Bertlmann
et la nature
de la réalité

Fondation Hugot
juin 17 1980

pink →

not
pink →

Fig. 1.

What is Fig. 1? My socks? Bertlmann ruffled through the pages and found, appended at the end, a little line sketch of the kind John Bell was fond of doing. He read on:

> But when you see that the first sock is pink you can be already sure that the second sock will not be pink. Observation of the first, and experience of Bertlmann, give immediate information about the second. There is no accounting for tastes, but apart from that there is no mystery here. And is not the EPR business just the same?

Bertlmann imagined John's voice saying this, conjured up his amused face. *For three years we worked together every day and he never said a thing.*

Ecker was laughing. "What do you think?"

Bertlmann had already dashed past him, out the door, down the hall to the phone, and with trembling fingers was calling CERN.

Bell was in his office when the phone rang, and Bertlmann came on the line, completely incoherent. "What have you done? *What have you done?*"

Bell's clear laugh alone, so familiar and matter-of-fact, was enough to bring the world into focus again. Then Bell said, enjoying the whole thing: "Now you are famous, Reinhold."

"But what is this paper about? Is this a big joke?"

"Read the paper, Reinhold, and tell me what you think."

A tigress paces before a mirror. Her image, down to the last stripe, mimics her every motion, every sliding muscle, the smallest twitch of her tail. How are she and her reflection correlated? The light shining down on her narrow slinky shoulders bounces off them in all directions. Some of this light ends up in the eye of the beholder: either straight from her fur, or by a longer route, from tiger to mirror to eye. The beholder sees two tigers moving in perfectly opposite synchrony.

Look closer. Look past the smoothness of that coat to see its hairs; past its hairs to see the elaborate architectural arrangements of molecules that compose them, and then the atoms of which the molecules are made. Roughly a billionth of a meter wide, each atom is (to speak very loosely) its own solar system, with a dense center circled by distant electrons. At these levels—molecular, atomic, electronic—we are in the native land of quantum mechanics.

The tigress, though large and vividly colored, must be near the mirror for a watcher to see two correlated cats. If she is in the jungle, a few yards' separation would leave the mirror showing only undergrowth and swinging vines. Even out in the open, at a certain distance the curvature of the earth would rise up to obscure mirror from tigress and decouple their synchrony. But the entangled particles Bell was talking about in his paper can act in unison with the whole universe in between.

Quantum entanglement, as Bell would go on to explain in his paper, is not really like Bertlmann's socks. No one puzzles over how he always manages to pick different-colored socks, or how he pulls the socks onto his feet. But in quantum mechanics there is no idiosyncratic brain "choosing" to coordinate distant particles, and it is hard not to compare how they do it to magic.

In the "real world," correlations are the result of local influences, unbroken chains of contact. One sheep butts another—there's a local influence. A lamb comes running to his mother's bleat after waves of air molecules hit each other in an entirely local domino effect, starting from her vocal cords and ending when they beat the tiny drum in the baby's ear in a pattern his brain recognizes as *Mom.* Sheep scatter at the arrival of a coyote: the moving air has carried bits of coyote musk and dandruff into their nostrils, or the electromagnetic waves of light from the moon have bounced off the coyote's pelt and into the retinas of their eyes. Either way, it's all local, including the nerves firing in each sheep's brain to say *danger,* and carrying the message to her muscles.

Grown up, sold, and separated on different farms, twin lambs both still chew their cud after eating, and produce lambs that look eerily similar. These correlations are still local. No matter how far the lambs ultimately separate, their genetic material was laid down when they were a single egg inside their mother's womb.

Bell liked to talk about twins. He would show a photograph of the pair of Ohio identical twins (both named "Jim") separated at birth and then reunited at age forty, just as Bell was writing "Bertlmann's Socks." Their similarities were so striking that an institute for the study of twins was founded, appropriately enough at the University of Minnesota in the Twin Cities. Both Jims were nail-biters who smoked the same brand of cigarettes and drove the same model and color of car. Their dogs were named "Toy," their ex-wives "Linda," and current wives "Betty." They were married on the same day. One Jim named his son James Alan; his twin named his son James Allen. They both liked carpentry—one made miniature picnic tables and the other miniature rocking chairs.

The correlations that Bell's theorem discusses are so obviously twin-like, so blatantly correlated, that the natural thing would be to imagine that they, like these lambs and Jims, have something approximating DNA. And that is where their mystery lies: for what the theorem shows is just how strange and nonlocal—"spooky"—those "genes" would have to be.

The person who most clearly presented the intellectual puzzle of Bell's theorem to nonphysicists was a solid state physicist at Cornell named David Mermin, and he first became aware of it in 1979, from a *Scientific American* article by Bell's friend Bernard d'Espagnat. Mermin hailed from an opposite corner of the physics world from Bell, studying slow atoms chilled to a few degrees above absolute zero. But soon Bell's hobby became his hobby, too. He boiled Bell's theorem down "to something so simple that I could convey the argument using no mathematics beyond simple arithmetic and no quantum mechanics at all."

From these musings arose "something between a parable and a lecture demonstration," centering around a cartoon version of a three-part machine like the one that Bell described in "Bertlmann's Socks." This machine can be viewed in two ways. It is a reified, more visual way to talk about the equations of quantum mechanics and their predictions and results. It is also an abstraction of an apparatus that, these days, may be found in any quantum optics lab. In the center is a box that, at the push of a button, emits a pair of particles, sending them in opposite directions. On either side of the box, and far from it, sit two detectors. Each has a lever or crank on one side that allows a person to realign its internal apparatus so that it measures the particle along a different axis. We can turn the crank from the "normal setting" (which measures the particle head-on) to the "vertical setting" to the "horizontal setting." Each detector also has a light on top that, upon receipt of a particle, flashes either red or green.

Mermin invites us to imagine that we have just come upon this machine with no further information. Tinkering, we press START, and shortly thereafter each detector flashes red or green. Garnering as much information as possible, we crank the detectors between the three settings, all the while pressing the button and noting which lights come on.

Over several hours, we accumulate thousands and thousands of apparently random results. But the results are not random. They are precisely what quantum mechanics predicts for certain two-particle situations.

This is what a sample of our results would look like (where H is horizontal, V is vertical, and * is normal):

LEFT DETECTOR SETTING	RIGHT DETECTOR SETTING	LEFT DETECTOR RESULT	RIGHT DETECTOR RESULT
H	*	**GREEN**	**RED**
H	**V**	**GREEN**	**RED**
*	**V**	**RED**	**GREEN**
*	*	RED	RED
*	**V**	**GREEN**	**RED**
H	H	GREEN	GREEN
V	V	RED	RED
*	*	RED	RED
V	V	RED	RED
V	*	**GREEN**	**RED**
*	**H**	**RED**	**RED**
*	**H**	**RED**	**GREEN**

Looking the results over, we can divide the data into two cases:

Case (1): When both detectors are on the same setting, they always flash the same color.

Case (2) (in bold type): When the detectors are on different settings, they flash the same color not more than 25 percent of the time.

"These statistics," Mermin remarks, "may seem harmless enough, but some scrutiny reveals them to be as surprising as anything seen in a magic show, and leads to similar suspicions of hidden wires, mirrors, or confederates under the floor."

Consider the case when the detectors are on the same setting. The same lights always flash. "Given the unconnectedness of the detectors, there is one (and, I would think, only one) extremely simple way to explain" this behavior, Mermin writes. "We need only suppose that some property of each particle (such as its speed, size, or shape) determines the color its detector will flash for each of the three switch positions." They share some kind of gene. Twin particles make twin lights flash.

This is such a reasonable explanation that it is disheartening to realize that the very same data prove it to be dead wrong.

If the hypothesis of genes is true then we can write out a prediction for further results. Here is an example of such a prediction, showing all the possible permutations for a series of pairs of particles all bearing "flash red for normal, flash green for horizontal or vertical" genes:

LEFT DETECTOR SETTING	RIGHT DETECTOR SETTING	LEFT DETECTOR RESULT	RIGHT DETECTOR RESULT
*	*	RED	RED
*	**H**	**RED**	**GREEN**
*	**V**	**RED**	**GREEN**
H	*	**GREEN**	**RED**
H	H	GREEN	GREEN
H	**V**	**GREEN**	**GREEN**
V	*	**GREEN**	**RED**
V	**H**	**GREEN**	**GREEN**
V	V	GREEN	GREEN

But particles such as these could never produce the results we actually got. Notice the cases where the settings are different (in bold type). The same lights flashed twice out of these six times: 33.3 percent of the time, not 25 percent.

This is the kind of result known as "Bell's inequality." It lay hidden for so long partly because no one, until Bell, thought to solve the equations of quantum mechanics for the situations in which the detectors were not aligned, and compare these with the predictions for particles with predetermined attributes. More than forty years after Bell's discovery, the completely mystifying unanswered question remains: if there are no connections between the detectors, and if what is arriving at them is not a pair of particles bearing identical "genes," what in the world causes identical lights to flash when the detectors are on identical settings?

In a sense Bell's argument in his theorem is really simple—to Bell it was, certainly—but there's something about it, as he said, that nobody follows originally. Because of this, Bell himself restated it in many ways, from his original five-line mathematical proof in 1964 to several formulations that rely on analogy—more than one of which are contained in "Bertlmann's Socks."

His friend Bernard d'Espagnat, for example, humorously gave this analogy to Bell's inequality:

The number of young nonsmokers
plus the number of women smokers of all ages
is greater than or equal to
the total number of young women, smokers and otherwise.

It is a logical statement this trivial—a tautology!—that quantum theory flouts. The problem is not the logic, but the premises: that it makes sense simultaneously to assign gender, age, and tobacco use to a person. Instead, in quantum theory, the whole seems to be greater than the sum of its parts.

The recoil of one ram from the head-butt of another; the lamb trotting up to the call of his mother; the arrival of the coyote and departure of the sheep: for all these correlations there is a cause and an effect; they all take place in time.

The ram's hard head moves as fast as the ram's muscles and hooves can carry him, covering perhaps ten meters in a second (roughly twenty miles per hour). The call of the mother sheep travels both faster and farther: about a kilometer in three seconds (nearly seven hundred fifty miles per hour) on a chilly spring day. The speed of the diffusion of the telltale smell of the coyote is slower and more arbitrary. And, even more than the ewe's call, the diffusion of this musk is at the mercy of the air: local contingencies of temperature, pressure, and any changes of these that form a wind all speed or slow its arrival in the quivering nostrils of a sheep.

The signal to the eye is the fastest one, since it occurs near the speed of light, almost three hundred thousand kilometers per second (compare this to the speed of sound, one-third of a kilometer per second). This speed is almost inconceivably fast—a ray of light could circle the earth seven times in a second—but it is, like the other local influences, a speed. It is not instantaneous.

With the possibility ruled out that Bell's strangely connected particles start out complete with their synchronized instructions, we might wonder if instead there is some kind of signaling going on. Upon reaching the detector with the green and red lights, one particle might somehow communicate with the other so they can coordinate the results. On a visit to Paris in 1979, John Bell explained the problem with this idea by telling a whimsical story about French TV.

"It has been feared that television is responsible for the disturbing decline of birth rate in France." He was speaking to a group of physicists at the University of Paris-South in Orsay, who must have wondered what this had to do with quantum physics. "It is quite unclear which of the two main programs"—Bell used the word in its old-fashioned sense as "channel"—"(France 1 and 2, both originating in Paris) is more to blame. It has been advocated that deliberate experiments be done, say in Lille and Lyon, to investigate the matter. The local mayors might decide, by tossing coins each morning, which one of the two programs would be locally relayed during the day." Bell noted that after enough time, one could get a pretty good sense for the number of conceptions in Lille and Lyon, respectively, after exposure to one or the other channels (a joint distribution of probability, involving both towns).

"You might at first think it pointless to consider such a *joint* distribution, expecting it to separate trivially into independent factors," Bell continued. "But a moment of reflection will convince that this will not be so. For example, the weather in the two towns is correlated, although imperfectly. On fine evenings, people do not watch television. They walk in the parks, and are moved by the beauty of the trees, the monuments, and of one another. This is especially so on Sundays." The investigators must recognize such extraneous factors affecting both towns, and remove their effects from the analysis.

It would be remarkable if the towns were still correlated after the investigators had dealt with these extraneous causal factors. It would be even more remarkable "if the choice of program in Lille proved to be a causal factor in Lyon, or if the choice of program in Lyon proved to be a causal factor in Lille"—if the result of showing France 1 in Lyon produced a spike in pregnancies in Lille. "But according to quantum mechanics, situations presenting just such a dilemma can be contrived. Moreover"—he came to the center of the problem—"the peculiar long-range influence in question seems to go faster than light."

And that is impossible, as the theory of relativity shows. Space and time are not constant realities unaffected by anything. Space, Einstein said, is merely what we measure with a ruler; time is what we measure with a clock. And as it turns out, the faster an object goes, the more it compresses and the slower ticks any clock it might carry (for example, its heartbeat). In fact, it compresses just enough, and its clock slows down just enough, that it can never reach the speed of light. Every day, accelerators like the ones that John and Mary Bell helped design—full of particles moving close to the speed of light—bear out in precise detail Einstein's amazing predictions: 299,800 kilometers per second is the universe's absolute speed limit.

Two years after the Orsay talk, as Bell was publishing "Bertlmann's Socks," a young experimental physicist there named Alain Aspect (with an auspiciously Hercule Poirotesque mustache) was about to test if such a long-range influence really must be faster than the speed of light. Rigging up something very much like the Bell-Mermin machine, he found that the mysterious correlations remained, unfazed no matter how fast he switched the settings on the detectors. Physical signals traveling at mere light speed could not explain the results.

So, there are no genes and there is no signaling. The world is entangled in a beautiful and mysterious way. At the beginning of the twenty-first century, after three-quarters of a century of coexistence with the idea, there is still no clear explanation of its magic. But a change is in the air.

One of the early people to think of trying to build something out of entanglement was Richard Feynman, in 1981 the greatest and most famous physicist alive. He had actually been thinking about Bell's theorem along the same lines as Mermin had. ("One of the most beautiful papers in physics that I know of is yours in the *American Journal of Physics* . . . ," Feynman wrote to Mermin in 1984 when he had read it. "All my mature life I have been trying to distill the strangeness of quantum mechanics into simpler and simpler circumstances," he explained, which had led him to craft a similar but twice as complicated gedanken demonstration "when your ideally pristine presentation appeared.")

Feynman had computers on his mind, and he had instantly seen that Bell's theorem prohibited them from simulating nature on the quantum level. Characteristically he viewed this as an opportunity rather than a problem. At the Massachusetts Institute of Technology that same year, as Aspect tweaked his machinery, Feynman brought Bell's inequality before a meeting of the best computer scientists in the world. He challenged them to produce a new kind of computer. This might hardly be recognizable to us as a computer (in fact, the first one made, at the end of the twentieth century, was a liquid of specially engineered molecules in a tiny vial), but in whatever form it takes, the person using it to compute will do so by manipulating the states of quantum particles. Most important to Feynman, a quantum computer would use the magic of entanglement, and bring us to an understanding of it in the process.

Not long after Feynman gave this speech, a few brilliant minds proved some of the things such a computer might be able to do. Most significant to nonphysicists, a quantum computer could crack all the codes that the security of our banks, government, and the Internet is based on.

As experimental quantum physics groups, in physics departments all over the world, turn their attention to the building of a quantum computer, entanglement remains a mystery. But it is a mystery that gets better known all the time. Physicists are beginning to look at the magical correlations as something as fundamental—and worth fathoming—as energy or information. Famously, it was through machine-building that scientists began to understand both these fundamental ideas. In the nineteenth century, advances in understanding energy were inseparable from the building and running of steam engines; in the twentieth century, the rise of computers was inextricably intertwined with the rise of information theory; in the twenty-first century, hands-on work on the quantum computer and quantum cryptography—another entanglement-based miracle—will make us both more comfortable and more awed at entanglement.

But it was the twentieth century that first encountered entanglement, and to tell the story of entanglement is to tell much of the story of quantum physics itself. The story of entanglement begins near the start of the century with a suspicion of the spookiness of the quantum theory. For centuries, physics had seemed to be on a relentless march to total understanding of the world. The dawn of the twentieth century brought the news that the deeper we delved into both matter and light, the more mysteries we would find.

The Arguments

1909–1935

Albert Einstein and Paul Ehrenfest, ca. 1920

2

Quantized Light

September 1909–June 1913

During the autumn in Salzburg, a hot, dry wind called the *Föhn* sweeps down the slopes of the Alps and through the chilly air of the city. It evaporates all haze and fog so that faraway things appear suddenly clear. But the atmosphere is heavy and unpleasantly unseasonal; people blame the *Föhn* for their headaches and irritability.

Albert Einstein, a rumpled and large-eyed thirty-year-old in a straw hat, about to change his official career from patent clerk to professor, came to Salzburg in late September of 1909 for his first physics conference. With its pale stucco facades and copper-roofed towers, the city is known for its particularly beautiful light, and light was what was on Einstein's mind.

Rooster tail feathers and hummingbirds; the pearly inner chambers of shells and the wing cases of beetles; soap bubbles and oil slicks; thick glass and dappled sunlight through gaps between leaves—all these, examined closely, show light as a wave. It does not rain down on us like the *Schnürlregen*, or "straight-string-rain" downpours of Salzburg. It ripples and interferes.

While the actual interference is invisible, its calling card is striking: precise bands of darkness in places that a "string-rain" of light would illuminate, and precise bands of light in places that the imagined downpour of light drops would leave in shadow. Color, moreover, is purely a wave phenomenon—it is the number of times a light wave rises and falls per second—and each color bends differently when it strikes liquid or striated surfaces, producing the iridescence of the soap bubble or beetle. The understanding of all these phenomena was the climax of centuries of study of electromagnetic radiation (light is merely the visible stretch of the electromagnetic spectrum—ranging from radio waves bigger than a house to X-rays and gamma rays smaller than an atom).

Then Einstein, in 1905, came across the beginning of a great mystery: sometimes light, so clearly a wave, seemed to behave like drops or particles. This mystery was as strange as the discovery of unicorn horns on Arctic beaches. Such a discovery encourages some to dismiss it entirely while others proclaim the existence of magic horned horses; only a few will search for the narwhal, the Arctic whale to whom the horn belongs. Einstein was going to Salzburg to tell his fellow physicists, whom he had never met, that light was made of neither waves nor particles, but some currently incomprehensible fusion of the two.

But the real narwhal is a strange beast—stranger than the mythical unicorn. Einstein was embarking on a quest to understand each particle of light (and matter) as possessing its own independence, disentangled from all the others; its own local reality; its own unique, separable state. This fifty-year search for separability led him again and again to its nemesis, entanglement. And the clarity with which he reported his unwanted results made this unsuccessful quest one of the most fruitful investigations of the twentieth century.

The year before the conference, Einstein's focus on light had reached an intense pitch. "I am incessantly busy with the question of the constitution of radiation," he wrote in 1908, engaging in arguments by letter with H. A. Lorentz of Leiden, Holland, and Max Planck of Berlin—the two preeminent theoretical physicists in the world, both more than two decades older than Einstein.

He found Planck "an utterly honest man who thinks of others rather than himself" (and because of this, he would show Planck more steadfast loyalty than he gave either of his own wives). "He has, however," Einstein decided in 1908, "one fault: he is clumsy in finding his way about foreign trains of thought." Lorentz, on the other hand, seemed to Einstein in 1908 "astonishingly profound," and by 1909 he was writing, "I admire this man as no other, I would say I love him."

Einstein and Lorentz believed that Planck, in the last month of the nineteenth century, had turned physics on its head. It had all started with light in a box with reflective inner walls (an image to which Einstein would return repeatedly). In 1900, after years of work, Planck had come up with a formula that predicted, for any given temperature, the energy carried by each color of light inside a box. (The "box" could be any shape, but it had to be black, absorbing all light that fell on it.)

In order to get the correct formula, Planck had to count energy in "quanta" of a certain size, symbolized by $h\nu$. *Quantum* was nothing mysterious yet—in German, it just means "amount"—while ν stood for the color (i.e., frequency) of the light, and h was a new, tiny, constant number.

Then came Einstein's discovery in 1905 that the data for bluer, high-frequency light showed that Planck's discovery was not just a counting device. Ultraviolet light, X-rays, and gamma rays behaved, in or out of the box, as if actually built from "mutually independent energy quanta"— atoms of light.

The man whose mathematical analysis they were all using to attack this problem was Ludwig Boltzmann of Vienna, who had done more than anyone to demonstrate logically that *matter* was built of atoms. But he lost himself in his ever-present depression and committed suicide in 1906 at the age of sixty-two. As that tragedy unfolded, his student, a twenty-five-year-old named Paul Ehrenfest, showed that Planck's new formula had introduced something completely new that could not be derived from, or even harmonized with, any physics that had gone before. A pure wave theory led to wildly wrong predictions for high-energy, high-frequency light. Ehrenfest dubbed the situation "the ultraviolet catastrophe."

"This quantum question is so uncommonly important and difficult," Einstein wrote in 1908, "that it should concern everyone." At the time, it hardly concerned anyone aside from himself, Planck, Lorentz, and Ehrenfest.

Early in May of 1909, Lorentz wrote Einstein a long and thoughtful letter criticizing Einstein's particles of light. "I must have expressed myself unclearly in regard to the light quanta," Einstein replied late that month. "That is to say, I am not at all of the opinion that one should think of light as being composed of mutually independent quanta localized in relatively small spaces." What strange things would mutually dependent or nonlocal quanta be? Einstein very much hoped that the truth, with further digging, would set him free of such chimeras.

On the way to Salzburg from his home in Switzerland, he had stopped in Munich to see the only schoolteacher who had really inspired him, a Dr. Ruess, who taught fourth- and fifth-grade languages and history. The perpetually amused, stubborn-as-a-mule patent officer standing before his old teacher had already accomplished much of what would later make him the most famous man in the world. He had already shown that though motion is relative to a reference point, the speed of light and the laws of physics are not. He had already shown that energy and matter were transmutable ($E = mc^2$). But Dr. Ruess saw only Einstein's frayed clothes, thought he must be there to ask for money, and sent him on his way, an inauspicious prelude to his first conference.

"It is my opinion," Einstein explained, looking out at the assembled physicists in Salzburg, "that the next phase of the development of theoretical physics will bring us a theory of light that can be interpreted as a kind of fusion of the wave and particle theories." The purpose of his speech, he explained, would be "to justify this opinion and to show that a deep change in our conception about the nature and the constitution of light is *indispensable.*" He showed an increasingly skeptical and bewildered audience that Planck's formula actually required light to be a particle as well as a wave.

Planck immediately stood up after the mystified clapping died down. He had come from a family of theologians on both sides, and he himself, with his serious eyes, dignified mustache, and thin frame, was in many ways the pastor of German physicists. "I will restrict myself naturally to the points where I have a different opinion from that of the speaker," he said. He felt that Einstein had made "a step that is in my opinion not yet necessary. . . . In any case, I think that we should first try to transfer the problem of quantum theory into the area of *interaction* between matter and radiating energy."

The belligerent Johannes Stark stood up, in pince-nez and a sweeping mustache that dwarfed his handsome face. He was a thirty-five-year-old Bavarian experimentalist, and, along with Planck, he had merited one of the few footnotes in Einstein's light-quantum paper of 1905. Einstein, who was not sensitive enough to be bothered by difficult people, seemed to have made a friend of Stark (who usually made only enemies). For a decade he would be Einstein's only light-quantum follower. But as Einstein's fame grew, Stark's jealousy became a mania. When Hitler rose to power, Stark would lead the cry for Einstein's blood, and for the removal of "Jewish physics" from Germany.

In 1909, Stark recognized Einstein's valid point. "Originally, I too was of the same opinion," he said to Planck. "However, there is one phenomenon which forces us to think of electromagnetic radiation as *separated* from matter and concentrated in space." Röntgen rays, as X-rays were then called, "can still act *concentratedly* [with their full force] on a single atom, even at up to ten meters": the opposite of a wave diffusing out into the world in ever-widening circles.

"There is something unique to Röntgen rays," Planck agreed. "Stark has mentioned something in favor of the quantum theory; I wish to add a remark against it," he continued. Most light behavior—most strikingly, interference—can hardly be explained by particles. "If a quantum interferes with itself," Planck said, "it must have a spatial extension of hundreds of thousands of wavelengths." How could a string-rain of particles produce those neat bands of light and shadow? Interference demands a wave explanation.

Stark responded confidently: "The interference phenomena would probably be different with very low radiation density." It would be three-quarters of a century before this reasonable statement could be shown to be false, when experiments would demonstrate that a single light-quantum could indeed interfere with itself.

Einstein stepped in, describing what he would later call his *Gespensterfelder*—ghost waves. An electron, he noted, is surrounded by an electric field, and each quantum might similarly emanate a field through which these ghost waves would ripple. On this hopeful note the meeting ended.

In the ensuing years Einstein tried to make this separable, one-wave-per-particle description work. The reason for the failure of this idea would slowly emerge over the decades with ever-greater clarity: entanglement, where two particles are not really separable because a single wave describes them.

"I have not yet found a solution of the light-quantum question," Einstein wrote that New Year's Eve. "All the same I will try to see if I cannot work out this favorite problem of mine." By December 1910, things were no better: "The enigma of radiation will not yield." In the spring of 1911, he wrote to his best friend, the engineer Michele Angelo Besso, "I do not ask anymore whether these quanta really exist. Nor do I attempt any longer to construct them, since I now know my brain is incapable of advancing in that direction."

He had spent three and a half years exclusively on light quanta and felt that he had gotten nowhere. In June 1911, he turned to a new problem, which would bring him his greatest success. "At present," he wrote in 1912, "I occupy myself exclusively with the problem of gravitation."

But the nonseparable (and hence uncountable) quanta remained in the back of his mind. Even as he repeated that "one cannot believe in the existence of countable quanta, since the interference properties of light . . . are not compatible with it," he yearned for the straightforwardness of separability: "I still prefer the 'honest' theory of quanta to the . . . compromises found so far to replace it."

It was a warm June evening in Zurich in 1913. Einstein was sitting in the garden of a coffee shop, an empty mug before him on the table. With him were two of his lifelong friends, Ehrenfest and Max von Laue.

They had just come from a colloquium on von Laue's spectacular discovery in Munich the year before. A crystal of blue copper sulfate with its regular rows of atoms, von Laue realized, could diffract X-rays: that is, the gaps between the atoms would cause the X-rays to fan out in concentric ripples on the other side, as waves do. These ripples, emerging from between each row of atoms in the crystal, then interfered with each other, wave peak joining with wave peak to make a wave bigger than the sum of their heights, or trough canceling out peak, producing nothing where two waves were before. Von Laue had immediately sent the photographs to Einstein, who wrote back congratulating him "on your wonderful success. Your experiment is one of the finest things to have happened in physics." How Einstein's quanta fit into this picture was still a mystery.

Handsome, thoughtful, and upright—even, later, in the midst of the maelstrom of the Third Reich—von Laue was the son of a peripatetic Prussian officer whose career had sent Max from town to town, school to school, throughout his childhood. In 1903, Laue had received his Ph.D. under Planck, whose first act was to introduce Laue to relativity. The enthralled student then spent his summer vacation hiking through the Swiss mountains to meet its author at the patent office in Bern. After he got over his shock that Einstein was a twenty-four-year-old like himself, the two talked for hours over cigars so awful that Laue let his "accidentally" fall in the river. Already famous in Berlin for the high speeds at which he flew down the city streets on his motorbike, in 1911 Laue published a book on relativity that Einstein regarded as "a little masterpiece."

Ehrenfest, on the other hand, had faced the quantum head-on ever since hearing a 1903 speech of Lorentz's. Although often suffering from depression, Ehrenfest usually hid it with his exuberance, and his brilliant criticism would make him "the conscience of physics." He was an unconventional Viennese who had settled in St. Petersburg with his wife, Tatiana, a Russian physicist. But in chaotic prerevolutionary Russia, much as he loved it, there were no university positions open to him as a foreigner, a Jew, and an atheist. In 1912, just as he had decided to move to Zurich to be near Einstein (which is what Laue had done the same year), Lorentz called him to be his successor at Leiden University, a place he would make supremely homelike to hundreds of visiting physicists while never feeling at home himself. Short and broad-shouldered in his Russian high-necked belted tunic, his deep-set dark eyes sparkling under his bristling black hair, he would emphasize the crux of complicated scientific concepts with words like: "*That's* where the frog jumps into the water!" or (in homage to Einstein), "There's the patent claim!"

Laue had warned Ehrenfest the first time he met Einstein, "You should be careful that Einstein doesn't talk you to death. He loves to do that, you know." But Ehrenfest could hold his own. On this, his second visit, Ehrenfest and Einstein had been talking for five days straight, walking alone in the hazy hot hills. A few days before, an eager crowd of physicists had joined them on a local mountain while Einstein explained his work, Ehrenfest asking ever sharper questions until it made sense. His jubilant shout, "I have understood it!" still rang in von Laue's ears.

Ehrenfest was now describing his struggles with the quantum theory to Einstein, who nodded sympathetically: "The more success the quantum theory has, the sillier it looks."

Ehrenfest turned to von Laue: "Did he tell you about the park in Prague?"

Von Laue shook his head.

"Well, when I came to visit Einstein in Prague a year ago—Einstein, you tell it."

"My office had a really quite nice view of a park with trees and gardens," said Einstein. "And people would walk in it, some deep in thought, others gesticulating intensely at each other in groups." Einstein grinned. "The strange thing was, it was only women in the morning and only men in the evening. And when I asked what was this place, they told me, 'It's the Bohemian Insane Asylum!' "

Ehrenfest turned to von Laue: "And he said, 'Those are the madmen who do not occupy themselves with the quantum theory.' "

3

The Quantized Atom

November 1913

Niels Bohr

THE LATE NOVEMBER FOG accompanied Max von Laue and Otto Stern most of the way up the Uetliberg. It was five months after Ehrenfest's visit. They were hiking the little mountain on this gloomy day because they knew that its peak was above the clouds, in the sun. That morning in the fog-bound streets, Stern and von Laue had seen yellow placards reading "UETLIBERG HELL"—Mt. Uetli light.

Laue and Stern had only recently become locals, Laue a professor at the University of Zurich and Stern a lecturer at the E.T.H. (the letters stand for Eidgenössische Technische Hochschule—the Swiss Federal Institute of Technology—but it is always known by its initials, the "*A. tay hah*"). Like the Alsatian Laue, Stern had grown up on the much disputed fringes of Bismarck's booming German empire, in Silesia (the as-yet unknown and ordinary town of Auschwitz sat on its border).

Stern was a decade younger than von Laue—chunky, long-chinned, radiating good humor. He would brilliantly occupy the border between theory and experiment, but he was so clumsy that he would do anything to avoid handling the more breakable equipment once he became a professor with assistants to do it for him. Even if it was toppling over, "you do less damage if you let the thing fall than if you try to catch it," he would explain, gesturing with his cigar. Spending his own money to join Einstein in Prague in 1911, he had been Einstein's sole confidant there; the memories of those "beautiful days" would bring tears to his eyes when Stern was an old man exiled by the Nazis on the other side of the world.

On the top of the Uetliberg, taking deep breaths of the cold air while sweat chilled between their shoulder blades, they looked out across the white sea of fog. The city behind them had vanished. Instead they saw the thunderous Alps: the snowy trinity of the Eiger, the Monk, and the Virgin, and the precipitous dark steeple of the Finsteraarhorn. The two men made a funny silhouette against this view: the short and semispherical Stern beside the tall and angular von Laue.

They were talking about the atom, as everyone they knew seemed to be. Ever since it became clear in 1911 that the atom was something like a tiny solar system (its sun the positively charged nucleus, exerting a constant electrical pull on its planets, the negatively charged electrons), there were problems. A charged object like an electron is inseparable from the electric field it emanates; if it moves, it generates a magnetic field, too; and if it changes its speed (speeding up, slowing down, or turning), that change makes a wave in the electric and magnetic fields surrounding it. This electromagnetic wave is what we call light. The electrons, charged and orbiting the heart of the atom, should be making constant light-wave ripples in their electromagnetic fields, leaking energy with each wave until every atom in the universe goes flat like a punctured tire. This, of course, does not happen.

In 1913, a few weeks before Stern and von Laue climbed the Uetliberg, the theoretical problem of the inexplicably stable atom had been declared solved by an unknown twenty-eight-year-old Danish physicist, Niels Bohr, just returned home to Copenhagen after a year in Manchester, England. It had taken him seventy-one dense pages, and his explanation was illogical. Rather than radiating light all the time, Bohr said, the electrons would emit light only when they made a kind of ineffable transition (the famous "quantum jump"). These jumps or leaps were nothing like the smooth bounding of a cat. They were a baffling, quantized, all-or-nothing disappearance in one orbit and emergence in another—like Earth suddenly materializing in the orbit of Mars.

The quantum leaps alone were like nothing that had ever happened in a physical theory before, but just as unsettling was the *frequency* of the light that the leaping electron emitted. We perceive the frequency of light as color, but the concept of frequency applies to anything that cycles, from literally round objects like wheels to things like the seasons that recur in a cyclical way. The frequency of a carousel, for example, is the number of times per minute the spotted pony with your little sister on it passes while you stand and wave. Inside the carousel a record player cranks out a tinkly band-organ tune. The frequency of the record is its number of rotations per minute (rpm), which is directly related to the frequency of the sound produced. If the operator happens to set the turntable at a slow 33⅓ rpm to play a 45-rpm record, the tinkling sound will deepen into a tired pizzicato, while the opposite mistake will produce a manic speed jingle. But in Bohr's atom, the frequency of the electron's orbit is not the same as the frequency of the light it emits. Incredibly, it is a single, pure frequency, equal to the difference between the energies of the starting and ending orbits (divided by Planck's constant, h)—as if the electron, radiating the light, already knew where it was going to stop.

"It's just absurd. I don't think it's physics at all," said von Laue finally, turning from the view. "He has simply stabilized the atom by fiat—"

Stern grinned: "The dictator!"

Von Laue laughed a little in his frustration.

They sat staring again at the peaks in the distance. Then they both turned to each other and said, "Have you talked to Einstein?"

"At this last colloquium when they presented Bohr's theory," von Laue said, "I stood up at the end and said"—he looked at Stern with a practical expression on his face—" 'but this is nonsense! If an electron is going in a circular orbit it has to radiate light.' "

Stern nodded.

"But *Einstein*—he said, 'Very curious. There must be something to it.' " Von Laue gave Stern a quick glance. "He said that he didn't believe it was pure chance that the Rydberg constant could be predicted so accurately in terms of basic constants."

The Rydberg constant had been for thirty years an unexplained number in the equation predicting the colors of light each element of the periodic table might emit. Bohr's preposterous theory had accidentally and effortlessly yielded this heretofore arbitrary number, producing meaning where before there had only been mandate.

"Well," von Laue finally said, "when he really thinks about it, he won't like the Bohr theory."

"I agree," said Stern. "This dictatorial decision, laying out flight paths and ordering inexplicable jumps for the electrons—it might seem successful right now, but it's not *physics*."

With a sardonic expression, von Laue commented: "*Someone* should stand up and stop this nonsense."

Stern, with a mock-elegiac tone in his voice, harked back to the fabled fourteenth-century birth of Swiss democracy: "The lone man with two arrows . . . where is Wilhelm Tell? Where are the men of the Ruetli Oath?"

Getting into the spirit of Schiller's famous play, von Laue quoted from it: " 'No, there is a limit to the tyrant's power.' " He and Stern were unwilling to accept a dictator, no matter how benevolent.

They both were laughing now. "Do you solemnly swear, Max Laue"—Stern grinned and corrected himself—"Max *von* Laue, to give up physics if Bohr turns out to be right?" (Max's father had been bestowed with hereditary nobility that year, hence the "von.")

Von Laue grinned widely. "Absolutely. I wouldn't be able to stand it. And you, Otto Stern, do you so swear?"

"How does the Ruetli Oath go?" asked Stern. "Something about the 'inalienable and indestructible stars . . .' "

"No," said von Laue, stretching out his hand: they were on the Uetliberg, after all. "We need new words," he said, beginning to grin.

Stern caught on: "For the *Uetli* Oath!"

"Swear by the atom," said von Laue.

The atom, ever since Greek philosophers in the fifth century B.C. first postulated it, has been one of the great mysteries. Is all matter—you, the chair you're sitting in, the air you're breathing—ultimately composed of the same building blocks? What would such an ultimate building block be like? People fumbled through philosophy trying to find an answer. Then, in the mid–eighteenth century, an inquisitive Scotsman, Thomas Melvill, burned table salt and looked through a prism at the light produced, as Isaac Newton had once looked through a prism at white light and seen the rainbow spectrum. Melvill did not see a rainbow spectrum: he saw only a pair of yellow-orange stripes, surrounded by darkness.

Sixty-two years later, Joseph von Fraunhofer looked through a prism at the sun while calibrating surveying lenses for a military supply company. He noticed for the first time that there were dark lines across Newton's rainbow spectrum. The rainbow was, in fact, missing two chunks of warm yellow—as if it were the salt spectrum turned inside out.

This coincidence went unexplored for almost half a century until Gustav Kirchhoff, a dapper little physicist who walked about the medieval halls of the University of Heidelberg on crutches, deduced why that chunk of yellow was missing: it had been absorbed by the sodium gas swirling about the sun (table salt being sodium chloride). Coming to his aid was his closest friend, the gentle giant Robert Bunsen (who twenty years earlier had lost the sight in one eye due to a sliver of glass from an exploding petri dish). "At present Kirchhoff and I are engaged in a common work that doesn't let us sleep," Bunsen explained to a friend in 1859. "Kirchhoff has made a wonderful, entirely unexpected discovery in finding the cause of the dark lines in the solar spectrum."

Kirchhoff's discovery meant that gases, even from great distances, could be identified by their characteristic spectra. The composition of the stars was suddenly an open book. Burned in a sufficiently hot flame that would not produce spectra of its own, earth elements—even in minuscule quantities—also would declare their names. Bunsen conceived of the necessary burner, while Kirchhoff jerry-rigged an accurate "spectroscope"—nothing more than a prism protected by a blackened cigar box and viewed through throwaway telescope ends. "Hitherto unknown elements" began to appear. Bunsen wrote: "I have been very fortunate with my new metal. . . . I shall name it cesium* because of its beautiful blue spectral line. Next Sunday I expect to find time to make the first determination of atomic weight."

Helium ("from the sun") was discovered this way, and a cascade of new elements followed. A small brotherhood of scientists was making spectroscopy its obsession. The first volume alone of the *Handbuch der Spectroscopie,* a list of spectral lines published in 1900, was eight hundred pages long. No one knew what mechanism was behind them, and Bohr remembered (in his idiosyncratic English) that the spectra seemed "marvelous, but it is not possible to make progress there. Just as if you have the wing of a butterfly then certainly it is very regular with the colors and so on, but nobody thought that one could get the basis of biology from the coloring of the wing of the butterfly."

Born in a mansion in Copenhagen that buzzed with the welcoming late-night talk of his father, a physiologist several times proposed for the Nobel Prize in medicine, and his father's three best friends, a linguist, a philosopher, and a physicist—all intellectuals renowned in Denmark—Bohr led a charmed childhood. He was an impulsive and wild child who did not know his own strength and in scuffles could leave his friends black and blue, but he was also a boy with an enormous smile, almost

---

*In their paper "On a New Alkali Metal," Bunsen and Kirchhoff explained that the name came from the Latin word *caesius,* which "the ancients used to designate the blue of the upper part of the firmament."

entirely without malice. His kindness and modesty, coupled with his sometimes insensitive strength (wielded intellectually when he was no longer a boy), would be noted throughout his life.

He had arrived in England, the land of subatomic exploration, in 1911 when he was twenty-six years old. He came to study with J. J. Thomson, the discoverer of the electron a decade or so earlier, and leader of the cadre of bright young experimenters among the cobwebs and neo-Gothic hallways and leaky ceilings of Cambridge's great Cavendish Laboratory. One of those bright young experimenters, the New Zealand–born Ernest Rutherford, was at the forefront of research into the mysterious phenomenon of radioactivity. Bohr sought him out at his new professorship in Manchester, where Rutherford had just that year discovered the atomic nucleus.

The newly fatherless Bohr became friends almost instantly with Rutherford, fourteen years his senior. They were both gregarious natural leaders (Rutherford beloved for his loud but mostly tuneless rendition of "Onward, Christian Soldiers" to inspire his atomic investigators); they were also both outdoorsmen and football players (Bohr's brother Harald was a silver-medal Olympian).

In the summer of 1912, his English year almost up, Bohr described his atomic hypothesis for Rutherford. It "is chosen as the only one which seems to offer a possibility of an explanation of the whole group of experimental results, which gather about and seems to confirm conceptions of the mechanismus of the radiation as the ones proposed by Planck and Einstein." Rutherford, who believed that a theory is not complete "unless you can explain it to a barmaid," thought that Bohr needed some editing.

Bohr was not at all interested in being edited. Instead, back in Copenhagen, he stumbled across a spectroscopic equation, published in 1885, the year he was born. Its author was the late Johann Jakob Balmer; at the time Balmer wrote his equation, he was a sixty-year-old teacher at a girls' school in Basel, Switzerland. The equation predicted the locations of yet-unseen spectral lines, and in the intervening thirty years, it had proved uncommonly accurate. In Bohr's hands, this equation describing the "butterfly wing" colors became a tool to see inside the atom. Fraunhofer's partially censored rainbow was the absorption spectrum: the censored colors were the frequencies of light that the atom could absorb, the leaps that its electrons were allowed to make outward into higher-energy orbits. Melvill's burning salt spectrum was just the reverse—the emission spectrum, with the electrons leaping inward toward the lowest orbit ("the ground state"), losing energy by the emission of light.

That fall of 1913, when Bohr published his paper, and von Laue and

Stern vowed to give up physics if he was right, Einstein was visiting Vienna. There he chanced upon Georg von Hevesy, a Hungarian experimentalist who was a good friend of Bohr's.

"I asked him about his view of your theorie," Hevesy wrote Bohr in his characteristically erratic English. "He told me it is a very interesting one if it is right and so on" ("faint praise," remarks Einstein's friend and biographer Abraham Pais, "as I know from having heard him make that comment on other occasions").

Hevesy then told Einstein that Bohr had been able to explain a mysterious series of spectral lines in starlight as belonging to helium. Bohr's theory had produced a result "in exact agreement with the experimental value"—an unheard-of feat in the field of spectroscopy. "The big eyes of Einstein looked still bigger," Hevesy reported to Rutherford. "He was extremely astonished and told me"—with his excitement, Hevesy's spelling grows even more scattershot—" 'Than the frequency of the light does not depand at all on the frequency of the electron . . . and this is an *enormous achiewment*. The theory of Bohr must be then wright.' "

He confided to Rutherford, "I felt very happy hearing Einstein saying so."

But the reaction of von Laue and Stern was far more common than the reaction of Einstein. "Bohr's work on the quantum theory . . . (in the *Phil. Mag.*) has driven me to despair," wrote Ehrenfest to Lorentz. "If this is the way to reach the goal," he continued, "I must give up doing physics." He liked to congratulate students with, "And now you've pulled the whole rat out of the soup!" Bohr, he felt, had left a lot of rat in the soup. He proceeded to ignore what he referred to as the "completely monstrous" Bohr model.

Before most physicists had absorbed the abstractions of the Bohr atom, and in the midst of the shortages and hardships of the First World War, in 1915 Einstein produced general relativity, science's greatest work of art. The earth turns about the sun; but according to general relativity, the sun also turns about the earth. There is no reference frame that is the correct one from which all the spinning of the worlds might be viewed; no "fixed stars," no stationary observer. General relativity is necessary to explain not only galaxies in their courses, but also the tiniest subatomic particle hurtling down from the sky. It is famously not amenable to quantization, and if Einstein had to pick between the two, he would choose relativity.

Ehrenfest wrote in 1917 that he only wanted "a general point of view which may trace the boundary between the 'classical region' " (which covered nonquantized physics, including relativity) "and the 'region of the quanta' "—a yearning that in the ensuing years Bohr would be glad to satisfy. Einstein, on the other hand, wanted a single unified physics, and would not be happy with such a truce.

In 1919, Bohr came to Ehrenfest's Leiden University. Speaking earnestly, with a "soft voice, indistinct pronunciation, and involved sentences, carefully qualified as to exclude possibilities which it was often unduly flattering to suppose one had considered" (as J. J. Thomson's son, G. P., described it in classic British understatement), Bohr lectured on his atom.

Accompanying his elliptical explanations were intricate, hypnotizing drawings of electron orbits crisscrossing about a central dot: "the Bohr atom." It was to be both the first and last quantum icon, before all such images vanished into clouds of abstraction. Bohr explained that these orbits were not to be taken exactly literally; but few could hear, while everyone could see the beautiful planetary drawings. Everyone could also see the successes of the Bohr theory, even if they couldn't understand it.

Ehrenfest was won over by the theory and the man himself. "Ehrenfest writes enthusiastically about Bohr's theory of atoms; he is visiting him," Einstein wrote in 1921 to his lifelong friend and colleague Max Born, whom he had met at that speech in Salzburg. "If Ehrenfest is convinced, there must be something in it, for he is a skeptical fellow."

4

The Unpicturable Quantum World

Summer 1921

Werner Heisenberg

Do you honestly believe," asked Werner Heisenberg, sitting in the grass beside his bike one late summer afternoon, "that such things as electron orbits really exist inside the atom?" He took a bite of cheese and looked over at Wolfgang Pauli, who was lying in the grass like the dead. Otto Laporte was drinking the sweet drafts of the really thirsty, holding his canteen above his head.

"Pass the cheese," said Pauli, without moving.

Heisenberg was nineteen. Pauli was only a year and a half older but had just completed his Ph.D. in Munich under Max von Laue's friend and former colleague Arnold Sommerfeld, with whom Heisenberg was also studying. Laporte was about to turn nineteen, and had come to Munich only a semester earlier, from Frankfurt am Main, where his family had lived until the German army took over their house.

Laporte and Pauli had met as "fellow sufferers" in the eight-hour-long old-fashioned experimental physics class of the dignified Nobel laureate

Wilhelm Wien. The three boys had escaped on a bicycle trip, in a whir of enthusiasm: "probably the only time Wolfgang dared enter my world," the outdoorsman Heisenberg wrote of his urban friend.

He passed the cheese, and squinted at the dusty road that led to the top of the Kesselberg. A noise came up from the grass.

Pauli, refortified, but still lying down, said, "I know, the whole thing seems a myth." He sat up with a huge effort, his heavy-lidded eyes almost closed in the midday sun—*he has a secretive face,* thought Heisenberg when he first met Pauli, the year before. Pauli and Heisenberg on first meeting could hardly have seemed more different. Heisenberg was blond, thin, "like a simple farm boy" (as his professor Max Born thought when he first met him). Pauli, dark-haired, already a little overweight, constantly oscillating and rocking, spent his extracurricular time in coffeehouses and nightclubs.

The physics brought them together. They were already the rising stars of their profession. In 1920, Pauli had written a monumental two-hundred-page review explaining all of general relativity—mathematically forbidding even to experts—which had impressed Einstein himself. And in the confused infant field of quantum theory, both of them were starting to produce original ideas, under the inspired (and unusually laissez-faire) guidance of Sommerfeld in Munich.

Laporte managed not to be overawed. Heisenberg liked his directness, the matter-of-fact expression behind his big, clunky, dark-rimmed glasses, his easy smile, and his interest in everything.*

"The talks we began during that tour and continued in Munich," Heisenberg remembered later, "were to have a lasting effect." The difficulty of picturing the quantum world as made of either waves or particles would lead Heisenberg and Pauli to deny the usefulness of striving for pictures at all, while Bohr would advocate holding two contradictory pictures in mind. For pictures are always simpler than what they describe, and oversimplicity can mislead. But just as deceptive is a repudiation of pictures: nothing is better than language for drawing an intricately vague veil over truth. It would turn out that what was obscured by an image-free description of quantum mechanics—or by Bohr's cubist description that welcomed contradictory pictures—was entanglement.

"You know," Pauli continued, "Bohr has succeeded in associating the strange stability of atoms with Planck's quantum hypothesis—which has

---

*A decade down the road, as a guest lecturer at the University of Tokyo, he would become fluent in Japanese, winning a prize for a haiku, and as a sideline to physics, he would specialize in the botany of cacti.

not yet been properly interpreted either—but I can't for the life of me see *how*, since he is unable to get rid of the contradictions."

"Well," said Laporte, "we should only use such words and concepts as can be directly related to sense perception."

Pauli's eyes almost closed. "Ahh, Mach. He always sounds so plausible, just like the devil." Ernst Mach was one of the main influences on nineteenth-century German physics, famous for his belief in positivism, according to which only observables had any meaning. Pauli's eyes opened. "He's my godfather, actually."

"What!" said Heisenberg.

"He's my godfather," said Pauli, nodding rhythmically. "Evidently he was a stronger personality than the priest and the result seems to be that . . . I am baptized antimetaphysical instead of Catholic. His apartment was chock-full of prisms, spectroscopes, stroboscopes, electrifying machines. When I visited him he always showed me a nice experiment . . . for the purpose of correcting thought processes, which are always untrustworthy and cause delusions and errors." Pauli grinned. "He always presumed his *own* psychology to be generally valid. But his positivistic doctrine is a waste of time."

Laporte, annoyed, said, "Didn't Einstein arrive at his theory of relativity by sticking to Mach's doctrine?"

Heisenberg nodded.

"I think," said Pauli, gesturing with his wedge of cheese, "that that is a crude oversimplification."

"What's wrong with sticking to observables?" asked Laporte.

"Mach did not believe in atoms because he could not observe them," said Pauli, with a firm sideways look at Laporte. "He was led astray by the very principle you're defending, and, as far as I am concerned, this was not by chance."

Heisenberg's brow furrowed.

Laporte said, "Mistakes are no excuse for making things more complicated than they are."

"Well, you're right about that, and my opinion is, these orbits should be the first thing to go. But Sommerfeld loves them. He relies on experimental results and atomysticism," Pauli continued. For an instant he raised his eyebrows.

"Atomysticism?" said Laporte.

Pauli laughed, rocking a little more. This was a word Sommerfeld's students had coined to describe what was, by 1921, called the Bohr-Sommerfeld model of the atom in acknowledgment of all the improvements Sommerfeld had made to it. With each improvement, the results grew more accurate, but the whole theory needed to be taken on faith. "You'll find that Sommerfeld talks about 'the atomic music of the spheres,' and believes very deeply in numerical links."

"Success sanctifies the means," Heisenberg told Pauli with defiance.

Pauli's mouth turned down in amusement. He took a bite from his salami, reclined against the grass, and then said, "Sometimes I think *I* will be the one to come up with the next step." His eyes were almost closed, like a Buddha. "But then"—the eyes snapped open—"perhaps it's easier to find one's way if one isn't too familiar with the magnificent unity of classical physics: you two have a decided advantage there." With a malicious grin, he added, "But then, lack of knowledge is no guarantee of success."

Heisenberg did not dignify this elaborate jibe with anything but the most oblique of comebacks. "Alright, I think it's time we hit the road again," he said, in his youth-leader voice. "Don't you, Otto?"

"Absolutely," said Laporte with a grin.

Pauli looked up at the steep wooded hill ahead and muttered, "How I let myself get dragged 'back to nature' by a disciple of St. Jean-Jacques Rousseau, I don't know." As an aside to Otto, he said, "Did you know he sleeps in a *tent*?"

No, Laporte didn't.

"He sleeps in a tent, gets up at some hour I don't even want to *think* about—"

"Wolfgang's usually rising and shining at about the crack of noon," said Heisenberg parenthetically.

"He gets up while the stars are still out," Pauli continued, "and hikes for an *hour*. But does this get him to class?" Laporte was grinning, Pauli getting deep into his exegesis. "No, no: *then* he has to board a *train*, which he rides into civilization, and finally, our *Wandervogel* arrives at Sommerfeld's lecture, which, as you know, starts at 9 a.m.—"

"—or so rumor would have him believe," said Heisenberg. "He hasn't ever *experimentally* verified that that's when Sommerfeld starts lecturing."

"I'm a theorist," said Pauli. "I leave that to other people."

Heisenberg, laughing, swung one leg over his bike and started again up the hill, and the world, as it always did, slid away from him. As a child he had been the sickly, hurt, and withdrawn little brother of the more favored son, Erwin. Werner's father (a professor of Greek at the University of Munich by the time Werner was eight) had fostered constant intellectual competition between the two boys. The young Werner had learned both to force himself to excel in things that did not come naturally and to escape into the woods when the effort of it all became too much. The lonely little brother who had once almost died of a lung infection became a mountaineer, skier, and the trustworthy and protective confidant for a circle of Boy Scout friends.

Pauli's childhood, on the other hand, had been secure and "always boring." All the excitement had happened two years before his birth, when his grandfather died and his father—like Bohr's, a famous medical scientist—reinvented himself: converting from Judaism to Catholicism, changing his name from Paschales to Pauli, moving from Prague to Vienna. Pauli's mother was a vivid, vocal, and ultimately despairing feminist and socialist writer, and Pauli learned from her to loathe the initially popular Great War (but he never took up newspaper reading). In a small high school class that would produce two Nobel Prize–winners, two famous actors, and a host of professors, Pauli was the class clown, organizing elaborate pranks and imitating his professors with deadly accuracy. As a college student, he would sharpen his wit on evenings out on the town and, upon his late-night return to his desk, produce his best work.

For Heisenberg, it was hikes and wanderings that gave him the best chance to think. And now, toiling up the hairpin turns of the Kesselberg, he was mulling over the news from Frankfurt of a fascinating experiment that had been in progress when Laporte left for Munich. Eight years earlier, with von Laue, Stern had sworn to give up physics if Bohr was right. Instead of fulfilling his vow, he was putting Bohr's atom to the test.

Surrounded by the magnetic fields produced by its circulating electrons, an atom acts like a magnet. Sommerfeld had shown that it would respond in a quantized way to an external magnetic field. Turning a deaf ear to those who told him not to take these things too literally, Stern decided to investigate this prediction, and he was lucky enough to have, as the head of the department in Frankfurt, Max Born.

Born was a self-sacrificing team player, a decorous revolutionary. Like Stern, he was from Silesia, a motherless child with a sad, distant father (like Bohr's and Pauli's, a medical researcher) and overbearing, wealthy maternal grandparents. With his sensitive, boyish face and eyes defensively squinted against the world, he struggled with—and sometimes overcompensated for—a lifelong insecurity, hardly shored up by his marriage to Hedi, an unstable playwright. But when they moved to Berlin at the beginning of the Great War, a shockingly confident and free-spirited presence entered their lives in the form of Einstein, who became their "dearest friend." Living alone nearby, he took to stopping in for conversation and music. Born later described that "dark, depressing time . . . with much hunger and anxieties" as "one of the happiest periods of our life because we were near Einstein." Einstein declared prophetically in 1920, when Born was trying to decide whether or not to move to Frankfurt, "Theoretical physics will flourish wherever *you* happen to be—there is no other Born to be found in Germany today."

With a knife-shaped magnet designed by the young experimentalist Walther Gerlach—an unofficial member of Born's lively, cutting-edge theoretical department—and financing, in those grim, money-starved days, from an American philanthropist and from Born's sold-out lectures on relativity, Stern was observing one of his trademark molecular beams. This was a ray of hot, gaseous silver atoms that passed through the magnetic field created by Gerlach's magnet to strike a screen on the other side.

Classical (pre-quantum) physics would predict a single smeared clump of silver on this screen as the result. Each atom on its flight approaches the magnetic field tilted at a slightly different angle, which affects its response to the big magnet, and thus each would land on the screen in a slightly different place, not too far from the middle. But Sommerfeld, trusting his quantized calculations, said that the silver atoms should hit the collecting screen in three neat piles, and he could predict the distance separating them.

But what Stern and Gerlach found, no one had predicted. No silver atoms ended up in the center. The atomic beam split neatly into *two* discrete beams (exactly as far apart as Sommerfeld had predicted, though). The response of the atoms to the field was even more quantized and less classical than Bohr and Sommerfeld had expected. The atoms had made a yes-or-no, up-or-down, either-or response to the field.

The Stern-Gerlach experiment was a sensation among physicists when it was published in 1922, and it was such an extreme result that many who had doubted the quantum ideas were converted. Bohr saw "the apparent contradictions inherent in quantum theory" emerging more strongly. A paper by Einstein and Ehrenfest, in which they had tried to understand what Stern's silver atoms did as they passed through the two halves of Gerlach's magnet, rang an alarm bell for him. As they "exposed so clearly, [the Stern-Gerlach experiment] presented with unsurmountable difficulties any attempt at forming a picture of the behavior of atoms in a magnetic field."

Einstein and Bohr would draw different morals from this difficulty of forming a picture of atomic behavior. Bohr would soon say it couldn't be done. The behavior of atoms, and their insides, were irreducibly unvisualizable.

Einstein would say there was something wrong with a physics that would come to this conclusion.

One thing was sure: quantization occurred; there was no real explanation beyond that fact, much as Sommerfeld might decorate it with musings on atomic harmony. As Heisenberg remembered, "This peculiar mixture of incomprehensible mumbo jumbo and empirical success quite naturally exerted a great fascination on us young students."

It would take three years for a solution to emerge—started, as soon as they got back to Munich, in a fit of inexplicable inspiration by Heisenberg, and clarified, corrected, and tied to reality by Pauli (a pattern that the friends would repeat many times). Heisenberg made the math work by introducing a half-quantum into Sommerfeld's equation, horrifying his professor's mystical leanings: "That is absolutely impossible! The only fact we know about quantum theory is that we have *integral* numbers, and not half numbers." Drily, Pauli suggested that his friend would next introduce quarter-quanta, then one-eighth-quanta, and in no time "the whole quantum theory will crumble to dust in your capable hands."

But eventually Pauli found that the half-quanta described something real about the electron. Real, but impossible to visualize: the electron seems to be spinning, but this is no spin anyone has ever seen; what should be a complete rotation is only halfway around to an electron. It must "spin" two full times to reach its starting point again. No one understands what this means. But the electrons that "spin" in one direction are attracted to one pole of the Stern-Gerlach magnet, and electrons that "spin" in the opposite direction are drawn the other way: two piles.* (Because 360 degrees is only halfway around for electrons, they are called by the ungainly name of "spin–½ particles.")

These deeper mysteries were still in the future when Heisenberg, on his bike, reached the saddle of the steep mountain road, where it prepared to plunge precipitously down to the Walchensee, a cup of blue among the mountains. He was light-headed from the exertion and the view. Laporte, with the sound of tires braking on the dirt, joined him in silence.

"This was Goethe's first view of the Alps," said Heisenberg finally.

---

*A technical footnote: Sommerfeld was wrong only because neither he nor anyone else yet knew about spin. This, along with electron angular momentum, affects an atom's behavior in a magnetic field. Not knowing about spin, Sommerfeld based his calculation only on the angular momentum. Like spin, it is quantized. But unlike spin (which has exactly two allowed values, up or down), angular momentum has an odd number of allowed values, one of which is always zero. Silver has a binary response to the magnetic field due to the spin (up or down) of a single unpaired electron in its outermost energy level. But if Stern and Gerlach had used a different atom—one in which the spin had added to zero (ups canceling downs), but the angular momentum did not—Sommerfeld might have been correct to predict a beam splitting into an odd number of piles.

After a while, Pauli appeared, muttering to himself. Then the breeze, like a brook running through the still day, met him, and he saw the view. He stood, rocking gently, gazing with those mysterious eyes out over the Walchensee.

Finally he broke the spell: "Now I am going to teach you two some physics." And he took off down the mountain road.

Laporte and Heisenberg, laughing, leaped on their bikes, trying to catch him, but soon their wheels were turning faster than they could pedal. They swerved down switchbacks, wide grins spread across their faces. The whirring of the tires, the air thrumming through the hair on their arms and tearing tears from their eyes, the flash of a sail on the lake in the sun—everything became an ecstatic blur, each boy inside his own blurry bubble, quivering to his own heartbeat.

Pauli's greater mass had carried him farther than the other two—the giddy descent done, he was still coasting while they pedaled again to catch him. The lakeshore beneath the Alps was striped with the gray-skinned trunks of knobby beeches leaning out over the water.

Heisenberg was still exhilarated. He loved the medieval towns, the little bucks that leaped about on cloven hooves up the cliffs, and these valleys, these lakes, these ever-present mountains: *How could anyone live anywhere else but Bavaria?*

He looked at his two companions and knew that they did not feel the same. Not Pauli, the child of the Viennese coffee shops, the cobblestone streets, the electric light; nor Laporte with his dreams of America.

But for Heisenberg the beauty had sometimes been the only thing keeping him going, through the years of the Great War (nearly starving in the turnip winter, and then the catastrophic, unexpected defeat) and the civil war afterward (the Red Terror, the White Terror). Like many others all over Germany, he had driven oxen, carried guns, and sneaked through enemy lines in the dark to get food for his family. There were five years of horror and upheaval, while his elders repeatedly, frighteningly proved that they had no idea how to live or run a country. A generation of intellectual German boys, Heisenberg among them, decided to ignore politics, with its commonness and militarism, to find a higher order. And many of them did and said nothing when Hitler gained power.

In a decade, these three boys' lives would be completely changed. In Ann Arbor, Michigan, Laporte would be an American citizen; the half-Jewish Pauli would flee because of the racial laws—first to Zurich, and finally to Princeton; and Heisenberg would remain, trying to save German science and his own skin from the barbarians.

"A saying occurs to me that ends thus—*but if it sinks in shining splendor, still it shines a long way back,*" Heisenberg wrote to his mother in the waning months of 1930. "I believe as long as we ourselves are in this world, we must be satisfied with feeling this shining-back. . . . About ten years ago," he told his mother, "that was the most beautiful time of my life."

5

On the Streetcar

Summer 1923

NIELS BOHR is riding the streetcar to pick up the world-traveling (and, after the astronomical confirmation of general relativity in November 1919, suddenly world-famous) Einstein at the port of Copenhagen. With him is Arnold Sommerfeld, visiting from Munich. Einstein has just given a speech in Sweden, making up for his nonattendance at his Nobel Prize ceremony in December 1922.

Von Laue, hearing that Einstein was planning a trip to Asia that year, had written him in barely cryptic fashion the September before the prize was awarded: "According to information I received yesterday and which is certain, events may occur in November which might make it desirable for you to be present in Europe in December. Consider whether you will nevertheless go to Japan."

The year after von Laue's oath with Stern on top of the Uetliberg, he himself had become a Nobel laureate for his beautiful interference experiment, showing X-rays to be waves. Seven years later, the Nobel committee was recognizing Einstein for suggesting that they were particles.

Einstein's fame had soured somewhat. On June 24, 1922, his friend Walther Rathenau, the German foreign minister, was assassinated, the latest of more than three hundred prominent Jews assassinated in the postwar search for scapegoats. Einstein knew he might well be next. "It is no art to be an idealist if one lives in cloud-cuckoo-land," he wrote in a eulogy for Rathenau. "He, however, was an idealist even though he lived on earth, and knew its smell better than almost anyone else."

Einstein himself only barely lived on earth, but enough so that he saw it was time for him to disappear for a while. The faithful von Laue took his place at a lecture overrun with antirelativity and anti-Semitic demonstrators (including Stark, who had received his own Nobel Prize in December 1919, but almost no recognition for it, as all eyes were on Einstein). Von Laue understood when, in December 1922, despite the forthcoming Nobel Prize, Einstein sailed as far from Europe as he could.

Independent from a young age, Einstein had renounced his German citizenship at sixteen. His father, a repeatedly hopeful but mostly unsuccessful small businessman (who had long before suppressed his own mathematical inclinations owing to the prohibitive expense of university), had moved the family from Ulm across Bavaria to Munich; to Pavia, near Milan; and then to Milan itself. Through all the entrepreneurial disappointments he remained kindly; the Einstein family was a warm, small circle. Albert and his younger sister, Maja, continued both the family closeness and its wanderings throughout their lives. The Einsteins' succession of houses were full of music from Albert's pianist mother, who encouraged the children to play. But upon the move to Italy, they left Albert alone in Munich to finish his friendless and uninspiring school. Science was entirely something extracurricular: a "miraculous" compass his father showed him when he was four and a "holy geometry book" he received at age twelve.

Now, in the summer of 1923, after belatedly paying his respects to the Nobel committee, Einstein was returning to Berlin, the heart of all he had once renounced, and for the last decade his home. But before that, he would make a stop in Copenhagen.

Einstein and Bohr, destined to spend their lives struggling together over the soul of the quantum theory, had first met three years before, when Bohr stayed with Planck in Berlin. The trolley cars stilled by a strike, Einstein walked nine miles to Planck's door in the suburb of Dahlem to escort Bohr back to his house for dinner—such as it was with the postwar food shortages. Bohr brought Einstein and his family— which now consisted of his second wife, Elsa, and her two daughters— food from "Neutralia, where milk and honey still flow," as Einstein described it in his thank-you letter.

"Not often in my life has a person, by his mere presence, given me such joy as you did," wrote Einstein in 1920, in that first letter to Bohr. "I now understand why Ehrenfest loves you so. I am now studying your great papers and in doing so—especially when I get stuck somewhere— I have the pleasure of seeing your youthful face before me, smiling and explaining. I have learned much from you, especially also about your attitude regarding scientific matters." ("What is so marvellously attractive about Bohr as a scientific thinker," Einstein wrote not long afterward, "is his rare blend of boldness and caution; seldom has anyone possessed such an intuitive grasp of hidden things combined with such a strong critical sense.")

Somewhat awed, Bohr wrote in reply, "To me it was one of the greatest experiences ever to meet you and talk with you. . . . You cannot know how great a stimulus it was for me to have the long-hoped-for opportunity to hear of your views on the questions that have occupied me. I shall never forget our talks on the way from Dahlem to your home."

The speech that Bohr gave on that trip to Berlin in 1920 was actually the first time he mentioned publicly his grave doubts about one of those views, the light-quantum. He did so in only the most polite and elliptical way, looking out on his new friend in the audience: "I shall not . . . discuss the familiar difficulties to which the hypothesis of 'light quanta' leads in connection with the phenomena of interference"—phenomena which, after all, the wave theory has shown itself to be "so remarkably suited" to explain.

Einstein, hardly concerned to learn of one more skeptic on this subject, wrote to Ehrenfest, "Bohr was here, and I am as much in love with him as you are. He is like an extremely sensitive child who moves around in this world in a sort of trance." To Lorentz he wrote, "It is a good omen for physics that prominent physicists are mostly also splendid people."

So here is Einstein on the platform of the ferry station in Copenhagen. And there is Bohr, with his wide, easy, lopsided smile, his broad woodcutter's shoulders and athletic stance, and Sommerfeld, squinting genially beside him, with his straight back and waxed mustache ("Doesn't he look the typical old Hussar officer?" Pauli had whispered to Heisenberg in class).

If Sommerfeld was hatless, you could see the long scar across his forehead from his youthful days in a drinking and dueling society at Königsberg on the Baltic Sea. With a lively, intellectual mother and a much older medical father whose pockets always held a beetle, shell, or chunk of amber to show his son, Sommerfeld became one of physics' greatest teachers, the first to include in his classes relativity (he was Einstein's constant correspondent during the gestation of general relativity) and the quantum theory. He spent an underappreciated decade teaching pure mathematics to students of mineralogy, mining, and engineering, until in 1906 an institute for theoretical physics at Munich was created around him. "He had the rare ability to have time to spare for his pupils," Max Born remembered years later—on skiing trips, at coffee shops, with money provided if they were broke. "Such a beautiful relationship between teacher and students is surely unique," Einstein wrote to Sommerfeld in 1909.

The three of them, hats tilted at different angles, three long coats flapping behind them, step out into the sunlight, Bohr and Sommerfeld bearing Einstein's suitcase and heavy bag of books and papers, while Einstein carries his violin.

"Einstein! It is so good to see you," says Bohr as they sit at the tram stop near the half-timbered clock tower of the ferry building.

"Tell us about Japan," says Sommerfeld.

"Probably much more interesting than the Nobel ceremonies, at any rate," says Bohr.

As Einstein was officially informed via telegram a hemisphere away, he had received the deferred 1921 prize,* and Bohr the 1922 prize that year. Bohr wrote him on November 11, 1922, the same day he found out: "For me it was the greatest honor and joy . . . that I should be considered for the award at the same time as you. I know how little I have deserved it, but I should like to say that I consider it a good fortune that your fundamental contribution in the special area in which I work as well as contributions by Rutherford and Planck should be recognized before I was considered for such an honor."

A month later, when the Nobel ceremonies occurred in Stockholm, Bohr threw down the gauntlet to the absent Einstein and his "fundamental contribution." "The hypothesis of light-quanta," Bohr had punned in his Nobel speech, ". . . is not able to throw light on the nature of [electromagnetic] radiation." He was serious, and he would be ready to say so to Einstein's face the next time he saw him.

Einstein, sitting on his deck chair sailing to Singapore, wrote back on the ship's stationery in January 1923:

Dear or rather beloved Bohr!
Your cordial letter reached me shortly before my departure from Japan. I can say without exaggeration that it pleased me as much as the Nobel Prize. I find especially charming your fear that you might have received the award before me—that is typically Bohr-ish. Your new investigations on the atom have accompanied me on the trip and they have made my fondness for your mind even greater.

As for Einstein's own mind, he continued conversationally, "I believe that I have finally understood the connection between electricity and gravitation"—a heroic, perhaps impossible, quest for unification that he would unsuccessfully pursue for the rest of his life. At that moment, the

*It seems that the prize was deferred, at least in part, because of a stalemate resulting from the Academy's skepticism over relativity, on one hand, and the growing pressure from theoretical physicists for Einstein to receive the prize, on the other.

bread-hungry and bloodthirsty madness of Berlin seemed very far away. "A sea voyage . . . is like a cloister. Warm rain drips lazily down from the sky, engendering peace and a plant-like state of semi-consciousness—this little letter attests to it. . . . Yours in admiration, A. Einstein."

"Look—here is the streetcar," says Einstein, standing to greet it. "Where are we going?"

"To the institute! Number 15 on the Blegdamsvej!" says Bohr over his shoulder as he climbs on, paying the three fares. Blegdamsvej (pronounced *Bly-dams-vai*) is the wide boulevard running past what will become the long green lawn of Bohr's newborn institute, only three kilometers from the ferry terminal. Five young physicists from five different countries, temporarily reposing in the future library and laboratory, have already come to work with Bohr. The first paper with the byline "The Institute for Theoretical Physics, Copenhagen," came out months before the building was completed, from two of Bohr's students too excited to wait.

The streetcar lurches to a start, its wheels shrill against the tracks. The three physicists, making their way down the aisle, are the only people speaking German, which emerges from the mouth of Bohr with a generous sprinkling of English and Danish for good measure.

"So, my dear Bohr," says Einstein, "I hear that you have predicted an element."

"Oh, yes, *that*," says Bohr.

Sommerfeld raises his eyebrows at this uncharacteristically brief statement. Einstein catches his glance and looks ironic. It was a new triumph for the Bohr atom that it was able to make sense, in terms of number of electrons, of the beautiful and hitherto inexplicable periodic patterns of the famous table of the elements.* Bohr had even described the properties of the (missing) element with seventy-two electrons.

"And then Hevesy, working right there in your new institute, *found* it,"

---

*The periodic table, first laid out in the 1860s and 1870s by the visionary Dimitri Mendeleev, is a way of organizing the elements into rows (roughly) from lightest to heaviest: each column holds a group of elements with similar properties. Bohr's solar-system atomic model explained these properties internally—for instance, through the number of electrons in an atom's outermost orbit. For example, neon (with ten electrons) is totally inert, while sodium (with eleven) is extremely reactive. Ten electrons completely fill neon's two atomic orbits, giving the atom a "smooth" outer surface—a full orbit leaves nothing for another atom to grab. It is the opposite case with sodium; its eleventh electron is alone in the atom's otherwise empty third orbital ring.

prompts Sommerfeld. Hevesy, though a Catholic nobleman, had been fired from his position in Hungary at the end of the Great War because of his Jewish ancestry; as a result, the friends Bohr and Hevesy were reunited in Copenhagen. In Hevesy's charming broken English, he wrote Rutherford in 1922, while he was searching for the new element: "Bohr reads the language of spectra like other people goes through a magazine."

The story begins to pour out from Bohr: "We quite innocently dropped into the most terrible muddle of postwar nationalism; we had never dreamt of any competition with the chemists in the hunt for new elements, but wished only to prove the correctness of the theory. There was Alexandre Dauvillier at the de Broglie lab" (Maurice de Broglie was a well-known experimentalist working on X-rays in his mansion in Paris) "trying to claim priority for it as 'celtium' for France—with de Broglie's younger brother Louis backing him up—and then a Brit emerged, saying he'd found it before anyone else and it should be 'oceanum' for the British navy . . . no one paying any regard to the important scientific discussion of the properties of the element."

Einstein can't resist: "You did not quite avoid nationalism, yourself, did you?"

"Right," says Bohr, beginning to laugh, "and when we couldn't decide whether to call it 'hafnium'—for the Latin name of Copenhagen—or 'danium,' we got a letter from the editor of *Raw Materials Review* in Britain saying the papers claimed we'd discovered *two* elements, was it true? And someone in Canada suggested 'jargonium.' " Einstein roared with laughter, so that people on the streetcar turned around, wondering what the joke in German could be, and Sommerfeld chuckled heartily into his big mustache.

"Oh, it's good to see you doing so well," says Einstein.

Bohr shakes his head, smiling: "My life from the scientific point of view passes off in periods of over-happiness and despair . . . as I know that both of you understand . . . of feeling vigorous and overworked, of starting papers and not getting them published"—his face is earnest—"because all the time I am gradually changing my views about this terrible riddle which the quantum theory is."

"I know," says Sommerfeld, "I know."

Einstein's eyes almost close; he is nodding. "That is a wall before which I am stopped. The difficulties *are* terrible." His eyes open. "The theory of relativity was only a sort of respite which I gave myself during my struggles with the quanta."

"But, you know, everything is so exciting," says Sommerfeld. "This crazy model my young Heisenberg has thought up—"

Bohr interjects (in what Heisenberg's biographer noted were "the strongest words he would ever use in criticism"), "Heisenberg's paper is very interesting, but it is difficult to justify his assumptions."

Sommerfeld nods. "Everything works out but remains in the deepest sense unclear." He begins to smile wryly and looks over at Einstein. "You know I can only contribute to the technology of quantum theory—you have to create its philosophy."

Bohr's brow furrows. "The question is not only the development of the interpretation of experimental facts," he says cautiously, "but just as much to develop our deficient theoretical conceptions. And here the hypothesis of light-quanta cannot really help us."

"I no longer doubt the reality of light-quanta," Einstein says, "although I still stand quite alone in this conviction."

Sommerfeld says, "I thought so, too, Bohr, but I have amazing news." While Einstein was in the Far East, Sommerfeld was touring the Far West. "The most interesting thing that I experienced scientifically in America," he says, "was the work of Arthur Holly Compton in St. Louis. It was so interesting that even though I wasn't certain he was right, I lectured about it wherever I went." Compton had shown that X-rays and electrons collide like billiard balls. Von Laue, while working with Sommerfeld in Munich, had proved that X-rays were waves; now Compton demonstrated that they were particles.

Sommerfeld, excited enough to forget about his friend von Laue for a moment, turns to Einstein with excitement. "After this experiment, the wave theory of X-rays will become invalid," he says.

But Einstein responds with a smile, "Well, that might be jumping to conclusions."

"I should say so!" bursts out Bohr, his voice a little strained. "You can understand my concern—for a person for whom the wave theory of light is a *creed*. The theory of light quanta can *obviously* not be considered as a satisfactory solution of the problem of light propagation. It excludes in principle the possibility of a rational definition of the *frequency,* since it is defined by interference experiments which demand a *wave* constitution of light."

"That's why I wanted to call your attention to the fact that eventually we may expect a completely fundamental and new insight," says the practical Sommerfeld.

"Well, yes," says Bohr. "But I can hardly imagine it will involve *light quanta.* Look, even if Einstein had found an unassailable proof of their existence and would want to inform me by telegram, this telegram would only reach me because of the existence and reality of radio waves."

Einstein's booming laugh rings out. His entire career—from patent office to Princeton—would be spent postulating ideas that no one else initially believed in; sometimes the consensus of physics would eventually come his way, sometimes not. He is equally serene in the face of Bohr's disbelief as in that of Sommerfeld's new enthusiasm.

"Einstein—" says Bohr; at the same time Sommerfeld, looking out the window, says, "Bohr—"

"Yes?" says Einstein.

"Yes?" says Bohr.

"Where are we?" asks Sommerfeld.

Bohr looks around and then starts laughing, which is infectious, with his wide mouth and big teeth and eyes shut. "We have missed our stop," he says, still laughing. "By, ah, about twelve stops, we have missed our stop."

Sommerfeld pulls the wire, the streetcar eases to a stop, and they emerge onto the street. It is still a beautiful day, and so it does not seem to be such hard luck as the three settle themselves on a bench beside the elm-bordered boulevard.

"Einstein, if you *should* be really able to prove that light is particles more or less in the commonly accepted sense," continues Bohr without a glance at his surroundings, "do you really believe that one could imagine seeing a law passed which would make it illegal to use diffraction gratings?"*

"Or conversely," Einstein fires back, "if you could prove that light is exclusively wavelike, do you then think that you could get the police to stop the use of photocells?"†

"Yes, but must I say"—a phrase famous around Bohr's institute, resulting from his idiosyncratic Danish-to-German translation—"that at the present time we are entirely without any real understanding of the interaction between light and matter."

Einstein is not interested in hiding one mystery (light sometimes behaves as a particle) behind another (the interaction between light and matter). "There are now two theories of light, both indispensable," he says, "and—despite twenty years of tremendous effort—without any logical connection. We must persist in our helpless attitude toward these separate results—those of the diffraction gratings and those of the photocells—until the principles have revealed themselves. But as long as the

---

*Diffraction gratings bend light so that it interferes with itself in classic wave form.

†Photocells (short for photoelectric cells) make up solar panels, which work by the photoelectric effect. Einstein's Nobel was for his explanation of this effect by light quanta.

principles capable of serving as starting points for the deduction remain undiscovered, individual facts are of no use to the theorist."

Bohr begins to speak, then falls silent, looking out at the canal. The ducks float into his field of vision and float out, dragging their undulating reflections and leaving long streamers of interfering ripples behind. James Franck, the beloved experimentalist, described Bohr deep in thought "in the early years": "One thinks of Bohr when one had a discussion with him. Sometimes he was sitting there almost like an idiot. His face became empty, his limbs were hanging down, and you would not know this man could even see. . . . There was absolutely no degree of life. Then suddenly one would see that a glow went up in him and a spark came, and then he said: 'Now I know.' . . . I am sure it was the same with Newton too."

The metal tracks in the road begin to hum again, giving Bohr a start, and all a-clang, a streetcar arrives going back the way they came. They climb on, sitting this time a little nearer the front.

"Well . . ." says Bohr as he sits down, "I suppose that during a stage in science where everything is still in ferment, it cannot be expected that everybody has the same views about everything."

Einstein smiles at this. "No, that is something that cannot even in the best of circumstances be expected."

Leaning against the side wall of the streetcar looking back at his friends, Bohr says, "But Einstein, I cannot quite understand what it is you hope to prove about the quantum theory." He leans forward, his arm gesturing over the back of the seats to explain. "I am thinking about your famous papers of 1916 and '17 where you showed that when an atom emits light, there is no telling *when* it will happen or *where* it will go. You seemed to think it was too crazy. . . ."

"I have full confidence in the route I have taken," replies Einstein. "But it is a weakness of the theory that it leaves the time and direction of such an elementary process to chance."

"To me," continues Bohr, "it was one of the most brilliant strokes of genius, it was almost the decisive point. What if causality does not actually hold for the quantum world?" This was an extreme position to take. The principle of causality—that everything has a cause—is the foundation of science; the purpose of science is to find those causes. But Bohr's intuition about causality in the atomic world would prove prophetic: inexplicable spontaneity remains one of the most characteristic quirks of the quantum theory. "I am inclined," Bohr says with half a smile, "in this case, to take the most radical or mystical views imaginable."

"Bohr," says Sommerfeld, bemused, "what *possible* calculations could have led you to think that causality might not be ultimately true?"

"Sommerfeld . . ." Sommerfeld is more than fifteen years older than he, but on this point, Bohr feels he needs to be taught: "You know how I feel about physics as merely 'mathematical chemistry'—everything does not always boil down to *calculations*. It could easily be that we do not yet even possess the mathematics with which we can talk about these things." Bohr has actually achieved most of his amazing successes not through mathematics but through an intuition, his "analogy principle," from which almost no one but he can extract any results. It involves building a quantum theory out of features that would average out, on a large scale, to produce this world we see. In 1920, Bohr renamed it "the correspondence principle," which made it no easier to operate or fathom. "A magic wand," said Sommerfeld in the most recent edition of his indispensable and religiously updated textbook of quantum theory.

Bohr's correspondence of large-scale quantum effects with classical physics would guide the young quantum theory for years. But his faithfulness to the spirit of this principle would also prevent him from discovering the possibility of long-range entanglement. For him, a concept like that—quantum correlation between two separated particles—would belong solely to a shadowy netherworld of subatomic semireality, which could never emerge into the bright Newtonian day of large sizes and distances. Instead, the skeptical Einstein would be the one to repeatedly show the way to the strange beauties of entanglement over large distances and of quantum effects writ large, which by the turn of the century were to become two of the most vibrant branches of all of physics.

"Bohr," says Einstein, "I greatly admire the sure instinct which guides all your work. I have learned before that you are brave enough and intuitive enough to try things that others pass by. But I have to say, abandonment of causality as a matter of principle should be permitted only in the most extreme emergency."

"Einstein," says Sommerfeld with half a grin, "I admire Bohr's sure instinct in the realm of physics as well. But on the streetcar I would have to say I consider his instinct less than sure. Look where we are."

Einstein and Bohr both look out the window, Bohr slapping his hand to his forehead, Einstein laughing and pulling the wire. They are almost back to the ferry station.

"We rode back and forth in the streetcar," Bohr reminisced much later, "because Einstein was really interested at that time. We don't know whether his interest was more or less skeptical—but in any case we went back and forth many times in the streetcar and what people thought of us, that is something else."

After disembarking, Sommerfeld wrote grandly to Compton, "Your work"—showing X-rays behaving as particles—"has sounded the death knell of the wave theory." Bohr, who (as Heisenberg wrote to Pauli back in January) "cares more for general theoretical principles than agreement with experiment," began work on a quantum theory designed to avoid light-quanta.

Compton's work did mark the beginnings of acceptance of light-quanta, strong enough that by 1926 the light-quantum would merit a name of its own: the *photon*. This word—approximate Greek for "light-being"—came from G. N. Lewis of Berkeley, the white-mustached "father of physical chemistry" (who also suggested that the time that light takes to travel a centimeter be called a *jiffy*). The next year, Compton would receive his Nobel Prize. But neither particles nor waves alone would be able to describe the quantum world, and the wave-particle fusion Einstein had been predicting since 1909 lay just around the bend, in several unexpected guises.

6

Light Waves and Matter Waves

November 1923–December 1924

Y OU KNOW THOSE DIFFICULTIES about not knowing whether light is old-fashioned waves or Mr. Einstein's light particles . . . or *what*," a twenty-three-year-old American named John Slater, touring Europe after finishing his doctorate at Harvard, wrote home to his parents in November 1923. "Well, that is one of the topics on which I perpetually puzzle my head; and about a week and a half ago I had a really hopeful idea on the subject. . . .

"It is really very simple. I have *both* the waves and the particles, and the particles are sort of carried along by the waves, so that the particles go where the waves take them, instead of just shooting in straight lines, as other people assume."

This idea was related to Einstein's "ghost waves," but it allowed entanglement to remain (unnoticed at the time) by not insisting on a single particle per wave. It was fated to have a strange history. After Slater, it was more fully proposed twice more: four years later by Louis de Broglie, and after World War II by another American, named David Bohm. Each time, the theory dramatically and impressively failed to attract a shred of interest. Its downfall stems from its lack of aesthetic appeal to physicists. But it was this thrice-rejected solution to the problem of waves and particles that would most clearly highlight the mysteries of entanglement to its most perceptive sleuth, John Bell.

At Christmastime in 1923, Slater brought his doomed theory to the center of it all, Bohr's institute in Copenhagen. There he talked with Bohr's right-hand man, the kind and ironic Dutchman Hans Kramers (only six years older than himself), and found that Kramers was excited by some of the mathematical elaborations of his idea. Even more thrilling, Kramers thought that Bohr—whom Slater had only seen looking exhausted and overworked—would also be interested.

When he got a chance to talk with Bohr, the great man seemed "decidedly excited." Slater spent his entire weekend in nonstop conversation with the two most important members of an institute bursting at the seams with people and ideas (a local newspaper reporter came by while Bohr, Kramers, and Slater were deep in their discussions and wrote that "five to six people sit at one table and compute").

There had, however, been warnings of what was to come. "Prof. Bohr wanted me to write it down for him to see," a bemused Slater wrote his parents, "and he told Dr. Kramers not to talk to me about it until that was done, and then went and spent all the time talking to me himself." It was a shock to realize what Bohr thought of light-quanta. And not just Bohr: "The theory of the light-quanta," said Kramers, "may be compared with medicine that will cause the disease to vanish but kills the patient."

Slater reported to his parents, "Of course they don't agree with it all yet. But they do agree with a good deal and have no particular argument, except their preconceived opinions, against the rest of it." Innocently he thought that they "seem prepared to give those up if they have to." But light-quanta dimmed under the full Bohr wattage ("You see, Slater, we agree much more than you think"). Bohr's last-ditch attempt at a quantum physics without light-quanta needed an idea like Slater's guiding wave, which Bohr rechristened a "virtual field" (the electromagnetic field carries energy; this field carried only guidance). The virtual field, rather than mapping paths for the light-quanta, would guide the interactions between the Bohr atom and ordinary light waves.

Pauli, when he heard about it, scoffed at the resulting "virtualization of physics." Without light-quanta, Slater's theory lost two related pillars of physics: causality, and conservation of energy—the principle that energy cannot be created or destroyed, but merely changed from form to form. For Bohr, this was a small price to pay. And as the years went by, causality's hold on the quantum theory would indeed appear more tenuous and statistical. (The same would not prove true for conservation of energy; still, every subsequent crisis in physics saw Bohr suggesting that energy might, this time, not fully be conserved.)

In his desperation to prove that light-quanta were unnecessary, it took only three weeks for Bohr to dictate to Kramers (his favorite form of paper writing) what is known as the Bohr-Kramers-Slater paper—by far the fastest-written paper of his life. "I haven't seen it yet," Slater reported to his parents a fortnight before its publication. He was pleased when he did: by its rejection of light-quanta, the theory obtained a "simplicity," he thought, which "more than made up for the loss involved in discarding conservation of energy and rational causation."

Slater was not the only twenty-three-year-old taken with this purported simplicity. Pauli visited Copenhagen in April, and briefly fell under the spell: "You succeeded by your arguments in silencing my scientific conscience, which much revolted against this perception," he wrote to Bohr about the Bohr-Kramers-Slater paper a few months later, after his scientific conscience had regained its voice.

The twenty-two-year-old Heisenberg also came to Copenhagen in the spring of 1924. He was overwhelmed by the brilliance of Kramers and the host of physicists surrounding Bohr, and was touchingly grateful when Slater and the other two Americans there—Frank Hoyt and the slightly older Harold Urey, a future chemistry Nobelist—took him on a trip to the Danish main peninsula, Jutland. But still better was when Bohr himself invited Heisenberg to take a walking tour that spring. Rucksacks on their backs, they followed the Zealand coast from Copenhagen north to Hamlet's Elsinore, and along the beach where the North Sea meets the Baltic Sea, flinging stones and talking all the time.

Famed for his instinct about physics, Bohr also had an instinct about people. When they returned from the trip, he told Slater's friend Hoyt, "Now everything is in Heisenberg's hands—to find a way out of the difficulties."

Difficulties certainly still remained. Einstein's reaction to the Bohr-Kramers-Slater paper came as a surprise to no one. "Bohr's opinion about radiation is of great interest," he wrote to Max Born that April. "But I should not want to be forced into abandoning strict causality without defending it more strongly than I have so far. I find the idea quite intolerable that an electron exposed to radiation should choose *of its own free will,* not only its moment to jump off, but also its direction." This was how the Bohr-Kramers-Slater paper would describe the Compton effect without light-quanta. "In that case, I would rather be a cobbler, or even an employee in a gaming-house, than a physicist. Certainly my attempts to give tangible form to the quanta have foundered again and again, but I am far from giving up hope. And even if it never works," he remarked in vintage form, "there is always that consolation that this lack of success is entirely mine."

The newspapers all over Europe began to run pieces on the conflict between Einstein and Bohr. "I was in Copenhagen and talked with Mr. Bohr," a friend of Einstein's wrote to him later in 1924. "How strange it is that the two of you, in the field where all weaker imaginations and powers of judgment have long ago withered, alone have remained and now stand against each other in deep opposition."

In April 1924 came word that the city of Copenhagen was going to give the overfull institute some of the land around it upon which to build, and in May, the first large donation of money arrived from the United States. Construction began a month later, and the Bohrs had a ground-breaking party on the institute's lawn.

Slater found himself talking with Mrs. Kramers (known to her friends as Storm). They were standing on the grass beside a little table with an ice chest holding bottles of Carlsberg, agreed among Bohr's students to be "the best beer in the world" (from the beginning, Carlsberg was the institute's biggest supporter). "Mrs. Kramers, pardon me, but you seem so Danish," he said finally.

She gave him a big smile. "I *am* Danish. They say that Hans has started a tradition which all foreign students coming to study under Niels Bohr must do: marry a Dane." She looked mischievous. "Now, Mr. Slater, have you any prospects for a Danish wife?"

Slater blushed and looked uncomfortable and young.

"Well, you still have time," she said.

"Actually, ma'am," said Slater, "I'm leaving in a week."

"You'd better start looking harder," she said. "Leaving in a week, my my. So how have you liked your time at the institute?"

"It's been fine," said Slater, toeing the grass.

She looked at him sideways, a small smile on her face, her eyebrows together skeptically. " 'Fine,' has it been?"

"Yes, ma'am," said Slater.

"*I* think you've been having a hard time, young man," she said. "Professor Bohr, much as it is like a religion around here to love him, can be very difficult if you disagree with him." She looked over in Bohr's direction, where her husband stood beside him. She nodded, and then turned back to Slater and said, "Let me tell you a story about my husband. About two years ago—yes, I think it must have been 1921—we had not been married for very long, at any rate—well, Hans came up with an idea, an idea about these light-quanta."

Slater looked up, surprised.

"Yes, he believed in them in those days, he thought they might actually be there, even though no one else thought so—except, of course, Professor Einstein. And especially Professor Bohr thought that these light-quanta were a very bad idea. Anyway, what Hans had come up with was very similar to your fellow American Professor Compton's idea—no *experiment,* but the idea was the same. And he was *insanely* excited. He was sure he was onto something very important, and he told Professor Bohr about it." The brisk spring wind blew her hair into her eyes, and she shook her head and raised her hand to pull it behind her ear.

"Well, immediately Professor Bohr started arguing with him, and daily, they would argue. These long terrible arguments, no holds barred. Hans would come home so weary and depressed he couldn't even eat his dinner. He was so confused and let down and torn between his idea and Professor Bohr, whom he loved and would have done anything for.

"Finally, he was so worn out that I had to take him to the hospital—no joke. He got better there, I think because the hospital was just too far away for Professor Bohr to come and argue with him." She smiled at Slater, but her eyes looked a little weary at the remembrance. "After Hans got better, he and Professor Bohr were *perfectly* in tune; when he talks to me about his physics, he never again suggests that Professor Bohr might be wrong. He made Professor Bohr's ideas and views his own ideas and views, and I have to say, it has been easier all round since."

Slater stood on the lawn, beside the green Carlsberg bottles glinting in the sun, shocked. He could hear in his mind Kramers intoning, "The theory of the light-quanta is like the medicine that will cure the disease but kill the patient." There was Bohr, so genial, chatting with the small group, telling one of his old jokes. Everyone was laughing.

"So you see, my dear," said Mrs. Kramers, "I understand what you are going through, and so does Hans. He tells me about you, he is very sympathetic, he has gone through it all himself. Now, you should run along—you only have a week to find a lovely Danish girl to marry. . . ." Then she smiled again. "Though I have a feeling you aren't looking for reasons to stay in Denmark. Hopefully some time you will return and you will have a better time of it, when you are older." She walked off to rejoin her husband, her white skirt flapping at her calves and her high heels poking into the lawn.

Slater now stood alone beside the beer, a tall, gawky American boy still shaking his head. Years later he would say in an interview, "I liked Kramers quite well . . . but he was always Bohr's yes-man." Everyone else around was in love with the great man, but "I completely failed to make a connection with Bohr. . . . I've never had any respect for Mr. Bohr since." He finished: "I had a horrible time in Copenhagen."

In the spring of 1924, as Slater was leaving Copenhagen, two unusual pieces of mail were slowly traveling toward Einstein's desk in Berlin, one from Paris, and one from Dacca, Bengal. The first came from Einstein's friend the mustachioed physicist and pacifist Paul Langevin, at the Sorbonne. It was the dissertation of a thirty-two-year-old student of Langevin's named Louis de Broglie, the younger brother of the experimentalist Maurice, the Duke de Broglie. Long ago, as the secretary for the first Solvay conference, in 1911, the older de Broglie had been impressed by Einstein's talk of light-quanta. In the next decade, war cut off communication between French and German citizens, but in the oak-paneled, tapestried rooms of de Broglie's lab (he was the first to bring X-ray equipment to France, with his footman as his technician), he showed his little brother how these X-rays sometimes behaved as waves, sometimes as particles.

Now the younger de Broglie, a thin, wide-eyed, bushy-haired aristocrat, declared that if Einstein had found a granular aspect to waves of light, then there should also be a *wave* aspect to grains of matter. He suggested that a stream of electrons passing through a small enough hole would diffract and interfere as if the electrons were waves. Langevin, "probably a bit astonished by the novelty of my ideas" (as de Broglie put it) and wondering what to think, sent Einstein a copy.

Einstein immediately passed it on to Max Born, saying, "Read it. Even though it might look crazy, it is absolutely solid." He wrote back to Langevin, "He has lifted a corner of the great veil."

Others were unimpressed. De Broglie had in the past few years written several papers forcefully advocating slightly half-baked ideas, delicately dismissed by Kramers's unmistakably Bohrian remark as "not consistent with the way in which the quantum theory at present is applied to atomic problems." With Alexandre Dauvillier—he of the "celtium" discovery— de Broglie had that January condescended to Sommerfeld in a paper that had already been disproved by experimental evidence, so he had few friends in Munich, either. The minutes of the star-studded Kapitza Club at Rutherford's Cavendish Lab recorded the unanimous opinion that the whole thing was nonsense.

It was, in fact, the first step toward the new quantum theory.

Satyendra Nath Bose, a lecturer at the brand-new University of Dacca, sent Einstein the other remarkable paper of that spring. Bose was a year and a half younger than de Broglie and approximately at the same stage in his education, but with no experimentalist older brother or renowned and well-connected professor at his side. Bose, too, started with the premise that light-quanta were real and derived startling results.

Like a scene from the Gospels in which total faith is spectacularly rewarded, Bose put his precious four-page paper, already rejected by

Philosophical Magazine, on the boat to Berlin as the heat broke and the monsoon rains began. "If you think the paper worth publication," Bose wrote to Einstein, "I shall be grateful if you arrange for its publication in *Zeitschrift für Physik*"—an astonishing thing for a total stranger to ask of one of the most famous men on earth, but (the genial and unflappable Bose continued) "we are all your pupils." He believed in Einstein's wisdom and goodwill, and had, moreover, a well-justified faith in his own paper's merits. "I am anxious to know what you think of it," he told Einstein, knowing he could not even hope for a reply before the rains ended in early September.

Einstein responded spectacularly. He personally translated Bose's paper from English to German, sent it to *Zeitschrift* with a strong recommendation, and in the subsequent six months followed it with three papers himself. He joined Bose's ideas to de Broglie's, applying the results that Bose had found in light-quanta to atoms, and seeing what this meant for the behavior of de Broglie's matter waves.

Bose's key idea was the complete indistinguishability of light-quanta, making meaningless any system that treated them as individuals. This indistinguishability meant strange things for quantum states. The likelihood of one of these fundamentally indistinguishable particles entering a given state is affected by the states of the particles that surround it.*

Bose's biographers tell a story about the statistics that result (known as Bose or Bose-Einstein statistics). Paul and Margit Dirac, visiting Calcutta in the mid-fifties, were in the backseat of Bose's car. Bose invited several students to sit in the front seat, which already held him and the driver. Alone in the back seat, the Diracs—whose thinness made a striking contrast with Bose's jovial rotundity—asked if they weren't too crowded. "Bose looked back and said in his disarming fashion, 'We believe in Bose statistics.'"

The differences—between Bose statistics and any other—"express indirectly" a certain hypothesis about a mutual influence of the particles," wrote Einstein in 1925, "for the time being of a quite mysterious kind." In their 2001 Nobel Prize lecture, the University of Colorado experimentalists Eric Cornell and Carl Wieman shared Einstein's wonder: "This mutual influence is no less mysterious today, even though we can readily observe the variety of exotic behavior it causes."

*A notoriously elusive concept, quantum *state* is perhaps best defined as what can be known about a quantum object. Bose-Einstein statistics allow amazing feats to be performed by unentangled particles in a definite quantum state. By contrast, entangled particles are in an *indefinite* state—neither here nor there, neither yea nor nay—until one is measured.

The exotic behavior that would still be resulting in Nobel Prizes three-quarters of a century later, Einstein described in a letter to Ehrenfest in November 1924. "From a certain temperature on, the particles 'condense' without attractive forces. . . . The theory is pretty, but is there also some truth to it?"

"Condensing without attractive forces" into the same state means taking on all the same attributes, so that the particles are not only indistinguishable but also moving in perfect harmony: together they become a quantum matter wave, visible to the naked eye. Soon to be known as Bose-Einstein condensation, this is what makes a laser's light so piercing, what allows a superconductor to carry a current of electricity forever, what makes a superfluid able to flow up walls.

Einstein had brought the quantum theory far past what it could handle in 1924 in his support and unification of de Broglie's waves of matter and Bose's sociable, indistinguishable particles of light. In the new edition of his textbook on the quantum theory, Sommerfeld wrote exhaustedly of waves and particles: "Modern physics is confronted here with two incompatible features and must frankly confess *non lignet*." (*Non lignet* is the archaic spelling of the verdict a jury gives "in cases of doubt," explains the Century Dictionary, "deferring the matter to another day of trial.")

"Your frank 'non lignet' is a thousand times more sympathetic to me," wrote Pauli to his old teacher in December 1924, "than the ad-hoc, artificial pseudo-solution of Bohr, Kramers, & Slater. . . .

"One now has the strong impression with all models," he continued, "that we are speaking a language that is not sufficiently adequate to the simplicity and beauty of the quantum world."

7

Pauli and Heisenberg at the Movies

January 8, 1925

Wolfgang Pauli, 1930

HEISENBERG AND PAULI were laughing until their insides ached. It was Heisenberg's second and Pauli's third viewing of Charlie Chaplin's *The Kid,* recently arrived in Germany after the war's import ban was lifted. Heisenberg had just turned twenty-three and Pauli would be twenty-five that spring. They were meeting in Munich, their old university town, after their Christmas vacations—Pauli's with his family in Vienna, Heisenberg's in the Bavarian Alps. The next day, Pauli would continue north to his new professorship in Hamburg (joining Stern, now professor of experimental physics there), Heisenberg to Copenhagen, where Kramers and he were furthering the Bohr-Kramers-Slater paper.

The plot of the movie revolved around a swindle (*schwindel* in German, one of Heisenberg and Pauli's favorite words) perpetrated by Chaplin, a figure with whom Pauli identified. Working in collaboration with him is the Kid, an innocent-looking little boy whom he adopts. The Kid breaks windows and then the Little Tramp comes around the corner, carrying new glass, offering his services.

"*Schwindlig* [giddy, dizzy] clown tricks," was the consensus of German film critics. The movie was "incredible *Unsinn* [nonsense]," it was "perfect *Unsinn*." Heisenberg and Pauli loved it. So did everybody else, despite the critics' cry that slapstick comedy was beneath this "nation of poets and philosophers."

The music stopped, the projectionist switched on the lights, the happy crowd began chattering and pulling on their coats. All these people, bludgeoned by inflation and hunger, were floating out of the theater in giddy groups. Heisenberg reached for the crutches lying in the aisle beside him, a token of a skiing accident a week earlier. With the faint pale lines around his eyes where his ski goggles had obscured the sun, he looked like the archetypal injured athlete.

Outside, a burst of cold hit their faces. Snow was swirling down through the glow from the curled street lamps. Heisenberg, the skier, smiled up into it. Opposite them, ghostly in the early dusk of January, stood the gate in the medieval city wall: two hexagonal towers with bricks mottled and softened with age, muted with ivy, now fringed with snow.

Heisenberg and Pauli started to make their way down the street to the tram stop. Behind them, people were already lining up before the *Lichtspieltheater* to buy tickets for the late show. The marquee glowed so bright it was hard to read the huge CHARLIE CHAPLIN JACKIE COOGAN it spelled out—the tramp giving the kid equal billing.

"So, tell me," said Pauli. "How's Pope Bohr? How's His Eminence Kramers the Cardinal?"

"You know, I don't think my physics career started until I met Bohr," Heisenberg said.

Pauli nodded. "It was the same for me."

"Bohr is more worried than anybody else about the *inconsistencies* of quantum theory," said Heisenberg, trying to describe the difference between his time in Copenhagen and his earlier experiences in Munich and Göttingen. Their college days with Sommerfeld in Munich had brought them deep into mysterious and disconnected mathematical attempts to describe the atom, untroubled by what it all *meant*. Göttingen, the center of mathematical learning in Germany, had become the third city in the quantum theorists' trinity, now that its theoretical physicists were led by the shy Max Born, with whom Heisenberg had studied in the winter of 1922 and spring of 1923. "You know how Sommerfeld is quite happy," said Heisenberg, "when he can apply nice complex integrals" (that is, mathematics involving fairly hard calculus). Pauli smiled at Heisenberg's comment; his eyes closed for a second. "And he doesn't worry too much," continued Heisenberg, "whether his approach is consistent or not. And Born in a different way is interested mostly in mathematical problems."

"It's true," said Pauli.

"Neither Born nor Sommerfeld really *suffers*, while Bohr can hardly talk of anything else. But Kramers . . . he is a bit strange to me. He makes jokes," said Heisenberg, "when I don't want any jokes made."

Pauli grinned broadly into the snow, thinking of his earnest friend working with the drily humorous Kramers.

"I don't know if you'll get the papal blessing with your new paper, though," said Heisenberg.

Almost simultaneous with Einstein's papers on Bose-Einstein condensation, Pauli had finished two papers. In the first, he made "a funny [*komische*] reflection" on the Stern-Gerlach experiment of 1922. He realized that its characteristic two discrete piles of silver atoms were due to a single electron in each atom—the outermost one—which, he wrote, "in a mysterious unmechanical manner . . . succeeds in running in two states with different momentum." These two exclusive options would soon be called spin up and spin down (though not by Pauli, who found talk of electrons as spinning balls "inadvisable").

The second was to become his most famous paper, on what their British friend P. A. M. Dirac would call "Pauli's exclusion principle" and Heisenberg and Ehrenfest called the *Pauli Verbot*—the prohibition. Pauli showed that, unlike the particles (i.e., light-quanta and many atoms) that Bose and Einstein were writing about at exactly the same time, electrons would never gather sociably in a single quantum state. The electrons liked making opposite pairs (spin up and spin down) but would not get closer than that.

Ehrenfest would explain, "Why is a crystal as thick as it is? Because the atoms are thick. Why are they thick? Because not all the electrons fall into inner rings. Why don't they do it? Because of the electric repulsion of the electrons against each other? No! For if it were only that, they would still be able to gather around the highly charged nuclei much more densely than they do.

"No, they don't do it for fear of Pauli!! And so we may say: Pauli *himself* is so thick, because the Pauli prohibition holds. The wonderful, the incomprehensible . . ."

Again the quantum theory encountered a "mysterious mutual influence"; again it moved forward not with models, explanations, or pictures, but with prohibitions, formal rules, and incomprehensibilities. "What I do here is *not a bigger* nonsense [*Unsinn*]," Pauli told Bohr, than anything else about the Bohr atom. "My nonsense is conjugate to yours. . . . The physicist who finally succeeds in adding these two nonsenses will gain the truth!"

Bohr showed the paper and the letter to Heisenberg, who, delighted, shot off a postcard to Pauli:

> December 15, 1924
> Copenhagen
>
> Today I have read your new work, and it is certain that I am the one who rejoices most about it, not only because you push the swindle [*Schwindel*] to an unimagined, giddy [*schwindlig*] height . . . and thereby have broken all hitherto existing records of which you have insulted me, but quite generally, I triumph that you too (*et tu, Brute!*) have returned with lowered head to the land of the formalism pedants; but don't be sad, there you will be welcomed with open arms.
>
> And if you think you have written something against the hitherto existing kinds of swindle, of course, this is a misunderstanding; for swindle × swindle does not yield something correct and therefore two swindles can never contradict each other.
>
> Therefore I congratulate!!!!!!!!
> Merry Christmas!!

Bohr, who always took longer than Heisenberg to do anything, wrote a week later to Pauli that he had *not* produced an *Unsinn* as promised but "complete *Wahnsinn* [insanity]." This was not necessarily a bad thing, but Pauli could tell that Bohr—with his careful phrases like "We are all enthusiastic about the many new beauties you have brought to light"— did not quite feel the same as the truly enthusiastic Heisenberg.

"I have the impression," wrote Bohr, "that we stand at a decisive turning point, now that the extent of the whole *Schwindel* has been characterized so exhaustively."

Years later, they would still be using the same joke words, including Bohr's famous "If you do not get *schwindlig* sometimes when you think about the quantum theory then you have not really understood it."

In Munich that January of 1925, Heisenberg and Pauli were struggling to understand the quantum theory, and the world. Heisenberg looked effortless on his crutches, and antsy, as he swayed himself through the snow down the street beside Pauli.

"He has 'no philosophy' and he doesn't care for clear formulation of the basic principles," Pauli had grumbled to Bohr a year before about Heisenberg. Despite this, he continued, "I consider him—apart from the fact that he is personally also very nice—as very important, even as a genius and I believe that some day he will produce major scientific advances. . . ."

Back in the mathematical citadel of Göttingen, Max Born was writing what many were thinking. Mechanics—physics that describes motion and the forces that make things move—had failed for the atom, and a new mathematical structure was needed. He called for a "*Quantenmechanik*," not yet knowing that he and Heisenberg would produce it.

"Could you perhaps," said Pauli to his friend on crutches, "send me page proofs of this paper you and Kramers are writing?" He started to grin. "Or will that be allowed by His Eminence? I am, after all, an infidel."

Heisenberg nodded. He could tell he was close. "I feel that we've—that Kramers and I have—now come one step further in getting into the *spirit* of the new mechanics. It seems—and Bohr thinks so too—that there *must* be some kind of wonderful new mechanics if we just push it a little further." His face glowed with excitement.

"I myself have discovered something new," said Pauli. "In fact, Stern is already calling it the Pauli effect."

"The Pauli effect? Does this have to do with your *Verbot*?"

"It has absolutely nothing to do with that. It has to do with what a good theorist I am." Pauli paused. "Well, so I guess in *that* sense it does."

"*And?*" prompted Heisenberg.

"You know how theoretical physicists can't handle experimental equipment—it breaks whenever they touch it. Well, I am becoming such a good theorist that it breaks when I walk in the room."

"Only you would turn a disaster into a compliment for yourself."

"Stern now will only shout questions through the closed door to his lab."

"You're talking as if you actually believe this happens!"

"I'm telling you, it *does*."

"I can't believe *Stern*—"

"Oh, you think Stern should be practical because he's an experimentalist? Stern told me a friend of his used to bring his apparatus a flower every morning, 'to keep it in good temper.' " Pauli grinned. "He himself 'had a somewhat higher method.' In Frankfurt, he would threaten the Stern-Gerlach apparatus with a wooden hammer to make it behave."

"You're all crazy."

"Well, listen to this: someone *borrowed* the hammer, and the whole thing stopped working until it was found." With raised eyebrows he nodded at Heisenberg.

"Well," said Heisenberg, laughing, "I'll tell Bohr that Hamburg is treating you well, even if you aren't returning the favor."

Pauli laughed too. "The astronomers have been particularly nice. I go to the observatory on full-moon nights—when it's too bright to make observations—and we drink wine."

"Wine?"

Pauli nodded seriously. "When I came to Hamburg I passed—under the influence of Stern—directly from mineral water to champagne. I have noticed," he continued, "that drinking wine agrees very well with me. After the second bottle of wine or champagne I usually adopt the manners of a good companion—which as you know I never have in the sober state—and then may at times enormously impress the surroundings, in particular if they are female."

"At the moment physics is again very much in disarray," wrote Pauli a few months later to his friend the American physicist Ralph Kronig, now studying in Copenhagen. "For me anyhow it is much too difficult and I wish I were a *Filmkomiker*"—movie comedian—"or something like that and had never heard anything of physics!

"Now I do hope that Bohr will save us all with a new idea. I urgently request that he do so."

8

Heisenberg in Helgoland

June 1925

Nord und West und Süd zersplittern,
Throne bersten, Reiche zittern:
Flüchte du, im reinen Osten . . .
North and West and South splinter
Thrones burst, realms tremble,
Flee you, in the pure East . . .
—Goethe, *West-östlicher Divan*

On June 7, 1925, Werner Heisenberg took the night train from Göttingen, where he was now assistant to Max Born. In the dawn, he boarded a ferry to the small North Sea island of Helgoland (the name of the island, pronounced *Hey-go-land*, means "holy land," because, surrounded by salt water, it is blessed with a spring). The island was German again after Danish and British interludes, but it flew its own cheerful flag—green and white surrounding the central red stripe of the German flag, instead of Germany's black and yellow.

Heisenberg's face was badly swollen. When he knocked on the door of an inn, the landlady whisked him to a room, saying, "Well, you must have had a pretty bad night." The room was on the second floor and the window looked out over the stony village, to the white dunes and the ocean, which Heisenberg could barely see from under his heavy eyelids. A thin white curtain was blowing from the open window into the room.

Heisenberg said with embarrassment, "It's just hay fever. But up here . . ."

She smiled in understanding: ". . . There are no flowers, no fields, and no fever."

Turning his face into the breeze from the window, he squinted out at the bare and beautiful view. Outside the day was windy and clear. He slept, ate, and went walking along the red cliffs. He stripped and dashed into the cold water. He spent the rest of the day reading Goethe's Persian idyll through his watery eyes, learning poems from *West-östlicher Divan* by heart.

> *Who the song would understand,*
> *Needs must seek the song's own land.*

Looking out at the view, he remembered Bohr on a different North Sea beach, trying to explain the charms of flat Denmark to the mountain climber: "I always think that part of infinity seems to lie within the grasp of those who look across the sea."

Alone on Helgoland, Heisenberg felt clean and focused. He wanted to see the quantum world on its own terms, to listen to the atom and hear what it was telling him. Back in Copenhagen, he and Kramers had given the Bohr-Kramers-Slater theory mathematical coherence. The month of April contained both the theory's triumph—the Kramers-Heisenberg paper—and the beginnings of its downfall, as ever more careful measurements by Compton and others showed that the principle of the conservation of energy had held, against Bohr's predictions. But "this mathematical scheme" that Kramers and he had come up with, Heisenberg remembered, "had for me a magical attraction, and I was fascinated by the thought that perhaps here could be seen the first threads of an enormous net of deep-set relations."

He lost himself in the colors of light emitted by the atom, thin flags of a still unknown realm. "A few days were enough to jettison all the mathematical ballast," he wrote. A clean, bare idea was arising in his mind, only dimly glimpsed before he left for Helgoland: it "results in a very peculiar point of view," he told his friend Ralph Kronig. His quantum mechanics would declare time and space nonexistent within the walls of the atom.

Or so it seemed at night on the island. As he worked and wondered, the North Sea sun began to set near midnight. The sea was red, the land dark; Heisenberg turned on the gaslight. His thin muscular back hunched over the desk, his shoulders curved near his ears, his feet hooked around the legs of his chair—he looked like a boy studying for an exam. Stars came out, the wind whined against the door to his little balcony. Much later Heisenberg was to say of the process of physics, "In science you just have to drill in very hard wood, and go on thinking beyond the point where thinking begins to hurt."

The clock struck three. He looked at what he had wrought: "I was deeply alarmed." As Max Born had predicted, a quantum mechanics would be very different from any description of forces and motion that physics had ever known before. "I had the feeling that, through the surface of atomic phenomena, I was looking at a strangely beautiful interior, and felt almost giddy at the thought that I now had to probe this wealth of mathematical structures. . . . I was far too excited to sleep."

He left his frenzied scribblings, and made his way through the dawning streets to the cliffs. "I had been longing to climb a rock jutting out into the sea." His hay fever gone, he could smell the salt water. The rough surface of the red sandstone felt harsher than usual, his trembling fingers felt newborn, his eyes more sensitive to the sunrise than on ordinary days, each wave carrying before its crest a pale petal of light. The spray splashed around him, on top of his rock, and he felt utterly triumphant, utterly empty, and filled with light.

On his way back to Göttingen he stopped in Hamburg for a few hours to present his quantum mechanics to Pauli, "generally my severest critic." But Pauli, ready for all models and visualizable descriptions of the quantum world to fail, had only encouraging words.

"My own work gives me almost no pleasure to write about because all is still unclear," he wrote to Pauli a week later. Space and time within the atom were gone—"In quantum theory it has not been possible," he noted in his paper, "to associate the electron with a point in space"—but he did not know what was left.

"He had written a crazy paper and did not dare send it in for publication," remembered Max Born of Heisenberg's return. "I read it and became enthusiastic . . . and began to think about it, day and night." Heisenberg's confidence ebbed and flowed, but of one thing he remained sure (he told Pauli in early July)—no trace of a space-and-time model, even one as weird as the Bohr atom, should appear. "My entire meager efforts go toward killing the concept of an orbit which cannot be observed *anyway*." With that, Heisenberg departed on vacation to Leiden, Cambridge, and Copenhagen, leaving Born with his paper.

"One morning about 10 July 1925 I suddenly saw light," remembered Born. "Heisenberg's symbolic multiplication was nothing but the *matrix calculus,* well known to me since my student days in Breslau." Well known to Born, a consummate mathematician, but unknown to most physicists less fond of math. The word *matrix* means "womb": an array of numbers arranged in rows and columns. A matrix, no matter how huge, is a single entity, fitting into an equation as a number would.

Born began to reframe Heisenberg's theory in terms of matrices, "and at once there stood before me the peculiar formula": $QP - PQ = \hbar i$, where

$\hbar$ ("h-bar") is equal to Planck's quantum constant, h (without which no quantum equation was complete) divided by 2 π, and i is the square root of -1; i, the first "imaginary number," is a common math tool, but this was (as David Wick writes in his book on the heresies of quantum mechanics) "the first use of the imaginary unit in *a seemingly essential* way in the history of natural science."

Mass is m, so, since the time of Newton, momentum has been p (for *impetus*), forcing position to be abbreviated by the counterintuitive q. When talking about a billiard ball, three numbers suffice to give its position. These are called the Cartesian coordinates, which on a pool table would be something like: x inches from the long side and y inches from the short side on a table that is z feet off the ground. Similarly, three numbers give the momentum: so many feet per second in each direction times so many pounds. But to describe the same aspects of an atom, an infinite array of numbers is required—none with any obvious connection to position or momentum.

In the equation Born found, **Q** and **P** are in bold capitals to signify that they represent these infinite matrices instead of an ordinary single number. Heisenberg, never a fan of the supermathematical sound of *matrix* and trying to keep things simple, called them "radiation-value tables," since they were essentially a list of all the characteristic frequencies that could be radiated by the atom in question.

These matrices were not like the attributes of a billiard ball in another way: the order of multiplication matters. The order does not matter when you multiply Cartesian coordinates: five times three is not different from three times five. But Born's equation shows that the rules change when we enter the quantum domain. **QP**, a position measurement following a measurement of momentum, is not equal to **PQ**.

Did measurement mean something different in the quantum world? Months of mulling provided no solution. "The worst is," Heisenberg wrote to Pauli, "that it does not become clear to me how the transition into classical theory takes place."

Neither was it clear to Born, nor to Pascual Jordan, a pale, shy, stuttering young student barely Heisenberg's age who "can think far more swiftly and confidently than I," Born told Einstein. Letters flew back and forth from Göttingen to Heisenberg in Leiden or Cambridge. The work, Heisenberg recalled, "kept us breathless for months."

It was thrilling and bewildering in equal measure. Only five days into this work, Born unbreathlessly wrote to Einstein, "I am fully aware that what I am doing is very ordinary stuff compared with your ideas and Bohr's. My thinking box is very shaky—there is not much in it, and what there is rattles to and fro, has no definite form, and gets more and more complicated."

Heisenberg, meanwhile, was lecturing in Cambridge—not on his inchoate quantum mechanics, but on the old familiar atomic spectroscopy on which it was based. In the audience was a thin, gangly, dark-haired young mathematical genius named P. A. M. Dirac. (When, soon afterward, a series of papers made this byline famous, many wondered what the initials stood for; people felt that the laconic Dirac cultivated the mystery, but it's more probable that the mystified never asked him point-blank. The answer would have been, in Dirac's high, precise Bristol accent, "Paul Adrien Maurice.") Too shy to talk to Heisenberg himself, Dirac eventually read a preprint of Heisenberg's paper. He set to work discovering the same idea that—unbeknownst to him or to Heisenberg—Born was simultaneously postulating on the Continent.

Born and Jordan wrote up their results, translating Heisenberg's original stroke of genius into matrices, but almost before they were done, Rutherford sent Born a reprint of a paper by Dirac. Born wrote, "This was—I remember well—one of the greatest surprises of my scientific life. For the name Dirac was completely unknown to me, the author appeared to be a youngster, yet everything was perfect in its way and admirable." In Göttingen, people started talking about *Knabenphysik,* the "kid physics": Dirac and Jordan were twenty-two, Heisenberg was twenty-three, and Pauli was twenty-five.

Among these wunderkinder, Dirac was unique. His paper of 1925 began a half decade in which mind-boggling papers—three in 1927 alone—steadily flowed from his pen, in careful, childlike cursive. He had been brought up in Bristol, England, by a tyrannical Swiss father who insisted that only French be spoken at his dinner table, silencing his young British wife. The result was that the young Paul spoke only when a direct question was posed to him or he had something important to say, and then in as few syllables as possible. These traits remained, in any language, for the rest of his life, producing a hundred humorous tales bandied about the international club of physicists, for whom he was "the purest soul," as Bohr described him. One of the most ethereal theoretical physicists of his age, he emerged from completely matter-of-fact schooling, attending the Merchant Venturers' Technical College—the school where his father taught French—and, in an attempt to please his father with a display of practicality, going on to higher education in electrical engineering. His only activities outside of physics were long solitary hikes, invariably in his formal black suit.

Copenhagen knew nothing of Heisenberg's breakthrough. Bohr's institute was convulsed with the death (at the hands of Compton and other experimentalists) of the Bohr-Kramers-Slater theory, or what Bohr and Franck—in a phrase borrowed from Pauli—had taken to calling "the Copenhagen putsch."

At the end of July, a month after Heisenberg's Helgoland breakthrough, Pauli wrote to Kramers. "I regard it as a magnificent stroke of luck," he started with characteristic lack of subtlety, "that your interpretation was so rapidly refuted by these beautiful experiments. . . . Every unprejudiced physicist can now regard light-quanta just as much (and just as little) physically real as electrons."

He greeted Heisenberg's new work "with jubilation":

> We have concurring opinions about almost everything, as much as
> this is at all possible between two people of independent
> thinking. . . . I now feel myself a little less isolated than I did for
> about half a year, when I felt quite alone between the Scylla of the
> number mysticism of [Sommerfeld's] Munich school and the
> Charybdis* of the reactionary Copenhagen putsch, which you
> propagandized with the excessive zeal of a true fanatic.

To the ever-confident Pauli it was much worse to be a convert like Kramers, than a missionary like Bohr: "Now I hope that you will no longer retard the reestablishment of a healthy Copenhagen physics, which in view of Bohr's strong sense of reality cannot fail to reemerge.

"Again all good wishes from your devoted Pauli."

Kramers already knew of Heisenberg's theory because he had visited Göttingen just as Heisenberg returned from Helgoland.

"You're too optimistic," he had told Heisenberg.

To Bohr on his return to Copenhagen, he had said nothing.

In a chatty letter to his friend Urey (Slater's friend who had the year before shown Heisenberg around Jutland), all he reported of the trip to Göttingen was that he saw Franck and Born and "others."

---

*In classical mythology, Scylla was a monster (later a perilous rock) standing across a narrow boating channel from the whirlpool Charybdis.

In the death of the Bohr-Kramers-Slater theory and the birth of the New Big Thing—and Heisenberg had only been extending the work that he and Kramers had done together—Kramers had seen his own star fall and a new one rise. Spiraling into depression, he (who for years had ignored the constant courting of other universities) accepted a position in his native Holland that September. For almost a decade Bohr's cardinal and scientific heir, Kramers now left Copenhagen for good.

Kept in the dark by an unsure Heisenberg and a miserable Kramers, Bohr could say only that the quantum theory "is in a most provisional and unsatisfactory stage," as he had written to Heisenberg on June 10. This letter would have greeted Heisenberg on his arrival back in Göttingen nine days later, but only at the very end of August did Heisenberg respond. "As Kramers perhaps has told you . . . I have committed the crime of writing a paper on quantum mechanics."

Einstein was also skeptical. "Heisenberg has laid a huge quantum egg," he wrote to Ehrenfest in September. "In Göttingen they believe in it (I don't)." When Pauli complained for the thousandth time of Heisenberg's breakthrough being smothered by "excessive Göttingen learnedness" (having already severely hurt Max Born's always tender feelings), Heisenberg fired back, "Your eternal ridiculing of Copenhagen and Göttingen is a flagrant scandal. You will have to admit that we are not seeking to ruin physics by malicious intent. If we are donkeys having never produced anything new, you are an equally big jackass, for you haven't done anything either!"

On Christmas Day in 1925, Einstein was writing to his best friend, Besso: "The most interesting recent theoretical achievement is the Heisenberg-Born-Jordan theory of quantum states. A real sorcerer's multiplication table, in which infinite matrices replace the Cartesian coordinates." With more than a bit of irony he added, "It is extremely ingenious, and thanks to its great complexity, sufficiently protected against disproof."

It was also extremely difficult to use, and only the superhuman Pauli was capable of demonstrating that it could mathematically describe the very simplest real-world application: the hydrogen atom. Pauli showed that all the ad hoc Bohr-Sommerfeld assumptions about the atom's size and shape were now unnecessary. These things emerged not by decree, but as a result of the new quantum mechanics. He submitted the resulting paper in January of 1926. Heisenberg, who had been attempting the same, rejoiced, and Bohr told Rutherford that he was no longer miserable: the longed-for quantum mechanics had really arrived.

Then, out of the blue and practically crossing paths in the mail, came a paper conveying a totally different quantum mechanics. It seemed to completely contradict Heisenberg's, yet it produced all of the same results, and—far from the thicket of matrices—did so using mathematics that any physicist could command.

The Schrödinger equation had arrived.

9

Schrödinger in Arosa

Christmas and New Year's Day 1925–1926

Erwin Schrödinger

ERWIN SCHRÖDINGER, THIRTY-FOUR YEARS OLD, had contracted tuberculosis. In the spring of 1922, just after he had started teaching at the University of Zurich, he "was actually so *kaput*," he told Pauli, "that I could no longer get any sensible ideas." With a Viennese chef and Anny, his wife of two years, he retreated to an Alpine health resort in Arosa, near Davos, where, at over a mile in altitude, the thin air would curb the tuberculosis bacteria. Einstein had brought his younger son, Eduard, who suffered from constant headaches and earaches, to the same hotel a few years before.

When the brisk air of autumn arrived, Schrödinger began reading Hermann Weyl. Weyl was a friend of Schrödinger's who had given a series of mathematical lectures on relativity during the war, at the E.T.H. in Zurich. These had been transformed into a famously daunting book, ubiquitous around physics departments, called *Space—Time—Matter*. Schrödinger noticed a point that Weyl made which, if followed through, suggested that the electron in an orbit behaves like a *standing wave* (whose crests and troughs oscillate up and down without advancing). Writing in his graceful small, slanted hand, he sat on the porch with a blanket over his knees plugging one equation into another, thinking fuzzily about matter and waves. These calculations resulted in a small paper, "On a Remarkable Property of the Quantized Orbits of a Single Electron," and Schrödinger gave no further thought to it.

Three years later, in 1925, Peter Debye at the E.T.H. (cigar clamped in his teeth) walked across the street to the University of Zurich to show Schrödinger Louis de Broglie's recent paper on waves of matter. He was entranced and, as Debye had hoped, gave a lecture on it at the next joint E.T.H.–University of Zurich colloquium.

This was on November 23, a week after Heisenberg, Born, and Jordan submitted their famous paper "On Quantum Mechanics" to the *Zeitschrift*. Felix Bloch, a twenty-year-old physicist soon to become Heisenberg's first pupil and then one of Pauli's first assistants, remembers that after the lecture, "Debye casually remarked that he thought this way of talking was rather childish. As a student of Sommerfeld he had learned that, to deal properly with waves, one had to have a wave equation" that described its undulation.

Schrödinger plunged into waves. In a paper on the Bose-Einstein ideas that he finished a few weeks later, he wrote (in stronger words than either Einstein or de Broglie had used) that he hoped to "take seriously the Broglie-Einstein wave theory of the moving particle, according to which the particle is nothing more than a kind of 'whitecap' on the wave radiation that forms the basis of the world."

The snow was falling on the steep, slanted roof of the hotel as Erwin Schrödinger, laughing, a woman by his side, made his way around the last kink of the road. He was back in Arosa again after four years, but now it was winter, and Anny and the Austrian chef were far away (no one knows anything about the woman beside him). It was almost Christmas. Anny was with Weyl, her lover and her husband's closest friend, whose book had inspired him the last time he was in Arosa to thoughts of waves (Weyl's wife, continuing the pattern, was the mistress of Pauli's friend the physicist Paul Scherrer).

Growing up in a cherub-festooned town house of pink stucco and marble in the heart of Vienna, within sight of its majestic Gothic Stephandsdom, Schrödinger had been a home-tutored only child treated to a constant stream of doting feminine attention from young aunts, maids, and nurses. His father, supported by an inherited linoleum business he despised, was an amateur botanist and landscape artist, his mother the frail and cheerful young daughter of a research chemist (whose second wife, in grand Viennese fashion, became Mahler's mistress for a decade). Schrödinger's grandmother—the research chemist's first wife—was British, and the little boy learned English before German. He grew handsome, cultured, charming, brilliant, and devoid of any sense that the world did not, in fact, revolve around him.

There, looking out through the snow-burdened pines to the Weisshorn, Schrödinger thought about Debye's comment: *The practical Dutchman! Of course we need a wave equation.* With pearls in his ears to dampen any distracting outside noise—the carols rising from the porch, the voice of the woman he brought with him—Schrödinger worked straight through Christmas, rapt at what he began to see unfolding on the page before him.

Willy Wien, the ferocious Munich experimental physicist who had flunked Heisenberg in college, was at his ski hut when he received a letter from Schrödinger, written two days after Christmas: "At the moment I am struggling with a new atomic theory. If only I knew more mathematics! I am very optimistic about this thing and expect that if I can only solve it, it will be *very* beautiful." Schrödinger followed this touching enthusiasm with the explanation that his theory was beginning to look as if it could produce the frequencies of the spectral lines of the hydrogen atom better than anyone had before—"and in a relatively natural way, not through ad hoc assumptions." Schrödinger was beginning to believe, looking down on the undulating whiteness all around, that the entire world was made of waves.

He came down the mountain on January 8 and went straight to Weyl, who helped him solve his equation. (A solution to the Schrödinger equation, describing the state or condition of a given quantum entity, is known as a wavefunction. The wavefunction is symbolized by the Greek letter psi, ψ.) When the E.T.H. and the University of Zurich met again for their fortnightly colloquia, Schrödinger stood up and declared in triumph, "My colleague Debye suggested that one should not talk of waves unless one has got a wave equation; well, I have found one!"

Schrödinger's assistant Fritz London (who a decade later became a pioneer in the theory of Bose-Einstein condensation) wrote jokingly to Schrödinger reminding him of that old paper:

Very Respected Herr Professor:
Today I must talk with you seriously. Do you know a certain Herr Schrödinger who described in the year 1922, a "noteworthy property of quantum orbits"? Do you know this man? What, you say you know him rather well, you were even with him when he wrote this paper and were implicated in the work? This is truly shocking. Hence you already knew four years ago. . . .

London went on to list the aspects of Schrödinger's 1922 paper that to him in 1926 seemed to lead, in a blindingly obvious way, straight to Schrödinger's wave equation.

He asked Schrödinger, "Will you now immediately confess that, like a priest, you held the truth in your hands and kept it a secret?"

"Now, hear that!" Einstein said, upon hearing that Schrödinger had so unexpectedly replicated the matrix results with a wave equation. "Until now we had no exact quantum theory, and now we suddenly have two. You will agree with me that the two exclude each other. Which theory is correct? Perhaps neither is."

What You Can Observe

April 28 and Summer 1926

Einstein stood by the window of his study in Berlin, his gray hair glowing in the last daylight. "You do not mention the path of the electron at all, Heisenberg," he said. "But yet when you look in a cloud chamber"—Einstein pointed his finger with squinted eye, as if he were following the jet trail of an electron—"the electron's track can be observed quite directly." (Electrons traveling through a jar of moist air leave trails like those of airplanes in the upper atmosphere.) "Don't you think," he asked, "that it's strange to say that there is a path for the electron in the cloud chamber, but there is no path for the electron in the atom? The existence of the path cannot depend merely on the *size* of its container."

Heisenberg's whole speech at the Berlin physics colloquium half an hour earlier had really been aimed at the gray-haired man in the second row. After all the people left, Einstein said, "Would you walk home with me, so we can discuss these things further?" Now they had arrived at Einstein's house, and Einstein was finished with the kind questions about Sommerfeld and Pauli, and was cutting to the meat of the matter. And Heisenberg was eager to convince him.

"But we have no way of *observing* the electron's path in the atom. What we actually record are frequencies of the light radiated by the atom, but no actual path," said Heisenberg. With a small flourish, he added, "And it is rational to introduce into a theory *only* such quantities as can be directly observed."

Einstein sat down in one of the big armchairs before the unlit fire. "Heisenberg," he said, "*every* theory contains unobservable quantities." Heisenberg looked up in surprise. "The principle of employing only observable quantities simply cannot be consistently carried out."

"But isn't that precisely what you have done with relativity?" asked Heisenberg.

Einstein smiled a little. "Perhaps I did use such philosophy earlier, perhaps I even wrote it, but it is nonsense all the same."

Heisenberg looked as if he were being told by his priest that he did not actually believe in God.

"You know, people talk about 'observation' all the time," continued Einstein, "but do they know what they actually mean by that? The very concept of 'observation' is itself already problematic." He searched his pockets for his tobacco pouch. "Every observation presupposes an unambiguous connection between the phenomenon to be observed"—he found the pouch, and began filling his pipe—"and the sensation which eventually penetrates into our consciousness." He lit a match, two fingers of his right hand holding the thin matchstick over the tobacco. It caught and smoldered, and Einstein blew the smoke out of his nose, leaning back in the armchair.

"But we can only be sure of this connection if we know the natural laws by which it is determined. If, however"—and here he looked straight at Heisenberg, who was leaning forward in his chair, his pale hair shining in the dim room—"as is obviously the case in modern atomic physics, these laws have to be called into question, then the concept of 'observation' loses its clear meaning." Heisenberg's quick mind was leaping about; Einstein's quiet voice was opening new doors all around his brain. "In that case," finished Einstein, "it is *theory* which first determines what can be observed."

Heisenberg felt as if a lightbulb with a filament made of fireworks had just lit up his brain. *It is theory which determines what can be observed.* Einstein was smoking and looking at him. *It is theory which determines what can be observed.* Part of Heisenberg wanted to immediately run out the door with this shiningly new concept and go ponder all these things; the other part of him wanted never to leave this chair, in the darkening room, listening to Einstein.

Einstein stood up and turned on a reading light near them, drawing them closer together, while the world outside went dark and their reflections bloomed in the windows. "Let there be light," said Einstein. He sat back down, a trail of smoke following him in the air.

"So, in your theory, we have an electron orbiting the nucleus and then suddenly it leaps—but wait, no, we don't really have that." Einstein smiled, catching himself: no orbits. "We have an electron. It's doing something, we don't know what, inside the atom." He looked at Heisenberg, who suddenly understood why people occasionally described Einstein's expression as impish. "Suddenly, it leaps into a different state, emitting a light-quantum"—one hand plucked an imaginary particle out of the air—"as it does so."

Einstein's brow furrowed, and he looked a little teacherly as his glasses slid down his nose. "But we also have an alternative idea, one that says the electron merely, like a little radio transmitter, beams out a wave motion in a continuous fashion, as the atom changes its state." Einstein puffed on his pipe. "Now, in our first scenario, we cannot for the life of us account for the interference phenomena which have so often been observed and can only come from overlapping waves; in the second, we cannot explain the fact of sharp spectral lines, each one a single frequency of light. So what do we do?"

Heisenberg thought of what Bohr would say. "Well, we are, of course, dealing with phenomena that lie far beyond the realm of everyday experience, and so we cannot expect these phenomena to be describable in terms of the traditional concepts."

"You are a good pupil of Bohr's," Einstein said. "But I am not. In what quantum state, then, do you suppose that the continuous emission of the wave is taking place?" That is, what list of characteristics might describe the atom while it emits the wave?

Heisenberg was a little taken aback, but, ever quick on his feet, said, "It could be like when you are watching a film, and often the transition from one picture to another does not occur suddenly—the first picture becomes slowly weaker while the second slowly becomes stronger, so that in an intermediate state we do not know which picture is intended. In the atom, too, a situation could arise in which—for a time—we just do not know what quantum state the electron is in."

"You are moving on very thin ice," said Einstein. "You are now talking of what we know about the atom and no longer about what the atom is really doing. If your theory is right, you will have to tell me sooner or later what the atom *does* when it passes from one stationary state to the next."

"I admit that I am strongly attracted," said Heisenberg slowly, "by the simplicity, and beauty, of the mathematical schemes with which nature suddenly spreads out before us—the almost frightening simplicity and wholeness of the relationships, for which none of us was in the least prepared. I know you have felt this, too."

Einstein sat smoking and nodding. Then he said: "Still, I should never claim that I really understood what is meant by the simplicity of natural laws."

"Do you believe in God, Herr Professor Schrödinger?" asked the blonde girl sitting beside him in the sand. It was the summer of 1926 and they were on the *Strandbad*, the bathing beach of the Lake of Zurich. She was a lovely fourteen-year-old, awkward and graceful, named Itha Junger.

Beside her sat her twin sister, Roswitha, who, like Itha, had long blonde braids. They wore matching convent school uniforms even though it was summer and they were at the beach. Both of them looked as if they were about to giggle, but the question was in earnest.

"I believe more in God the Father with the white beard than I believe in Nothing," said Schrödinger. He was lying on the sand in his bathing trunks. By his side was a small portable blackboard, dull and chalky in the brilliant sun flashing on the sand. The water lapped against the wet shore a few feet away.

"I thought scientists didn't believe in God," said Roswitha.

"What people who say that don't understand," said Schrödinger, "is that the world picture of science becomes accessible at the price that everything personal is excluded from it. A personal God cannot be encountered *there*. 'I do not meet God in space and time,' says the honest scientific thinker . . . and reproaching him for it are the same people whose catechism says 'God is Spirit.' "

Schrödinger had always liked teaching on the beach. "In summertime, when it was warm enough, we went to the *Strandbad* on the Lake of Zurich," remembered a recently graduated student of Schrödinger's. The students "sat with our own notes in the grass and watched this lean man in bathing trunks before us writing his calculations on an improvised blackboard which we had brought along."

Itha and Roswitha's mother was a friend of Schrödinger's wife, Anny, who had persuaded him to give them math lessons once a week during the summer, so that Itha, who had flunked math, could catch up with her class. Schrödinger asked Weyl what was taught to fourteen-year-olds, and the great mathematician, who that January had helped him solve the first Schrödinger equation, obligingly made him an outline.

Schrödinger was a good teacher, and Itha learned the necessary math, but along with that were many discussions on other subjects, including Schrödinger's equation. He told them that it seemed to suggest that everything was made of waves.

"What do you mean," Itha asked, "everything is a wave? Even water is little droplets when it's small."

Schrödinger nodded, explaining how much smaller than a water droplet were the "waves of matter" his equation described—the smallest water droplet imaginable, one dwarfed by a pinhead, would contain trillions of waves of matter.

"And matter is like light," he said, "and it diffracts." Squinting into the sun beyond Itha's head, he explained diffraction: "You know how small things with the sun behind them—a lock of your hair, a mote of dust, a spider's web—light up mysteriously; the single strand of hair itself becomes, as it were, its own source of light."

The bigger and slower-moving a quantum-mechanical particle is, the smaller its associated wave. "Just as a mote of dust diffracts a light wave, so the nucleus diffracts an electron. An atom in reality is merely the diffraction halo of an electron wave captured by the nucleus of the atom."

Itha was entranced to imagine she might be made of diffraction haloes, but Roswitha had had enough. "I'm going swimming," she said. Sand flew as Itha, tripping over her uniform, rose up to follow her. Both girls splashed in, clothed and screaming. Schrödinger followed, walking, watching them as they bobbed off the shore. He broke into a sprint, and in one smooth motion dived into the water.

The girls looked around, kicking in place, swirling their hands in the water, their skirts ballooning below their elbows.

"Well, where is he?" said Itha.

"I'll bet he's sneaking up on us," said Roswitha.

"Eek!" said Itha, her nerves humming.

Roswitha shivered with pleased expectation.

Itha looked a little blue in the lips.

Then she screamed. Her mouth was filled with water as she was pulled under by her leg. She emerged with the grinning professor, whose hair was plastered and peaked in funny ways on his head, his face looking years younger without the little round glasses and the presence of his wife. Itha was coughing, shivering, and laughing simultaneously. Roswitha shook her braids in a tremendous water-splattering circle, and then said, "It's too cold, I'm getting out."

She began swimming for shore but her twin didn't follow her quite yet. She asked the professor, "When you were under the water for so long, was it nice to be surrounded by waves? Was it like your equation?"

Schrödinger looked at her and then slowly smiled as they swam to shore.

Settled back on the sand beside her sister, Itha said, "I like your theory about the world being made of waves."

"The funny thing was," said Schrödinger, "a German called Heisenberg, with the help of my friend Max Born, had come up with another theory half a year before me. His theory had no space or time within the atom." He shook his head. "I don't know what that means. And because of the to me very difficult-appearing methods of transcendental algebra—much harder than what we're doing here!—and because of the lack of *Anschaulichkeit*, I felt deterred by it, if not to say repelled."

Anschaulichkeit—the physical naturalness of an idea, so that it can be pictured in the mind's eye—was one of Schrödinger's favorite things about his theory, and the area in which he felt it most obviously out-reached Heisenberg. As far as Heisenberg was concerned, Schrödinger could have it: "The more I think of the physical part of the Schrödinger theory," he wrote to Pauli that June, "the more abominable I find it." Mimicking one of those gentle Bohr phrases they both knew so well, he continued, "What Schrödinger writes about *Anschaulichkeit* 'is probably not quite right,' in other words I think it is *Mist*"—bullshit. He grumpily decided that "the greatest result" of the Schrödinger theory was that it could be used to extend his own where the math was too hard.

"But then"—Schrödinger leaned his thin frame back on his elbows against the sand, his face still showing the surprise of the discovery— "while playing around with his theory and mine I found that they were mathematically equivalent."

Heisenberg's quantum mechanics and Schrödinger's wave mechanics were two ways of saying the same thing.

Two decisive moments had come in late June, and they both involved how the wave mechanics dealt with more than one particle. First came a paper from Schrödinger published in *Annalen der Physik*. The wavefunction ψ, the solution to his equation, he knew, described an electron as a wave in three dimensions: perfect. But it described a *pair* of electrons as a *single* wave in six dimensions: nonsense. (Two waves in three dimensions would make intuitive, *anschaulich*, sense.) The wavefunction ψ "cannot and may not be interpreted directly in terms of three-dimensional space—however much the one-electron problem tends to mislead us on this point." He would spend the rest of his life trying to make a three-dimensional world built out of waves. Meanwhile, however, an eerie fact lay in the back of his mind: that two electrons together are somehow a single six-dimensional wave means they are tied together. Entangled.

Simultaneously came a paper from Born, published in *Zeitschrift für Physik*. Always a bit numinous and hazy, Schrödinger's waves under the gaze of Born were evaporating ever quicker. Born looked at how Schrödinger's equation described two particles colliding. He found that "one gets no answer to the question, 'What is the state after the collision?,' but only to the question, 'How *probable* is a specific outcome of the colli-sion?' " Born gave the wave its modern interpretation. It became a very useful tool, with a tenuous and undefined relation to reality—a mathe-matical sibyl in a cage that could be induced, with sufficient prodding and squaring, to prophesy probable fates of particles.

"Max Born betrayed us both," said Schrödinger with a half-grin. "He brought together particles, which I didn't want, and waves, which Heisenberg didn't want. And everyone has left Heisenberg and me behind." He laughed wryly. "But Born overlooks something. If the waves are there when you don't look, and particles when you do, it depends on the taste of the observer *which* he now wishes to regard as real."

"Like animals talking when we aren't around?" asked Itha.

Schrödinger laughed. "You don't believe that animals talk when you're not around."

Itha shook her head. "Only when I was younger."

"Maybe my waves are only as real as talking animals," he said as he rolled over to let the sun onto his sandy back. He was silent. Finally he said, "But the world, the physical world, can be much stranger than we tend to give it credit for."

Itha wrinkled her nose questioningly. "What do you mean?"

"Well, back during the Great War, I had to spend a lot of time on an observation post near the Italian border, and—"

"Did you have a gun? Did you kill anyone?" Roswitha, who had been lying on her back with her eyes shut, suddenly turned on her side to face Schrödinger.

Schrödinger looked at her in partial amusement and partial horror. "My father bought me two pistols, a small one and a large one, but luckily I never had to fire them at man or beast."

Itha said, "But you were going to explain about how strange the world is," giving her sister a withering look. Roswitha rolled her eyes and turned back to her supine sun-drinking position.

"Yes," said Schrödinger, smiling at her, "how strange the world is."

"You were on the observation post," prompted Itha.

"Right. We would be awake, under the stars, watching the mountain pass, you know."

"For the enemy," whispered Itha.

"Well, one night we saw lights moving up the dark Alpine slopes toward us, where there were no paths."

Ithi's eyes were wide. Even Roswitha, with eyes closed, turned her head slightly.

"And it was St. Elmo's fire," said Schrödinger, in a voice that reflected the wonder, with its underlying warm rush of relief, that he had felt. "Each barb on the barbed wire of the entanglements was haloed with fire. Only it's not even fire. The sharp barbs discharged the electricity in the air, melting the air into *plasma*—the same thing that makes up lightning and the aurora borealis and the sun and all the stars."

"So it doesn't even really belong on earth," said Itha.

"Yes, you feel that way when you see it. The Portuguese, in fact, call it '*corpo santo.*' "

"Holy body," translated Itha.

"Sacred body," said Schrödinger, nodding. "Sailors used to see it on their masts and think that it was a saint protecting them from beyond time and space. And Niels Bohr, another of the quantum physicists, wants to describe the atom in the same way. *He's* not surprised that the reality of the wave or particle might depend on *observation,* because he believes that a space-time description of atoms is impossible," he told her. "But this I reject *a limine.*"

She looked at him.

"From the doorstep," he explained. He smiled at her. "Sometimes I think Bohr has no idea what you yourself know perfectly well: that physics does not consist only of atomic research, science does not consist only of physics, and life does not consist only of science." She laughed, pleased to be talked to as an adult.

"What we cannot comprehend within space-time," he said, "we cannot comprehend at all." He smiled ruefully. "There *are* such things, but I do not believe atomic structure is one of them."

He was about to meet Bohr himself for the first time, and he had no idea what he was getting into.

11

This Damned Quantum Jumping

October 1926

Schrödinger lay coughing in the guest bed of the Bohrs' house in Copenhagen, his face red and sweaty from a fever. The Bohrs hovered over him, concern in their voices: Margrethe proffering tea and broth, Niels saying, "But, Schrödinger, you must admit that the quantum jumps *happen....*"

They had been arguing for three days now, ever since Bohr had picked Schrödinger up at the station. The whole day for Margrethe and Heisenberg, who was still living on the top floor of the institute, flowed around these two obdurate rocks. Every conversation, every meal, every walk was swallowed up by the continuous battering of Bohr against Schrödinger.

"Although Bohr was normally most considerate and friendly in his dealings with people," wrote Heisenberg, "he now struck me as an almost remorseless fanatic, one who was not prepared to make the least concession or grant that he could ever be mistaken. It is hardly possible to convey just how passionate the discussions were, how deeply rooted the convictions of each man—a fact that marked their every utterance."

Schrödinger turned uncomfortably in his bed. His voice hoarse, he said, for what felt like the thousandth time, "You must realize, Bohr, that the whole idea of quantum jumps is bound to end in nonsense." He went into detail, almost sitting up: "The jump from one energy level to another is either gradual or sudden—if it is gradual, how can we explain the persistence of sharp spectral lines? But if it is sudden, we have no idea how to explain what the electron might be doing during the jump."

Recall that the color of light and its energy are closely related by the fundamental quantum equation defining Planck's constant, $E = h\nu$ (energy equals Planck's constant multiplied by frequency). When a high-energy atomic electron falls to a lower energy level it emits the extra energy in the form of a telltale frequency of light—one of the narrow bands of color known as a "spectral line." If the electron moved from one energy level in an explicable fashion, the result would be a spectrum of color, a visual analog to the increasing hum of a car engine as the car accelerates from a standstill to sixty miles per hour.

The only explanation for the pure, clear, single frequency that actually occurred, however, seemed to be the inexplicable quantum jump, as if the car went from a standstill one moment to speeding the next, with nothing in between. Schrödinger's face, drawn and bare of his glasses, looked like someone else's. He sank down against his pillow. "The whole idea of quantum jumps necessarily leads to nonsense," he said.

Bohr, sitting at the end of the bed with one knee up on the blanket, looking calm except for his intense eyes, said, "What you say is absolutely correct."

Schrödinger looked at him warily, not turning his face on the pillow.

"But," said Bohr, "it does not prove that there are no quantum jumps. It proves only that we cannot imagine them, that the representational concepts with which we describe events in daily life and experiments in classical physics are inadequate when it comes to *describing* quantum jumps." Heisenberg, sitting in the corner, was nodding. "Nor should we be surprised to find it so, seeing that the processes involved are not the objects of direct experience." Bohr's forehead was creased with his effort to cover every nuance.

Heisenberg looked pleadingly at Schrödinger: *Can't you understand? Let go of the pictures.*

In the petulant voice of the invalid, Schrödinger almost whispered, "I don't wish to enter into long arguments about the formation of concepts; I prefer to leave that to the philosophers." One hand was fluttering around the edge of the blanket covering him, which he made into a fist. "*I wish only to know what happens inside an atom.*" Suddenly he looked straight at Bohr and said, reasonably, "I don't really mind what language you choose for discussing it. If there are electrons in the atom, and if these are particles—as all of us have believed—then they must surely *move* in some way. And it *ought to be possible in principle* to determine how they behave." Bohr's face was blank.

Schrödinger continued. "But from the mathematical form of wave mechanics or quantum mechanics alone, it is clear that we *cannot* expect reasonable answers to these questions." He drank from the cup of chicken stock that Margrethe had left by his bedside. "The moment, however, that we *change* the picture and say that there are *no discrete electrons*—"

Bohr shifted his position, looking troubled.

Schrödinger's voice strengthened. "The moment we say there are no point-particle electrons, only electron *waves* or waves of matter, then everything looks quite different. The emission of light is as easily explained as the transmission of radio waves through the aerial of the transmitter, and what seemed to be insoluble contradictions have suddenly disappeared."

Bohr said, "I beg to disagree." He sat up straight. "The contradictions do not disappear; they are simply pushed to one side." The very interaction between light and matter that had caused Planck, twenty-six years earlier, to introduce the concept of *quantization* requires discontinuity, he reminded Schrödinger.

"I cannot believe that the last word in describing atomic interactions is this picturesque pageantry of individual molecules swallowing or respewing whole energy parcels," said Schrödinger.

"Not to disagree, but merely to understand," said Bohr, "Schrödinger, I think you are far too wedded to pictorial ways of thinking."

"But then you, too, have so far failed to discover a satisfactory physical interpretation of quantum mechanics," said Schrödinger.

" 'Physical interpretation,' " said Bohr; "as I said, you place too much emphasis on things of that sort."

Schrödinger said firmly that there was no reason why he should not be able to ultimately come up with an explanation of the quantized interaction between light and matter: "an explanation that will admittedly look somewhat different from all previous ones."

"No," said Bohr. "There is no hope for that at all. We have known what Planck's formula means for the past twenty-five years," he stated with papal assurance. "And quite apart from that, we can *see* the inconsistencies, the sudden jumps in atomic phenomena, quite directly—when we watch flashes of light on a scintillation screen or the rush of an electron through a cloud chamber. You cannot ignore these observations."

Schrödinger was again lying flat on the bed. His eyes were closed. He said wearily, "If all this damned quantum jumping were really here to stay, I should be sorry I ever got involved with quantum theory."

Heisenberg looked at him with shock.

Bohr finally felt bad. Here was his guest, lying sick on the bed before him. "But the rest of us are extremely grateful that you did," he said. Schrödinger did not open his eyes. "Your waves have contributed so much to mathematical clarity and simplicity; they represent a gigantic advance over all previous forms of quantum mechanics."

Heisenberg's student and best friend, Carl Friedrich von Weizsäcker, in his beautiful reminiscence of Bohr in the thirties, described what it was like to be in an argument with Bohr over the meaning of your own ideas. Bohr, in the midst of exhausting work trying to help German Jewish physicists escape from Hitler to Denmark (and elsewhere), had asked the twenty-year-old to write a paper on an arcane subject.

When they met, Bohr "was late and looked infinitely tired. He pulled the paper from the pile and said, 'Oh, very good, very good; that is a very nice piece of work, now everything is clear. . . . I hope that you will publish it soon!'

"I thought to myself," remembered von Weizsäcker, "The poor man! He probably has had almost no time to read the paper.

"He continued: 'Just to clarify something, what is the meaning of the formula on page seventeen?'

"I explained it to him.

"Then he said, 'Yes, I understand that, but then the footnote on page fourteen must mean . . . so and so.'

" 'Yes—that is what I meant.'

" 'But then . . .' And so it went on. He had read everything.

"After an hour he was getting fresher all the time, and I came to a point where I had difficulty with an explanation. After two hours he was flourishingly fresh and complete master of the situation, full of naïve enthusiasm, while I felt that I was getting tired and was being driven into a corner.

"In the third hour, however, he said, triumphantly but without any trace of malice, 'Now I understand it. . . . The point is that everything is exactly the opposite of what you said—that's the point!'

"With due reservations about the use of the word 'everything,' I agreed that it was so."

Recovering at home, Schrödinger wrote about Bohr to his friend the crusty old experimentalist Willy Wien: "There will hardly again be a man who will achieve such enormous success, who in his sphere of work is honored almost like a demigod . . . and who yet remains—I would not say modest and free of conceit—but rather shy and diffident like a theology student. I do not necessarily mean that as praise, it is not my ideal of a man. Nevertheless this attitude works strongly sympathetically compared with what one often meets in stars of medium size in our profession. . . ." In spite of everything, "the relationship with Bohr, and especially Heisenberg—both of whom behaved towards me in a touchingly kind manner—was cloudlessly amiable and cordial."

He described how Bohr talked "often for minutes almost in a dream-like, visionary, and really quite unclear manner, partly because he is so full of consideration and constantly hesitates—fearing that the other might take a statement of his [i.e., Bohr's own] point of view as an insufficient appreciation of the other's (in this case, in particular, of my own work)."

In his approach to atomic problems, Bohr "is completely convinced that any understanding in the usual sense of the word is impossible. Therefore the conversation is almost immediately driven into philosophical questions, and soon you no longer know whether you really take the position he is attacking, or whether you really must attack the position he is defending."

Heisenberg's response to the debate was evangelical: "We in Copenhagen felt convinced toward the end of Schrödinger's visit that we were on the right track," he remembered, "though we fully realized how difficult it would be to persuade even leading physicists that they must abandon all attempts to construct perceptual models of atomic processes." He immediately whipped off a paper with "pedagogical" intent, he explained to Pauli, "against the Lords of continuum theory."

Bohr, too, was spurred into action. "We had great pleasure from the visit of Schrödinger," he wrote to a friend. "After the discussions with him it is very much on my mind to complete a paper dealing with the general properties of the quantum theory." This was hard, and a few weeks later, Bohr lamented to Kramers, "How little the words we all use are suitable in accounting for empirical facts, except when they are applied in the modest way characteristic for the correspondence theory."

Certain Bose-Einstein condensates, however—which so violate the spirit, if perhaps not the letter, of the correspondence principle—would, strangely enough, revive Schrödinger's original and thoroughly discredited interpretation of waves of matter and charge. When thousands of superchilled atoms or electron pairs are flowing in perfect, numinous union, matter really is made of waves: it is the particles that evanesce. Particles and probability reign only in the heat.

It was too early for this realization. Meanwhile, Max Born admitted to Schrödinger, "It would have been beautiful if you were right. Something that beautiful happens, unfortunately, seldom in this world." Pauli's take on Schrödinger's ideas was less elegiac: "Local Zurich superstitions." In November of 1926, this prompted an indignant reply from Schrödinger.

Pauli was kind and almost diplomatic in response. "Look on the expression as my objective conviction" that discontinuities are necessary to describe the quantum world. "But don't think that this conviction makes life easy for me," Pauli went on. "I have already tormented myself because of it and will have to do so even more."

Schrödinger appreciated this frankness. "We are all nice people, and are interested only in the facts and not in whether it finally comes out the way one's self or the other fellow supposed," he replied. ". . . Such capriciousness serves science better than uniformity."

Uniformity, however, was the direction quantum physics was headed in. Even Born, a good friend of Schrödinger's, ignored his pleas for open-mindedness. In fact, one of Schrödinger's letters struck Born as so obviously ridiculous, or insidiously dangerous, that he read it to his students on a walk in order to expose its perfidy, like a "commander-in-chief summoning his army."

On that military walk were P. A. M. Dirac and his polyglot American roommate, J. Robert Oppenheimer, who described the scene to the stuttering quantum mechanician Pascual Jordan. He thought that Born "put off some of the Americans by how sure he was that his way was the right way." That same November, in his characteristically brimful phrases, Oppenheimer gave the definitive description of Göttingen physics: "They are working very hard here, & combining a fantastically impregnable metaphysical disingenuousness with the go-getting habits of a wallpaper manufacturer. The result is that the work done here has an almost demoniac lack of plausibility to it, & is highly successful."

The demonically successful and implausible ideas wallpapering Göttingen at the time of Oppenheimer's letter were, of course, no longer matrices, but Born's waves of probability. "About me it can be told that physics-wise I am entirely satisfied," Born wrote to Einstein that same November, "since my idea to consider Schrödinger's wave-field as a *Gespensterfeld* [ghost wave] in your sense of the word proves more useful all the time." Einstein had never published the *Gespensterfeld* idea, but many physicists remembered it in some detail, including Bohr, who wrote of his first meeting with Einstein six years earlier: "Certainly his favored use of such picturesque phrases as 'ghost waves guiding the light quanta' implied no tendency to mysticism, but illuminated rather a profound humor behind his piercing remarks."

Whatever Einstein may have meant by *Gespensterfeld,* it was not the racetrack tote board that Born's interpretation of the Schrödinger equation represented, in which the odds for any given quantum event were all that could be known. In early December, Einstein wrote to the Borns, "Quantum mechanics is certainly imposing. But an inner voice tells me that it is not yet the real thing. The theory says a lot, but does not really bring us any closer to the secret of 'the old one.' I, at any rate, am convinced that *He* is not playing at dice." This, wrote Born, "came as a hard blow to me."

A few weeks later, in January 1927, Einstein dealt with the other side in a letter to Ehrenfest: "My heart does not warm to 'Schrödingery'—it is un-causal and altogether too primitive." And even his favorite hope—a field theory, where the fundamental idea is not waves or particles but smooth fields of force, like gravity or electromagnetism—was deserting him: "All attempts of recent years to explain the elementary particles of nature by means of continuous fields have failed," he wrote in a paper in the first days of 1927.

Max's wife, Hedi, wrote Einstein a few days later to thank him for his criticism of her recently completed play. Included in the letter was a drawing: "At the special request of the children, I enclose the result of a writing game we played yesterday. They are sketches which are made up as follows: one person draws the head, a second the trunk, a third the lower part of the body, and none of them knows what has been done by the others before him. Finally, a name is written underneath at random. You are going to be very pleased with your portrait."

Einstein wrote back, "What applies to jokes, I suppose, also applies to pictures and to plays. I think they should not smell of a logical scheme, but of a delicious fragment of life, scintillating with various colors according to the position of the beholder.

"If one wants to get away from this vagueness one must take up mathematics. And even then one reaches one's aim only by becoming completely insubstantial under the dissecting knife of clarity. Living matter and clarity are opposites—they run away from one another. We are now experiencing this rather tragically in physics."

12

Uncertainty

Winter 1926–1927

IT WAS LATE CHRISTMAS EVE in 1926. An almost transparent coating of snow had fallen on Blegdamsvej. In Bohr's institute, under its terra-cotta-tiled eaves, now fringed with frost, one light burned. Someone walking by would have seen two men standing in an attic room, far apart. But the snow on sidewalk and street was like an unwritten page. Everyone else from the institute had gone home days ago.

"What the word *wave* or *particle* means, one no longer knows—there are too many classical words for a quantum world," Heisenberg was saying doggedly. "But the mathematics, on the other hand, is now perfect: Dirac has made quantum mechanics as complete as relativity." (For Heisenberg, still squeamish about the word *matrix* and distrustful of waves, "quantum mechanics" *was* matrix mechanics—which Dirac and Jordan had massively extended at the beginning of December.) He sounded tired: "I just want to see where it leads." Almost quivering, he stood by the dormer window, while Bohr wandered around the room.

"Heisenberg," Bohr said, "the nicest mathematical scheme in the world won't solve the paradoxes we are up against. Classical words like *wave* and *particle* are all we have. This paradox is central. I first want to understand how nature actually avoids contradictions. Mathematical schemes like yours or Schrödinger's are just tools, and you shouldn't restrict yourself to just one tool. We must search for the deep truth—"

Heisenberg broke in. "I *don't* want to make any concession to the Schrödinger side that's not already contained in the quantum mechanics!"

Bohr's heavy eyebrows rose slowly, and he stopped in his tracks.

Heisenberg realized he was shouting, and admitted, "Perhaps this is just, psychologically, because I come from quantum mechanics." He looked up, his eyes still burning. "But at the same time I feel that whenever the people on the Schrödinger side add something to it then I always expect that it will probably be wrong."

He tried to laugh, apologetically, but Bohr was serious. "Heisenberg, you must understand the torture I am putting myself through in order to get used to the mysticism of nature," he said. Slowly pacing and slowly explaining, Bohr continued. He wanted to look at both approaches together, and look beyond them to their "epistemological lessons." His quiet voice pressed on, until Heisenberg felt he could stand it no longer, felt all the progress they had once made becoming disordered heat energy in his skull.

"I would try to say, 'Well, this is the answer,' " Heisenberg remembered. "Then Bohr gave the contradictions and would say, 'No, it can't be. . . .' In the end, shortly after Christmas, we both were in a kind of despair. . . . We couldn't agree and so were a bit angry about it." After a tense and exhausting month, Bohr went skiing in Norway, leaving Heisenberg in his attic room.

"So I was alone in Copenhagen."

It was a winter midnight. Heisenberg's chair scraped against the floor as he rose. Tiredly, he dragged his fingers along the desk, hit his pencil— sending it on an inevitable trajectory across the strewn paper, into an entirely deterministic free fall, accelerating at 9.8 meters per second squared, until it slapped gently against the floor and rolled into shadow.

With his face next to the windowpane, he could see through his own ghostly reflection the trees bowing over the entrance to Faelled Park behind the institute. The light was so low that the color-sensitive cones in the inner workings of his eyes could barely operate; what he saw were shades of glowing gray. The windowpane fogged with his breath. On the desk behind him lay dozens of false starts; electron trajectories sputtering into blank paper; Greek symbols that meant nothing; mathematical statements that had turned out to be lies.

The institute was quiet again at midnight, for the first time in weeks. Heisenberg stood by the window, thinking.

Before the impasse, while they were still constructing *gedanken* (thought) experiments in what Einstein called their thought-kitchens, he and Bohr had made progress of a sort. "Coming nearer and nearer to the real thing, only to see that the paradoxes . . . got worse and worse because they turned out more clearly—that was the exciting thing," Heisenberg recalled. "Like a chemist, who tries to concentrate the poison more and more from some kind of solution, we tried to concentrate the poison of the paradox. . . ."

In his mind he replayed the speedy, silent progress of an electron through a cloud chamber again and again. It was so simple, and no one could explain it, no one could describe it mathematically. The results of his attempts to do so lay on his desk, scribbled out and crumpled.

"When I realized fairly soon that the obstacles before me were quite insurmountable, I began to wonder," remembered Heisenberg, "whether we might not have been asking the wrong sort of question all along. But where had we gone wrong? The path of the electron through the cloud chamber obviously existed; one could easily observe it. The mathematical framework of quantum mechanics existed as well, and was much too convincing to allow for any changes."

He felt trapped. He wondered if Einstein ever felt this brain pinch. His frustration was so palpable it was like another presence in the room.

Don't you see, Heisenberg? It is theory which first determines what can be observed.

Einstein's words rang out in his cramped mind, and then, with incredible relief, the boards that bound it began to fall away. *Don't you see?*

Heisenberg almost ran out of the little attic room, down the stairs of the institute, the door slamming behind him as he emerged into the fresh cold air and emptiness of the park. He walked between the bare trees at the gate, his booted feet leaving dark footprints, melting the frost that fringed the grass. *It is theory, it is theory, it is theory which first determines what can be observed.*

"We have always said so glibly," Heisenberg told his frustration, or the trees, or Bohr, or Einstein, "that the path of the electron in the cloud chamber can be observed." He turned to see his footprints among the frost, and he thought of the electron shooting through the cloud chamber, leaving its footprints of dew behind, small condensed clouds. "But perhaps," he continued slowly, "what we really observed was something much *less*. Perhaps"—he was walking faster now, his breath, like the electron, leaving clouds behind him—"we merely saw a series of discrete and ill-defined *spots* through which the electron had passed. In fact, all we do see in the cloud chamber are individual water droplets which must certainly be much larger than an electron." He stopped again. "The right question should therefore be: Can quantum mechanics represent the fact that an electron finds itself *approximately* in a given place and that it moves *approximately* with a given velocity, and can we make these approximations so close that they do not cause experimental difficulties?"

Pauli had written him a letter in October which "continually makes the rounds in Copenhagen" with Heisenberg, Bohr, and Dirac all "scuffling" for it (as Heisenberg enthusiastically wrote back a week after receiving it). A phrase from this letter was now resonating in his mind:

"The first question is . . . why can't the p's *as well as* the q's be prescribed with arbitrary precision. . . . One can look at the world with the *p*-eye and one can look at it with the *q*-eye but when one would like to open both eyes, then one gets dizzy" (*schwindlig* again).

Why can't the *p*'s—momentum—and *q*'s—position—both be known at the same time? This consequence of the new mechanics was a drastic departure from the old Newtonian system, where knowing the *p*'s and *q*'s was the start to every solution. Once you knew the position of a billiard ball on the green felt, and its momentum, you could tell exactly what the ball would do next. Knowing the position and momentum of all the other billiard balls on the table, and the position and momentum of the cues, you could tell how the whole game would unfold. In its infinite wisdom (having pinned down the position and momentum of every particle in the universe) the Marquis de Laplace's hypothetical Demon knew the whole future this way.

This was a concept that had become wildly popular in the late eighteenth century, with the rise of science and the fall of kings. It was called determinism, and it was (then) thought of as the newly chic scientific attitude: the Great Clockmaker had wound up this great clock, the universe, and everything was playing out exactly as it was constructed to do. In these similarly unsettled times a century and a half later, the tide had turned against causality, and ever since Germany's sudden defeat in 1918, Weimar intellectuals, socialites, and back-to-nature Boy Scout troops like Heisenberg's yearned for an "irrationality" and "holism" that would transcend the mechanical chain of cause and effect. Fitting the spirit of the age, Heisenberg's quantum mechanics had no determined end. It questioned the Clockmaker and defied the Demon.

Heisenberg walked back across the frost to the door of the institute and up the stairs. He sat down at his desk, brushing all the false starts out of his way, and wrote down an equation almost immediately:

$$\Delta p \Delta q \geq h$$

". . . *one can look at the world with the p-eye and one can look at it with the q-eye, but when one would like to open both eyes, then one gets dizzy.* . . ."

This is the famous Heisenberg uncertainty principle. It states that the uncertainty in the exact momentum of a particle times the uncertainty in the exact position of a particle must be greater than or equal to the tiny number that is Planck's constant (the unchanging, mysterious number that appears in every quantum equation). It is paralleled by the same equation relating energy to available time. *i.e.* $\Delta p \Delta q \geq \hbar / 2$ ✓

Refined in later years so that *h* is now divided by 4π, the essential point is that, though small, the number on the right is not and never will be zero. And in the way of equations like this, if there is absolutely no uncer-

could
(is better here)

tainty in the measurement of the momentum of a particle, then there must be an *infinite* amount of uncertainty in the measurement of its position: it could be anywhere. (The concept of "measurement" here became a principle of physics—a weird, unwanted, and seemingly irreversible intrusion, like the moment the ship's rats fatally decamp on a South Sea island hitherto populated only by marsupials.)

When the particle is pinpointed in one place, its momentum has fuzzed out, becoming indeterminate, so that no one—not even Laplace's Demon, or God—knows what it is, because it *isn't*. Against Heisenberg's wishes, the wave nature of quantum subjects has here overwhelmed their particle nature. Waves do not have a specific position and momentum; these two attributes are linked in exactly the same way that Heisenberg's principle describes (a wave in a tiny box—a specific position—caroms off its walls in a wild and jumbled mess; a wave left to spread out in the world—a blurred position—has room to have a specific momentum).

A world wavelike and causal when unseen, making quantum leaps into particularity when seen: this was the dilemma of quantum mechanics, leading straight to entanglement. But Heisenberg did not want to think about waves at all. About particles, he wrote in his paper, "In the strong formulation of the causal law—'If we know the present exactly, we can predict the future'—it is not the conclusion but rather the premise which is false."

Before the end of February, he had written the whole thing up in a fourteen-page letter to Pauli. "The solution can now, I believe, be expressed pregnantly by the statement: the path only comes into existence through this, that we observe it." But was it all "old snow"? he asked Pauli. Slater, in his discussion of the relationship between time and frequency central to the Bohr-Kramers-Slater paper, had introduced the time-energy version three years before Heisenberg, though not with the same interpretation. Heisenberg needed Pauli's "relentless criticism": "To make anything clearer, I must write to you about it," he told Pauli.

In a somewhat wary note to Bohr, he wrote, "I believe that I have succeeded in treating the case where both p and q are given to a certain accuracy. I have written a draft of a paper about these problems" (did Bohr here expect him to write, "which I am about to send to you"?) "which yesterday I sent to Pauli."

Bohr spent that month in scarf and ski goggles, slowly cutting Christies down snowy Norwegian slopes. Bohr's new right-hand man, Oskar Klein, described him sympathetically: "He was very tired at that time, and I believe that the new quantum mechanics caused him both much pleasure and very great tension. He had probably not expected that all this would come so suddenly but rather that he himself perhaps might have contributed more at the time. At the same time he praised Heisenberg almost like a kind of Messiah and I think that Heisenberg himself understood that that was a bit exaggerated."

Bohr wished that Pauli had been there. Pauli would have said to Heisenberg, *Stop it, you are being an idiot,* and to Bohr, *Shut up!* He would have stormed around as the wrath of God personified, quivering throughout his whole rotundity; with his barbed wit he would have pierced every hope and desire until there was nothing left but unity, the way Bohr loved it to be.

Poles against his sides, Bohr turned in a smooth swoop. The snow flew from his skis, sparkling up to sting his face. He began to feel those past three months slide away behind him like his blue-shadowed tracks. Then the wind picked up and even his tracks were gone, and Bohr's mind felt clean and empty as if the snowy air had blown through it, too. The hill steepened, his turns quickened, his smile freezing to his face.

A long-dismissed thought was sitting in the back of his mind: *There exist particles* and *waves.*

Bohr's head came up and his eyebrows pulled together as he felt this thought emerging in a different way than before. As he skied on, the thought came again: *There exist waves and particles together.*

A turn. *There exist particles, and waves, both necessary, but not at the same time.*

A turn. *Why do we imagine that we can observe physical phenomena without disturbing them? There exist particles when we look for particles, and waves when we look for waves.*

A turn. *All our ordinary verbal expressions,* he thought, *bear the stamp of our customary forms of perception—particle, wave; space and time, causality . . . and to these customary forms of perception, quantization is an irrationality.*

A turn. *What if this is a fundamental limitation? What if the very nature of the quantum theory forces us to regard pairs of ideas as complementary but exclusive features of the description?* Bohr felt an overwhelming sense of peace and acceptance. *There exist waves and particles; they both exist in a complementary way.*

He slowly swooped down the mountain, scarf flapping behind: one warm, bundled human with dark goggles and red nose; one spot of ordered energy through all the vast disordered whiteness.

Glowing like Moses, he came down from the mountain with his new commandment. He returned just as Heisenberg was about to send off his paper. Bohr took a look at it and decided it was just a special case of his important new concept of complementarity. Besides, Heisenberg had tried so hard to avoid waves that—experimental physics never having been his strong suit—in his example he hadn't even explained correctly how a microscope works. Heisenberg should not publish.

Heisenberg was thoroughly irritated, and sent the paper off anyway. But Bohr, in his slow, intense way, hounded him, seconded by his assistant Klein (like all the other assistants, self-sacrificially devoted to Bohr's perspective, and a little jealous of Heisenberg). Bohr told Pauli that he would pay for the ticket if he would come to Copenhagen. (Pauli couldn't.) Heisenberg felt he was going crazy with frustration: "It ended by my breaking into tears because I just couldn't stand this pressure from Bohr."

In the end, Heisenberg's paper was published in May with the correct microscope example (and with thanks to Bohr for helping him with it) and the note that "recent investigations by Bohr have led to points of view that permit an essential deepening and refinement of the analysis of the quantum mechanical relationship attempted in this work."

Meeting Schrödinger on his own territory, it was titled "On the *Anschaulich*"—visualizable, intuitive—"Content of the Quantum . . . Mechanics." Bohr set to work on his own paper. "Bohr," wrote Heisenberg to Pauli, "wants to write a general paper on the 'conceptual basis' of the quantum theory, from the point of view that 'there exist waves and particles'—when one starts like that, then one can of course make everything consistent." Heisenberg preferred the way of Dirac and Jordan—"*unanschaulicher* and more general." At any rate, "There are essential differences in taste between Bohr and me about the word *Anschaulich.*"

Heisenberg's first draft of his uncertainty paper arrived at *Zeitschrift für Physik* late that March, a few days before the two hundredth anniversary of the death of Newton. This occasion did not mean much to Heisenberg, but had spurred Einstein into composing a pair of tributes—one for the German-speaking world, and one for the English. The German-language essay ended with bravado and uncertainty: "Who would presume today to decide the question whether the law of causation . . . must definitely be given up?"

Two weeks later, on April 13, Bohr sent Einstein a copy of Heisenberg's uncertainty paper, as Heisenberg had asked him to do. In his accompanying letter, Bohr emphasized how complementarity deepened the "significant . . . exceptionally brilliant . . . contribution" of Heisenberg. In words that amounted to a challenge Einstein would not be able to resist,

Bohr explained that Heisenberg had now reconciled particle with wave, "since the different aspects of the problem never manifest themselves simultaneously."

Bohr wanted to make it clear that, despite the mild sound of "complementarity," he was talking about a violent disjunction. We have "only the choice between Scylla and Charybdis"—the boat-crunching rock and the man-eating whirlpool: particle phenomena and wave phenomena— "depending on whether we direct our attention to the continuous [smooth] or the discontinuous [quantized] features of the description."

Heisenberg still saw the situation differently. As Einstein sat down to read the uncertainty paper for the first time, its twenty-five-year-old author, in the backseat of a Berlin taxi, was telling the solemn fifteen-year-old son of the German ambassador to Denmark, "I think I have refuted the law of causality."

Einstein responded speedily to these twin challenges. He delved into the Schrödinger equation to see if Max Born's statistical interpretation, coupled with Heisenberg's uncertainty principle, really was the last word. A month later, he was whipping off a relieved postcard to a skeptical Born: "I handed in a short paper to the Academy, in which I show that one can attribute quite *definite movements* to Schrödinger's wave mechanics, without any statistical interpretation. This will shortly be published in the minutes of the meetings. Kindest regards."

The short paper was titled "Does Schrödinger's Wave Mechanics Determine the Motion of a System Completely, or Only in the Sense of Statistics?" "As is well known," he began, "the opinion currently prevails that, in the sense of quantum mechanics, there does not exist a complete space-time description of the motion of a mechanical system." Einstein believed that in this paper he had shown the reverse, refuting Heisenberg's uncertainty.

After two months of fighting, Heisenberg had just acquiesced to Bohr's concerns and demands over the uncertainty paper, when Max Born relayed the news of this new attack. Days after sending off the final copy of the uncertainty paper, he wrote worriedly to Einstein of this "paper in which you . . . advocate the view that it should be possible after all to know the orbits of particles more precisely than I would wish." He was sure that something was wrong, and a month later was writing to Einstein in words that amounted to the quintessential exchange of views between the two men, using first Einstein's and then his own favorite phrases: "Perhaps we could comfort ourselves that the dear Lord could go beyond it and maintain causality. I do not really find it beautiful, however, to demand more than a physical description of the connection between experiments."

It was not Heisenberg's avowed conviction, but a question from the experimentalist Walther Bothe that had already scotched Einstein's new paper. Bothe was a meticulous and thin-skinned physicist who as a captured German cavalry officer had spent much of his young manhood as a prisoner of war in Siberia. He passed the time of his exile learning Russian, working on physics in his head, and courting and winning a Russian wife. He returned to prove the Bohr-Kramers-Slater theory wrong, and causality correct, with two different experiments (one with his mentor, Rutherford's assistant Hans Geiger) in 1925 and 1926, having "the singular good fortune of being able to discuss the problem constantly with Einstein."

In 1927, Einstein was writing a postscript to his paper. Bothe had alerted him to the odd behavior of coupled systems in Einstein's theory: their total motion did not seem to have anything to do with the motions of each component system—a situation that must be rejected, Einstein believed, "from a physical standpoint." He thought he might avoid such a fate for his theory with a few refinements.

As the Prussian Academy was printing up its minutes, however, Einstein called. Upon his request, his paper was removed.

13

Solvay

1927

In October, less than six months after Heisenberg published his uncertainty principle, thirty quantum physicists gathered in Brussels for what would become one of the most famous conferences ever. A few minutes of those history-making days were captured on film. Nancy Greenspan, Born's biographer, describes the jittery black-and-white video: "Here is Max Born exiting from an ornately grilled door; Niels Bohr animatedly conversing with a natty Erwin Schrödinger; Werner Heisenberg flashing a youthful, cocky grin; Paul Ehrenfest making faces; a rumpled Albert Einstein nodding in acknowledgment to the anonymous cameraman; and a boyish Louis de Broglie looking about. It is the Solvay Congress, Brussels, 1927. In what appear to be the first days, all are smiling, determinists and non-determinists alike." But a few days later the mood has changed. "The last frames of these same men leaving the conference and descending the stairs show a few pinched smiles and many aggravated looks."

Heisenberg would remember it as a great success for Bohr, Pauli, and himself, and the beginning of the spread of "*der Kopenhagener Geist*"— the spirit of Copenhagen: classical ideas used in complementary ways. But the apparent failures of the conference would, from the point of view of the history of entanglement, prove far more important. The heretical suggestions of de Broglie and Einstein in Brussels would become the direct intellectual ancestors of John Bell's surprising theorem four decades later.

De Broglie's speech came early in the conference. He described a theory by which a wave guided—or, in de Broglie's word, "piloted"—the particles, as Slater had tried to propose four years earlier. This meant that he added *position* for the particles in the wave, and those positions came to be called "hidden variables"—hidden from quantum mechanics. "The indeterminist school," de Broglie recalled, "whose adherents were mainly young and intransigent, met my theory with cold disapproval." In the other camp, "Schrödinger, who denied the existence of particles, did not follow me at all. . . ."

Then Einstein walked up to the front of the room, his stiff collar neatly standing up with the tips turned down in the best old-fashioned and out-of-date way. "I must apologize," he said, "for not having penetrated quantum mechanics deeply enough. Nevertheless, I would like to make some general remarks.

"Imagine we have an electron traveling toward a screen." He turned around to the blackboard and picked up a piece of chalk. In a few lines, the chalk skipping dustily, he had the path of the electron (dash-dash-dash) and the screen (slash). He half turned. "In the screen is a slit"—turning back and smudging the center of the straight line with a finger, "which will scatter the electron." He drew a series of half circles spreading from the smudged hole: the Schrödinger wave representing the electron will spread out from the little opening. "Beyond the slit is another screen"—Einstein's chalk made a second slash—"which will catch the electron."

Einstein looked like a magician about to pull a rabbit out of a hat, and at the same time like someone who did not believe in magic. "If I see *this one* electron arrive *here*"—he tapped a point on the top of the second slash—"then it is out of the question that it simultaneously arrived *there.*" He moved his chalk to another point along the screen.

His eyes narrowed. "But the wave—the Schrödinger wave interpreted as giving the probability for *this* particle to be situated at a certain place—the wave covers the whole screen, and not just one point on it," said Einstein, gesturing at the semicircular ripples billowing from the smudged gap.

His quiet voice pierced the room: "This interpretation presupposes a very particular mechanism of instantaneous action-at-a-distance to prevent the wave from acting at more than one place on the screen." (Recall that relativity denies simultaneity any meaning, declaring that no information can travel faster than the speed of light.) "It seems to me that this difficulty cannot be overcome unless the description of the process in terms of the Schrödinger wave is *supplemented* by some detailed specification of the location of the particle."

He looked over to where de Broglie sat, staring earnestly at his hero. Einstein nodded and said, "I think that Monsieur de Broglie is right in searching in this direction." De Broglie, who had received no encouragement from anyone, melted with gratitude. Einstein repeated, "If one works only with Schrödinger waves, the interpretation contradicts the principle of relativity." He sat down.

Everyone was silent; then suddenly everyone was crowding around him with something to say. Lorentz tried to keep order, but Einstein had struck a nerve. In the midst of the fray, Ehrenfest remembered the Tower of Babel, and wrote on the much-smudged blackboard: "The Lord did there confound the language of all the earth."

Bohr finally gained the floor. "I feel myself in a very difficult position," he said, "because I don't understand what precisely is the point which Einstein wants to make. No doubt it is my fault." Kramers, next to Bohr, was scrivening down the words from the master. "I would put the problem in another way," Bohr continued. "I . . . do not know what quantum mechanics is. I think we are dealing with some mathematical methods which are adequate for the description of our experiments. Using a rigorous wave theory we are claiming something which the theory cannot possibly give." In other words, the math of quantum mechanics—though indubitably a wave equation—could come far nearer to describing the quantum world than any theory involving actual waves, such as Schrödinger's or de Broglie's. "We must realize that we are—*away,*" said Bohr slowly, "from that state where we could hope of describing things by classical theories."

"Ominously," as Andrew Whitaker, John Bell's biographer, writes of this exchange, "Einstein and Bohr were already talking past each other." "Some mathematical methods adequate for the description of experiments" was hardly any physicist's motivation, even Bohr's—Bohr, who would come to believe that complementarity solved many of the intellectual woes of humanity.

Almost no one really followed Bohr through the thicket of words and ideas as he launched into a discussion of complementarity. "Naturally, once again the awful Bohr incantation terminology," Ehrenfest wrote home to his students. "Impossible for anyone to summarize. (Every night at one a.m., Bohr came into my room just to say ONE SINGLE WORD to me, until three a.m.)"

Even though Bohr could rarely turn a phrase neatly, or halt a run-on sentence, he still was able somehow to dominate through sheer conviction and charisma. When Ehrenfest described his overall impression of the conference, his tone was very different (with his whimsical capitals): "BOHR towering completely over everybody. At first not understood at all . . . then step by step defeating everybody."

Bohr could not, however, defeat Einstein's simple objection, which lives on unsolved as "the collapse of the wavefunction" or, more generally, "the measurement problem." The wave corresponding to any particle spreads out over the whole world, but the particle itself is discovered in one tiny discrete location, and nowhere else. The vast wave collapses down to the size of a single particle. Quantum mechanics does not describe this moment of discovery, but only gives the probability that the discovery will occur in the place it does. There is no explanation of the particle's doings while unobserved (or, alternatively, of its genesis, if the act of measurement somehow created a particle where before there was only a wave).

From a wide-ranging, matterless probability wave to an incarnation as one specific being—it is a postulate so nonintuitive as to sound like the tenet of a religion. Einstein, who did not believe in a personal deity either, found this strange. As if the moment some mystic had the idea of searching for God—*poof!*—he's not a huge, disembodied, omnipresent spirit-mind anymore, but just a little tyke in a barn in a specific hamlet in a specific countryside.

The measurement problem is one symptom of the nonseparability of quantum systems; another, to which the conversation naturally turned, is the uncertainty principle—uncertainty is what you get when you try to treat nonseparable things as separate.

Couldn't you follow the electron in its flight to the screen? Einstein wondered. What would repeated measurements of its state show us? Bohr suggested for this purpose another screen—this one with two slits—inserted between the original slit screen and the collecting screen. You could then measure which of the two slits the electron went through on its way to the final screen. As they elaborated on this thought-experiment, Bohr eventually showed that the very apparatus that might specify the location of the electron in its flight would itself suffer uncertainty, making that location unclear, just as complementarity and the uncertainty principle dictated.

"Like a game of chess, Einstein all the time with new examples," Ehrenfest described to his students the subsequent Bohr-Einstein debate, which filled all the extracurricular conference time. Considering his own distaste for both unclear speaking and smelly pipe smoke, Ehrenfest found apt metaphors at hand: "Bohr from out of the philosophical smoke clouds constantly searching for the tools to crush one example after the other. Einstein like a jack-in-the-box; jumping out fresh every morning. Oh, that was priceless."

One day, after Einstein had asked for the umpteenth time if Bohr really believed God played dice to determine the future, a smile dawned on Bohr's face. "Einstein," he said, "stop telling God how to run the world."

"After the same game had been continued for a few days," remembered Heisenberg, Ehrenfest looked over at Einstein in bemused semi-exasperation. "Einstein," he said, "I am ashamed of you; you are arguing against the new quantum theory just as your opponents argue about relativity theory." Heisenberg gave Pauli a look that said: *Finally someone's said it.*

Einstein looked at Ehrenfest with raised eyebrows and a little smile.

Ehrenfest, suddenly serious, said, "I will not be able to find relief in my own mind before concord with you is reached."

Back in Holland, he told his old student Samuel Goudsmit, "I must make a choice between Bohr's and Einstein's position." Goudsmit realized with a start that Ehrenfest was crying. Ehrenfest looked away and then looked at Goudsmit again: ". . . and I cannot help but agree with Bohr."

14

The Spinning World

1927–1929

Max Born

Einstein, ever the bird of passage, was going straight to Paris from Brussels, so de Broglie, riding the same train, got to spend a few more hours with "the idol of my youth." Among the crowds on the arrival platform of the Gare du Nord, their train panting and sweating beside them, Einstein said, "This exaggerated turn toward formalism which quantum physics is taking worries me." He looked at de Broglie. "I really believe that all physical theories—math aside—ought to lend themselves to so simple a description that even a child could understand them."

De Broglie smiled politely, but his raised eyebrows betrayed his skepticism.

Einstein, amused, saw this and ignored it.

They began walking together toward the exit, Einstein nodding to himself with a faint smile. "Almost twenty years ago now, Planck asked me what I was working on," he said, "and I described the bare bones of the general relativity theory which had just begun to emerge in my mind. He said to me, 'As an older friend, I must advise you against it,' "— Einstein, imitating Planck, looked solemn and waved his index finger— " 'for in the first place, you will not succeed; and even if you succeed, *no one will believe you.*' " Einstein's eyes twinkled. "So, as an older friend," he said, "I must advise you against the course you have taken." He gave de Broglie a half-smile. "In the first place, you will not succeed; and even if you do—"

De Broglie finished, laughing: "No one will believe me."

As they walked out of the station doors and went their separate ways, Einstein called out, "But carry on! You are on the right road!"

De Broglie watched Einstein turn and fade into the crowd on the Paris street. He smiled broadly, his odd face lighting up in a way that was so innocently elated that a few people bustling past him stopped to look at him again. He didn't see them: *You are on the right road.* Einstein's voice rang in his mind. *Carry on!*

A few months later, however, the discouraged and exhausted de Broglie converted to the unstoppable spirit of Copenhagen. This left only Einstein and Schrödinger standing before the juggernaut. Schrödinger was still a skeptic in late May of 1928, when he wrote to Einstein in frustration and confusion about a discussion he had had with Bohr by letter over the uncertainty principle. He was again wondering if "position" and "momentum" weren't just inapplicable concepts in the quantum world.

Bohr had responded by telling him that there was no need for any new understandings; the principle of complementarity removes all the difficulties. He asked Schrödinger to pass his reply on to Einstein, not realizing how incomprehensible his ideas were to them.

"What is it that we human beings depend on? We depend on our words. We are suspended in language," Bohr famously said years later. "Our task," he maintained, "is to communicate." His old friend Aage Peterson explained that "Bohr was not puzzled . . . by questions as to how concepts are related to reality. Such questions seemed sterile to him." For Bohr, there was no meaning beyond the fabric of classical words spun, weblike, over the quantum abyss.

Einstein was confined to bed, diagnosed with enlargement of the heart. He wrote back to Schrödinger the very next day,

Dear Schrödinger,
I think you have hit the nail on the head. . . . Your claim that the concepts p, q will have to be given up, if they can only claim such

"shaky" meaning, seems to me to be fully justified. The Heisenberg-Bohr tranquilizing philosophy—or religion?—is so delicately contrived that, for the time being, it provides a gentle pillow for the true believer from which he cannot very easily be aroused. So let him lie there.

But this religion has so damned little effect on me that, in spite of everything, I say

    not: E and ν

    but rather: E or ν;

    and indeed: not ν, but rather E (it is ultimately real).

But I cannot make head or tail of it mathematically. My brain is also too worn out by this time. If you would give me the pleasure of a visit from you again sometime it would be good of you and very fine for me.

Best regards from
A. Einstein

The relation $E = h\nu$ is fundamental to quantum mechanics: it defines Planck's constant h. When an electron falls from one energy level to another in a quantum leap, the energy it loses is carried away all at once, in a packet, and the energy (E) of this light-*quantum* is proportional to the frequency (ν) of a corresponding light *wave*.

Einstein had written to Schrödinger's friend Weyl a year earlier, "In the depths of my soul I cannot reconcile myself to this head-in-the-sand conception of the half-causal and half-geometrical. I still believe in a synthesis of the quantum and wave conceptions, which I feel is the only thing that can bring about a definitive solution." This synthesis was not Bohr's duality ("*E and* ν"). Einstein was searching, as in all his endeavors, for a unified view of the world, not for "complementary" ways of describing it.

"It is very important," he wrote to Weyl, "to know whether or not the field equations are . . . refuted by the facts of the quanta. One is indeed naturally inclined to believe this and most do believe it." But if Einstein could find a field that would subsume both waves and particles, then quantum theory would become part of general relativity, the most beautiful field theory of all.

In January of 1928, four months before, quantum theory had made its uneasy truce with special relativity (the part of relativity that deals with frames of reference moving at constant speeds), in an astonishing equation from Dirac. The trail to the Dirac equation had actually started four years earlier (before quantum mechanics), when Pauli and then Heisenberg unleashed their ridicule on the idea of the spinning electron. "This after all, has nothing to do with quantum mechanics or politics or religion—it is just classical physics," Pauli groused in a letter to Kramers. "I appeal to you as a significant scientist . . . to destroy this heresy." He signed the letter "*Geissel Gottes*"—God's Whip. "P.S. This title Ehrenfest has given me. I am very proud of it!"

The problem was that the electron was so small that if it spun on its axis, its equator would be moving much faster than the speed of light. It also mysteriously took two full rotations to get back to where it started. Pauli, in his "*komische* reflection" of 1925, had described the same phenomenon as a "peculiar, classically non-describable, suggestive two-valuedness"—a phrase that did not take the world by storm.

Instead, everyone was talking about the "spinning electron" or the "electron magnet" (since a magnetic field arises wherever an electric charge is in motion). Bohr was converted to "a prophet of the electron-magnet gospel" (as he joked to Ehrenfest) by a train trip during which at every stop he was met by a different pair of physicists intent on convincing him of opposite conclusions, from Pauli and Stern's protestations at the Hamburg station to Einstein and Ehrenfest's decisive reasoning on the Leiden platform: Einstein was able to show how the spinning electron could coexist fruitfully with relativity.

Heisenberg made a bet with Dirac that it would take three years to understand the spinning electron. Dirac bet that it would take three months. As it turned out, in three months, Pauli developed three matrices that described what he still deplored calling "spin": one matrix to describe the phenomenon in each of the three perpendicular Cartesian directions.

Dirac had not, however, made any money: Pauli's matrices were not compatible with the theory of relativity, and in fact, Pauli had become convinced that a relativistic spin theory was not three years in the future, but an impossibility. Another bet ensued, this one between Pauli and Kramers, who threw himself into a relativistic equation for the spinning electron.

Either fate, or a roll of the dice-throwing God, however, had sadly decreed that Kramers, throughout his long career, should always be scooped by some frighteningly brilliant young genius, and this time— just as he finished his horribly complicated relativistic theory of the spinning electron—it was Dirac. In 1926 and 1927, Ehrenfest was already telling Dirac that his papers were "crossword puzzles," while Einstein was complaining to Ehrenfest, "I have trouble with Dirac. This balancing on the dizzying path between genius and madness is awful." Dirac's equation of January 1928 raised the genius and madness to even giddier heights. In one stroke he transformed Pauli's matrices almost beyond recognition into a relativistic equation for the so-called spinning electron that solved all previous problems effortlessly.

But the equation had a second solution, a positive *antimatter* electron, which had never been heard of before and was clearly science fiction. "The saddest chapter of modern physics is and remains the Dirac theory," wrote Heisenberg to Pauli that summer. "The spinning electron has made Jordan melancholic." Pauli retired from physics, as he told Bohr in June, to await "a fundamentally new idea" and write a utopian novel. Pauli got as far as his title (*Gulliver's Journey to Urania*) and outline before Heisenberg threw aside all forms of magnetism to enlist his aid on a unified field theory. They would avoid the Dirac equation for the next several years.

That fall, Einstein proposed de Broglie, Schrödinger, or Heisenberg for a Nobel Prize: "a difficult case," he wrote. "With respect to achievement, each one of these investigators deserves a full Nobel Prize although their theories in the main coincide in regard to reality content. However, in my opinion, de Broglie should take precedence, especially because his idea is *certainly* correct, while it still seems problematic how much will ultimately survive of the grandiosely conceived theories of the two last-named investigators."

De Broglie won the next year, "for his discovery of the wave nature of the electron."

Slowly striding on long wooden skis through the Black Forest's blue shadows, Max Born is falling apart. Three years ago he explained Heisenberg's breakthrough in terms of matrices; two years ago he explained Schrödinger's breakthrough in terms of probabilities; now he has allowed the inexplicable outside world to distract him from physics. He feels as if he has been taking poison through his ears for the last month.

It is the last winter of the roaring twenties. It is the coldest winter of recent memory; you could walk the three miles across Lake Constance from Germany to Switzerland. The sanatorium by the frozen lake where he had gone to heal his frayed nerves had only frayed them further, with his fellow patients luxuriating in a chatter of hate and harsh laughter. They talked of the poor fatherland so trampled in the muck, and blamed the Jews. They talked of Adolf Hitler, who would bring back the greatness of Germany.

So Born left, and now has come to rest here in the heart of the Black Forest in a tiny town called Königsfeld. The trees (their bark ancient, Jurassic) are covered with snow, boughs bowed down; a few clumps of flakes fall down his neck, chill his spine. Out there, beyond this solitary peace, *things fall apart, the center cannot hold,* civilization is crumbling, the old polite veneer is cracking and flaking and under it is the ugly beast.

The early sun sets among the crests of the trees—with infinite wistfulness and lingering and yet, seeming to have happened all at once—the colors change from white to long purples. He is tired; he has not skied as long as perhaps he could have before, in the days when confusion was so much cleaner. Confusion in those days meant, *What does quantum mechanics mean?*, not *How evil will the future be?*

At the end of the trail, he unstraps his skis from his boots, pulls his mittened hands out from the loops of his poles, collects all the shafts of wood into one pile over his shoulder, and walks into town. The twilight makes the buildings seem to huddle together.

Ahead of him stands a tall and spacious white church. Its windows are dark but someone is playing an organ. Born leans his skis against the wall, climbs up the outside stairs, and opens a green door. He finds himself directly underneath a delirious blare of organ sound that falls from the choir loft like a mantle around him. He sits down, leans back, stretches his legs. A globe-shaped lamp on the wall by the organ glows on gilding and baroque swirls and the bent head and white hair of the man in the organ loft.

Sometimes he plays over and over again one series of bars, as if he is ironing the wrinkled music with smooth, hot strokes from an iron—and then plays on, the fabric of music cool and lovely, hung out on the line, flapping in the breeze. Born's eyes are almost closed; the rolling of the waves of sound seems to be warming the very foundation of the church, eddying in the arches of his feet, in the mazes of his ears.

When the the organist stops, Born is so far out on the waves of music that he almost does not know what has happened. He can hear, in the thick, reverberating silence, hands collecting sheets of paper, the scratching of a careful pencil, marking some detail of performance. Then, out of the darkness of the organ loft, a man with a bushy mustache and an unruly mop of hair descends. Born is not fully awake: *Ahh,* he thinks, *it's Einstein.*

"So, I had a one-man audience," says the man, walking toward Born.

Born gets up to shake the organist's hand. "I'm so glad that I happened to be walking by."

The organist smiles. His face is not Einstein's, but it is a face as well known and well beloved.

Born says, "It's such a pleasure to meet you. . . . Dr. Schweitzer, isn't it? I'm Max Born."

Albert Schweitzer smiles. "Max Born. Well, it is quite an honor to meet you, too. . . . Dr. Livingstone, I presume?" He laughs a little, a deep, organlike laugh. "I'm going home now. Might we walk together?"

Born and Schweitzer meet often in the ensuing days for long tramps in the snow. They talk about physics. They talk about Schweitzer's hospital where he lives in equatorial West Africa. He is raising money for it by playing Bach throughout Europe for six months. Born, finally, begins to heal.

In March of 1929, Einstein is alone as well. He has just built a little half-timbered house on a piney patch of land above the red-roofed village of Caputh, near Berlin on the river Havel. It has no telephone and is reachable only by a combination of train, bus, and foot. After a year spent slowly recovering from the weakness of his dilated heart, he is beginning to feel well again.

For his fiftieth birthday that month, his friends have given him a magnificent new sailboat, the *Tümmler* (tumbler, or dolphin). Sailing on one of the Havel lakes that curl around the city of Berlin, he is surrounded by the creaking of the boards, the sound of the wet sheets and metal rings against the mast, the slap of the water, the ripples creasing out from the prow. Elsa, his second wife, finds the ownership of land in these troubled times consoling. Yet Einstein knows his time in Germany is running out. What is consoling, what is in fact wonderful, is the meandering water of the Havel, rushing under his feet and the ribs of his boat, and its circuitous progress through its necklace of lakes and the tangle of Saxon rivers to the Elbe and finally the North Sea.

"Illness," he remarked in the March 23, 1929, volume of *Nature*, "has its advantage; one learns to think. I have only just begun to think."

By 1929, Bohr's institute is triumphant. Both de Broglie and Schrödinger are converted and a new generation of talented physicists is flocking to Copenhagen as the center of the physical universe. In Leon Rosenfeld's description of the Copenhagen conference that year, the overriding note is of joy and jokes. A chair breaks (suspiciously) spontaneously when Pauli is debunking the ideas of its occupant. George Gamow, a young practical-joking physicist, shouts out: "Pauli effect!"

"Oh, the Pauli effect," says Ehrenfest. "The Pauli effect is just a special case of the more general phenomenon that misfortunes rarely come singly."

Ehrenfest shines at the conference, teasing Bohr and keeping everyone honest with his searching questions and critical thinking. No one guesses how deeply to heart Ehrenfest still takes Bohr and Einstein's intellectual division, nor his anxiety over the encroaching Nazi movement (which most people believe will not really come to anything). "Of this inner tension there was no outward sign," Rosenfeld will remember. "To the last we saw him as cheerful, witty and warmhearted as ever."

No one knows that the previous August, Ehrenfest had written to his old student Kramers, begging him to "please help me. . . . Practically all new theoretical physics stands before me as a completely non-understandable wall and I am at a total loss. I no longer know the symbols, the language, and I no longer know what the problems are."

It is here at the 1929 Copenhagen conference that Rosenfeld—"a stubby young man with a serious round face . . . given to philosophical contemplation," as a friend remembered him around that time—first meets Niels Bohr, who explains to him the importance of the distinction between the classical measuring apparatus and the quantum object under observation. "It was one of the few solemn moments that count in an existence," Rosenfeld would later write, "the revelation of a world of dazzling thought, truly an initiation." He soon becomes Bohr's new assistant and amanuensis.

John Wheeler, another young physicist on his way to becoming one of the most influential teachers of quantum theory in the twentieth century, writes of those days: "Nothing has done more to convince me that there once existed friends of mankind with the human wisdom of Confucius and Buddha, Jesus and Pericles, Erasmus and Lincoln, than walks and talks under the beech trees of Klampenborg Forest with Niels Bohr."

15

Solvay

1930

$M_{AGNETISM}$ was the official subject of the Solvay conference that fall. Between the speeches, however, Einstein and Bohr were back at their old discussion, sitting companionably side by side, smoking their pipes.

Einstein said, "I have a new *gedanken* experiment for you, Bohr."

Neither man was ever very proficient with experiments in the lab, but they both loved thought (or *gedanken*) experiments, which could be performed entirely inside the head. Bohr raised his eyebrows in anticipation.

"What if we had a box filled with a certain amount of radiation," said Einstein, "equipped with a shutter," he explained, moving his hands, "which can open and close by means of a clockwork within the box." He leaned forward, the straight line of his trouser leg from knee to ankle fuzzy in the late sun from a window nearby, his hair in an attitude of static shock from leaning against the back of the couch. Smoke rose from his forgotten pipe, and billowed back and forth with a stray gesture from his hand.

"The shutter will open when the clock strikes a certain time, and then close again very swiftly, so that just one light-quantum is emitted." Einstein looked over at Bohr with barely suppressed excitement. "But we could weigh the box, and that would allow us to measure its energy, due to $E = mc^2$—"

Bohr's face looked blown open. "We have a box with a shutter"—his eyebrows came together, leaning so far out they looked as if they might topple off his face—"with a shutter which is controlled by a clockwork." Einstein was nodding, puffing on his pipe. Bohr brought the stem of his pipe to his lips. He found it, as he often did, no longer burning. "Einstein, do you have a match?" he asked abstractedly, patting his pockets.

Einstein delved into his pockets and produced his matches. Bohr slowly lit his pipe again. "We weigh the box before and after a photon is emitted," he continued as if there had been no interruption, while absentmindedly pocketing Einstein's matches, "and now, contrary to the uncertainty principle, we have both the energy emitted and the time of emission."

"I grant you the consistency of the uncertainty principle," Einstein said, laughing a little. "I'm not done with the experiment, Bohr. We can allow the light-quantum to travel to a sufficiently distant mirror—say, half a light-year away"—Einstein's hand with his pipe extended—"to ensure spacelike separation when we decide to learn about it by either weighing the box or checking the clock." (When even an influence traveling at the speed of light will not be able to connect two events, then they are "spacelike separated"—the closest thing relativity allows to "simultaneous.")

Bohr, who was still thinking about the uncertainty principle, nodded with a preoccupied look on his face. "This amounts to a serious challenge," he said quietly. "I will have to thoroughly examine the whole problem."

"*Wait*, Bohr," said Einstein. "The problem is not yet complete. If, while the light-quantum is half a light-year distant, we now check the clock, we will be able to predict exactly *when* the light-quantum will return to us, which gives us a sharp determination of its position. If we *weigh the box* instead, we will be able to predict exactly its *energy* and thus color."

Bohr, his face now devoid of expression, was deep in his own train of thought. "When the exigencies of relativity theory are taken into consideration," he said slowly, "does that indeed allow us to control the interchange of energy between the objects and the measuring instruments, even when they are serving their purpose of defining the *space-time* frame of the phenomena?"

"*Bohr!* Let me finish. Now, according to the uncertainty principle, the state of a light-quantum *cannot* involve *both* a sharp determination of its location in time *and* a sharp determination of its energy. But we can learn either characteristic from one or another measurement on the box.

"Here is my question," said Einstein. "Should we assume that the subsequent measurement we make on the box *physically* influences the fleeing light-quantum, now half a light-year distant? That would be a superluminal"—faster than the speed of light—"action-at-a-distance. Of course it is logically possible, but so very repugnant to my physical instinct that I cannot take it seriously.

"So the real state of the light-quantum is independent of whatever is done to the box." Einstein remembered his pipe, took a puff, and moved in for the kill. "But from that it follows: every characteristic of the light-quantum—every characteristic we could glean from a measurement of the box—*exists,* even if this measurement is not performed. Accordingly, the light-quantum has a definite localization and a definite color, and the quantum-mechanical description is incomplete."

"It was quite a shock for Bohr," wrote his new assistant, Rosenfeld. "I shall never forget the vision of the two antagonists leaving the club: Einstein a tall majestic figure, walking quietly, with a somewhat ironical smile, and Bohr trotting near him, very excited."

That evening, as everyone was sitting down to dinner, Bohr told first one person, then another—Ehrenfest, Pauli, Heisenberg—of Einstein's new *gedanken* machine, always stopping at the point that interested him, the possibility of measuring the photon's time of emission by reading the clock and, simultaneously, its energy by weighing the box.

That Einstein's thought-experiment did not require this, Bohr had hardly noticed. He would similarly pay no attention when Ehrenfest told him this in so many words nine months later. "He said to me," Ehrenfest wrote to Bohr, "that, for a very long time already, he absolutely no longer doubted the uncertainty-relations," and "had BY NO MEANS invented the 'weighable light-*blitz* [flash] box' . . . *contra* uncertainty relation." That is, Einstein's box was not supposed to be a weapon against the uncertainty principle.

For Bohr, a discussion about unobserved reality had no meaning, and he was focusing on the aspect of the experiment that, in his estimation, did. But now and later, Bohr's "resignation as regards visualization and causality," his eagerness to "consider this very renunciation as an essential advance," his modesty in trusting only "the demand of avoiding logical inconsistencies," had blinded him to perhaps the most important thing. From "irrationality" in the twenties to "wholeness" in the sixties, Bohr welcomed into physics the expressions of the zeitgeist. Shadowing the precision of entanglement under their capacious wings, these foggy ideas ensured that Bohr would never recognize its long-range possibilities, like a man who never walks on the land that he has bought.

The year before, Bohr had declared again, "Any observation necessitates an interference with the course of the phenomena." This "should not be regarded as a hindrance," he explained. "We must only be prepared for the necessity of an ever-extending abstraction from our customary demands for a directly *anschaulich* [visualizable] description of nature."

Instead of heeding Bohr's declarations and admonitions, Einstein had decisively sharpened his two decades of worry about separability into a picture of entanglement over half a light-year's distance. If the light-quantum was to be disturbed by the act of observation, as Bohr insisted, this disturbance would have to be nonlocal, spookily acting at an effectively infinite distance. But it would take two more presentations of this argument (one by Ehrenfest the next year, and one in the famous Einstein-Podolsky-Rosen [EPR] paper of 1935) before Einstein would be able to convince Bohr to focus on what was really bothering him.

Bohr would come to realize, as he wrote about a conversation with Einstein three years later, that "the difference between our ways of approach and expression still presented obstacles to mutual understanding." What bothered him and what bothered Einstein about this new *gedanken* experiment spoke volumes about the two men's "ways of approach and expression."

"Ehrenfest, it can't be true," Bohr said for the fifth time. And then: "It will be the end of physics if Einstein is right." His face looked more solemn and ponderous than anyone had ever seen it, and more in pain.

"Bohr will work it out," said Heisenberg to Pauli and Ehrenfest. He looked sideways at Bohr, now deep in discussion with Max Born. Bohr's face had lit up—but then Born was shaking his head, those small whimsical eyes squeezed into a squint, his mouth a regretful fist. Bohr deflated, and then just as quickly started on another angle. Born looked thoughtful, raised an objection. Bohr will work it out, Heisenberg had said.

"Won't he?" he added.

Ehrenfest shrugged his eyebrows with a trace of humor, to hide a feeling that was almost panic.

They all stayed late at the dinner table, and when they moved, they all moved together, an amorphous school of physicists, into the club room, where, under clouds of smoke, they clustered around Einstein and Bohr. Ehrenfest, trying to mediate between the two, felt his head splitting.

"The next morning came Bohr's triumph," said Rosenfeld.

Bohr strode into the meeting hall, its tall pilasters and mirrors pale and blinking in the morning light, up to the chalkboard, where he started drawing an elaborate apparatus. Everyone gathered around began to realize that it was Einstein's light-*blitz* box, but now it was realized in minute buildable detail (and in fact, the practical-joking Russian physicist George Gamow did build a mock version of it). It was the embodiment of all Bohr aspired to in his writing, usually at the costly expense of clarity—he had anticipated every contingency, every physical effect, bent close to the chalkboard as he drew in deep concentration. The box was now suspended by a spring (Bohr wrote, illegibly, "spring," with an arrow pointing at a chalky curlicue) from a sort of metal gallows, and had a pointer on it against a little ruler fixed to the vertical post of the gallows. Under the box was a hook for a weight to hang on.

Einstein came up to the blackboard and the two of them together worked out the solution to what was bothering Bohr: the movement involved in the weighing of the box meant that, ever so slightly, the clock would run more slowly, as explained by Einstein's own theory of relativity. This made the time of the shutter's opening to release the photon uncertain, and Bohr showed, writing the steps out on the blackboard, that one arrived back at the time-energy uncertainty relation that Einstein had seemed to refute.

Einstein, looking quizzical, pulled out his pipe, but could not, inevitably, find any matches. Bohr spotted this, and for once was able to find his own (Einstein's). He lit Einstein's pipe for him. Einstein nodded once, as if he were tipping his hat. Bohr slowly smiled.

Once more, again in Brussels, Bohr had shown that measurement is a messy process, a physical disturbance—of the object, the apparatus, or both. But he had not touched Einstein's pristine light-quantum, half a light-year away from the turbidity of measurement.

When Bohr finally turned his attention to that untouched, entangled quantum, the debate would have one more round. It would be, in fact, the last debate of the golden years of the quantum theory, as war closed in on the physicists and cast them to the wind like seeds.

Things Fall Apart

1931–1933

(The SPIN OF THE PHOTON, *dressed in Indian guise, slithers across the stage, accompanied by fugitive music)*

Attention again! Here's *The Spin of the Photon*
With some kind of Indian *sari* and coat on.
(It's clear that no modest, respectable *Boson*
Would traverse the platform without any clo'es on!)

P.A.M. *Dirac in* The Blegdamsvej Faust, *1932*

Man only loves destiny," wrote Eduard Einstein in an aphorism for his high school paper. "It is the worst destiny to have no destiny and be no destiny for anyone else. . . ." In 1931, two years after his high school graduation, the best of his aphorisms were published in the *Neue Schweizer Rundschau* (the *New Swiss Review,* a literary magazine). By then, the beloved and brilliant "Tetel" had dropped out of medical school in Zurich and abandoned his dreams of being a psychiatrist, rarely leaving his dark rented room and unraveling mind.

When Einstein separated from his first wife and their two sons in 1914, Tetel was four. An otherworldly child with luminous eyes and uncut hair, he eerily prefigured the beatified icon his father would become. He had been fearsomely precocious, diving into Schiller and Shakespeare so soon that his father cautioned him, age eight, to "save yourself some until you're grown up." As Einstein told his friend and confidant Besso, he had seen schizophrenia "coming slowly but irresistibly since Tetel's youth." Listening to the teenage Eduard absorbed in the piano moved everyone: his friends saw him without his shell of irony and what one of them described as his usual "slight sense of being absent"; his father heard "frantic" and wooden music, and felt foreboding.

In the early summer of 1930, Eduard's self-described "rapturous letters"—full of literary and philosophical attempts to catch the attention and love of the distant father he worshipped—turned hysterical, vindictive, and despairing. This got the attention of his father, who came to Zurich to find all of his fears for his beautiful son coming true. He tried to tell Tetel that this breakdown would help him become "a really good doctor of the soul," but privately felt only deterministic dread.

Leaving his son in Zurich in the hands of the psychiatrists, Einstein retreated to his house in Caputh. "He has always aimed at being invulnerable to everything that concerned him personally," Elsa Einstein said, trying to explain her husband to a friend. "He really is so, much more than any man I know. But this has hit him very hard." He aged visibly. When asked to summarize his philosophy that summer, he made (for him) a startling statement: "We know from daily life that we exist for other people—first of all for those upon whose smiles and well-being our own happiness is wholly dependent." But the absent father would not change his ways now. "The winter here can bring nothing but sadness," Elsa said at the end of 1930. "We intend to go away for a long time."

"It is well known," wrote Einstein in an Eden seven thousand miles from Zurich and Berlin a few months later, "that the principles of quantum mechanics limit the possibilities of exact prediction as to the future path of a particle." But what of a particle's past? Heisenberg, who believed that his principle had bearing only on what was yet to come, had four years earlier castigated Einstein for not treating a particle as a particle. But Einstein was beginning to suspect that the past histories of Heisenberg's particles were exactly as fuzzy as their futures. This was the question on which he mused on a warm Pasadena winter day early in 1931, his pipe smoke trailing behind him as he wandered across campus beside the tall, thin, neatly dressed Richard Chace Tolman, who was followed by his own aromatic cirrus of smoke.

Tolman was one of the first professors of the new California Institute of Technology, which had been founded a decade before by a trio of physicists—the physical chemist Arthur Noyes; the measurer of electron charge, Robert Millikan; and the first director of the Mt. Wilson Observatory, George Ellery Hale. Tolman, appropriately, was a physical chemist who had measured the electron's mass and now contemplated the heavens. He had just written a pair of books on two of Einstein's great loves, statistical mechanics and relativity.

Joining them on the verdant winter lawns was an affable-looking and awkward recent Caltech Ph.D., a Russian-American named Boris Podolsky, who had just spent his postgraduate time with Heisenberg in Leipzig. When Paul Ehrenfest, who was always nostalgic for his own Russian days, descended on Caltech a month before, he and Podolsky collaborated with Tolman on a paper called "On the Gravitational Field Produced by Light." Now Tolman and Podolsky turned their attention, with Einstein, to "Knowledge of Past and Future in Quantum Mechanics."

Podolsky had grown up in a little Russian village near where the Don meets the Sea of Azov, working in his aunt's grocery store, wrapping purchases with heavy twine that he had learned to break with his bare hands. He arrived on Ellis Island just before the First World War as a seventeen-year-old steerage passenger, and crossed the country by Greyhound bus to another aunt who lived in Los Angeles, arriving the same time as the aqueduct and the movie industry that would together create the future possibilities of the city. To pay for college, he worked for a plumber who refused to show him how to seal a toilet into its flange for fear that his quick-witted assistant would then know too much and start his own competing business. Instead, Podolsky went from a bachelor's degree in electrical engineering to a master's in math to a doctorate in physics. His second American job found him designing copper piping to bring power to Los Angeles from the (then) Boulder Dam, plumbing and electrical engineering on a grand scale. Even working for the electric company, he approached problems from unusual directions. He showed, for example, that the best indicator of the timing of the important spring melt was not the one the company was using—snow depth in the mountains above the dam—but high temperatures in Tokyo.

Einstein, with Podolsky and Tolman, had found a new use for his light-*blitz* box. "The purpose of the present note is to discuss a simple ideal experiment which shows," they explained, "that the possibility of describing the past path of one particle would lead to predictions as to the future behavior of a second particle of a kind not allowed in quantum mechanics." Einstein's perfect *gedanken* experiment had almost arrived.

"In a local pub, I met your son," wrote Max Born, visiting Zurich, to Einstein in Pasadena shortly after the paper with Tolman and Podolsky was published. "I liked him very much; he is a fine, intelligent fellow and laughs in exactly the same wonderful way you do. Well, what else is there to tell you? Things in Europe do not look pleasant, either politically or economically. . . . But things must surely improve in spite of Hitler and his consorts. I know all sorts of things about California, as I am just reading Ehrenfest's wonderfully vivid, descriptive letters about his travels which he has sent to Hedi. How well that fellow observes and describes his experiences."

Hedi added birthday greetings, writing, "I am always very amused to see and hear you in the weekly newsreel—being presented with a floral float containing lovely sea-nymphs in San Diego, and that sort of thing. The world has, after all, its amusing side. However crazy such things may look from the outside, I always have the feeling that the dear Lord knows very well what he is up to. In the same way as Gretchen sensed the Devil in Faust, so he makes people sense in you—well, just the Einstein. For none of them will ever be really able to *know* you—however thoroughly they may have studied the theory of relativity."

Also in America, Pauli—cavorting about Ann Arbor with Sommerfeld in the summer of 1931—suffered what Sommerfeld termed an "inverse Pauli effect," when (as Pauli admitted) "in a slightly tipsy state" at the Michigan home of his old bicycling friend Otto Laporte, he tripped on a stair and broke his shoulder. Writing to his assistant Peierls, Pauli explained, "In spite of the opportunity for swimming I here suffer much from the great heat. But under the 'dryness' I don't suffer at all because Laporte and Uhlenbeck are superbly stored up in alcohol (one notices the vicinity of the Canadian border)." But "the cigars Sommerfeld misses very much."

Sommerfeld accompanied Pauli to America because his old student desperately needed him—the only person Pauli would ever address with "*Ja, Herr Professor*" and "*Nein, Herr Professor.*" ("Why it was just you who managed to instill awe in me is a deep secret," Pauli wrote to him, "—one, no doubt, many others would have liked to learn from you, in particular, all my later bosses including Herr Bohr.") In the last few years of his twenties, Pauli's life had taken a horrifying turn for the worse, and he needed the stable, reliable guide of "those cheerful student days." First his vibrant, idealistic mother committed suicide while his father dallied with, and then married, a girl uncomfortably close to his son's age. In late 1930, Pauli's own new bride, a cabaret dancer, left him for a chemist after one year of marriage. He bravely managed one of his trademark double-edged insults ("If it had been a *bullfighter*—with someone like that I could not have competed—but such an *average* chemist!"), and spent the nights drinking.

Nineteen thirty-one was the year of the particle and the box. Einstein took his *gedanken* experiment everywhere—from its debut in Brussels to Pasadena to Berlin to Leiden, and back to Brussels. Everywhere Einstein went the experiment changed slightly as he explored its contours, pushed its envelope, and worked his way toward EPR.

"Without in any way interfering with the photon," Einstein wrote, "we are, thus, able to make accurate predictions pertaining *either* to the moment of its arrival *or* to the amount of energy liberated by its absorption."

It is theory that first determines what can be observed: Heisenberg set his student von Weizsäcker to analyze a variant of the light-*blitz*-box *gedanken* experiment in 1931. Einstein's musings on the same subject would lead him to the EPR paradox, paving the way for Bell's theorem and all the experimental magic of long-distance entanglement. But all von Weizsäcker and Heisenberg saw was a sterile and correct "exercise in quantum field theory," quickly laid aside, its buried treasure still buried.

Ehrenfest, like Einstein, thought it was more than an exercise, but did not know how to respond to it. He did what anyone would do: he wrote Bohr, on July 9, 1931, to tell him that Einstein would be in Leiden at the end of October. Could Bohr come too, he asked (with one of his characteristic explosions of capitalization) so that they might have a "QUIET" exchange of views? Ehrenfest described the experiment again to Bohr in the meticulous detail with which Einstein had written him (e.g., "Weigh the box during the first 500 hours and screw it firmly to the fundamental reference frame"). "It is interesting to get clear," explained Ehrenfest to Bohr, "the fact that the projectile, while flying around isolated on its own, must be prepared to satisfy very different non-commutative predictions"—the kind of predictions denied simultaneous meaning by quantum mechanics—"without knowing as yet which of these predictions one will make (and test)."

Ehrenfest mailed the letter to Margrethe, not Niels, beseeching her in a postscript to give it to her husband only if he was not too tired, and "there is *ABSOLUTELY NO NEED FOR AN ANSWER.*" Whether or not Margrethe handed the letter on to Bohr, there was no answer: Bohr persisted in misunderstanding Einstein's point, giving a speech that October in Bristol, England, in which he again demonstrated his utter incomprehension that the point of the light-*blitz* box was—in Schrödinger's word, coined four years later—entanglement.

Meanwhile, in Cambridge, Massachusetts, a graduate student of

Slater's named Nathan Rosen was investigating the structure of the hydrogen molecule, producing the first reliable calculation of its two bound-together hydrogen atoms. It is a strange object. Only the molecule as a whole has a quantum state; the component atoms are entangled. They have no states of their own, and, as far as quantum theory is concerned, a measurement performed on one instantly affects its twin. When the logical analysis of Podolsky married Einstein's light-*blitz* box to Rosen's entangled hydrogen atoms, the EPR paradox would be born.

That September, Einstein again nominated either Heisenberg or Schrödinger for the Nobel Prize. "Personally, I assess Schrödinger's achievement as the greater one," he wrote, "since I have the impression that the concepts created by him will carry further than those of Heisenberg." He noted, "This, however, is only my own opinion, which may be wrong." His recommendation went against the opinions of everyone else and threw the Nobel committee into such confusion that no physics prize was awarded for 1931.

Despite Ehrenfest's plea, Bohr did not show up in Leiden in October. Ehrenfest was unusually muted as Einstein again presented the light-*blitz* box, stressing, as Ehrenfest's assistant Hendrik Casimir remembered, that "after the light quantum has escaped we can still choose whether we want to read the clock or alternatively determine the weight of the box. So, without touching the light-quantum in any way, we can either determine its energy or the time at which it will return from a distant mirror." Could the state of the long-gone photon be dependent on actions made on the box?

Instead of answering Einstein himself, or asking one of his famously piercing and clarifying questions, Ehrenfest entrusted his twenty-five-year-old assistant "with the task of opening the discussion, and I explained to the best of my ability the Copenhagen views on such questions," Casimir remembered. "Einstein listened, perhaps slightly impatiently, and then he said—and I believe I now can vouch for the exact words: 'I know, this business is free of contradictions, yet in my view it contains a certain *Härte*' "—a certain hardness, or "unreasonableness," as Pais would have it. "I think," wrote Casimir, remembering the ironic look of the determined skeptic, " 'a certain *unpalatability*' comes closer to what Einstein had in mind."

On a liner heading to Pasadena in December of 1931, Einstein was watching the seagulls. "Today, I made my decision essentially to give up my Berlin position, and shall be a bird of passage for the rest of my life," he wrote in his travel diary. "Gulls are still escorting the ship, forever on the wing. They are my new colleagues."

As the pillars of European civilization vibrated in 1932, the mathematical foundations of quantum mechanics were made secure. The obstreperous twenty-nine-year-old Hungarian genius John von Neumann showed that the mathematical structure of quantum mechanics—be it waves or matrices—reduced to a pure and austerely abstract group of foundational mathematical statements, which mathematicians call axioms. With the power of this mathematical analysis, von Neumann was able to conclusively demonstrate that the acausal nature of quantum theory—Einstein's bête noire—was irremediable. Its mathematical structure would allow no fuller completion. Von Neumann's book was a tour de force of elegant and fearsome mathematics, the sort of thing most physicists never read, but its existence and conclusions were deeply consoling. The phrase "von Neumann has shown . . ." entered the quantum physicist's lexicon as a debate ender.

Younger than Pauli, Heisenberg, and Dirac, von Neumann had already given the same treatment to a famously paradoxical branch of mathematics (set theory), and also come up with the insight that would found a whole new discipline of mathematics and economics (game theory); in a year, he would become one of Einstein's few colleagues at the brand-new Institute for Advanced Study in Princeton, entertaining himself with pranks like putting the absentminded older man on the wrong train.

"The five years following the Solvay Congress [of October 1927] looked so wonderful that we often spoke of them as the golden age of physics," wrote Heisenberg forty years later. Pauli and Dirac—hesitantly, almost in desperation—each predicted a particle. Pauli's came first: a chargeless, massless particle he called the neutron would be the only thing that could explain a form of nuclear radiation that was mystifying the entire community of scientists. Dirac, meanwhile, was suffering with his theory, which predicted both positive and negative solutions for electron charge, almost as if there were a positive antimatter electron as well as a negative one that made ordinary matter. It couldn't be the proton, as his friend Oppenheimer pointed out; if it were, then all of matter would have collapsed into itself in a blaze of light only 10^{-10} seconds after the Big Bang.

Years later, theoretical physicists would predict particles with gay abandon, but at this point the only particle that had ever been successfully predicted was Einstein's photon, one of only three particles known at all. The other two—the electron and the proton—had been found experimentally. It seemed foolhardy in those days to say that one's equations demanded a particle that no one had found with a cloud chamber or a Geiger counter.

"Nobody took Dirac's theory seriously," said the Cavendish Lab's quintessential experimentalist, the crotchety P. M. S. Blackett, far too much of a disciple of Rutherford's to listen to theoretical moonshine about positive antimatter electrons. (Rutherford was overheard in the late twenties to say, "There is only one thing to say about physics: the theorists are up on their hind legs and it is up to us to get them down again.") But then an American experimentalist, Carl Anderson, seemed to have actually found this antimatter in a cloud chamber.

Bohr brought a picture of Anderson's experiment back with him from California to Denmark. With his sixteen-year-old son, Christian, he rejoined Heisenberg and two of his current students, Felix Bloch and von Weizsäcker, at the Oberaudorf train station in southern Bavaria. From there, they skied up to Heisenberg's cabin—Heisenberg nearly carried away en route by a small avalanche. The next day, lying exhausted on the roof in the sun, with snow like an ocean all around them and the Alps above, they pored over the picture, and argued over whether or not the curving trail of water droplets was the wake of a so-called positron.

Back in England, Blackett was shocked to get the same results. Rutherford found it "regrettable" that Dirac's *theory* of the positron had preceded its experimental finding. "Blackett did everything possible not to be influenced by the theory," Rutherford said proudly, "but . . ." still, there it was. "I could have liked it better if the theory had arrived after the experimental facts had been established."

Bohr and Heisenberg's ski party passed the chilly evenings playing poker at Heisenberg's cabin. On the third night, Bohr proposed *cardless* poker. His money was on Bloch or his son Christian to win, since they were the most convincing bluffers. With what Heisenberg remembered as "strong grog" to keep them warm, "the attempt was made."

"My suggestion was probably based on an overestimate of the importance of language," Bohr admitted a little while later, as the game dissolved into absurdity. "Language is forced to rely on *some* link with reality." Firelight leaped on his genially bemused face as they laughed at him. "In real poker, we can use language to 'improve' the real hand with as much optimism and conviction as we can summon up," he said with a sheepish grin. "But if we start with no reality at all, even Christian can't persuade me he has a royal flush."

Rutherford himself in 1920 had predicted a "neutron," a neutral twin to the (positive) proton, which would solve many of the mysteries of how atoms, especially the heavier ones, held together. James Chadwick, the assistant director of the Cavendish Lab, found this neutron twelve years later, in February 1932.

But it was not Pauli's neutron. He did not give up hope, and neither did Ehrenfest's student Enrico Fermi, who that year published his great theory of beta decay (an otherwise mysterious change of an atomic nucleus), made possible by what he called Pauli's "little neutron"—the *neutrino*. (Nearly a quarter of a century later, half a world away in the postwar Los Alamos labs in New Mexico, the neutrino was found; its finders fired off a telegram to Pauli telling him the good news.)

At the Cavendish, back in 1932, two more members of Rutherford's team—the laconic John Cockcroft, who was known to be working two and a half full-time jobs around the Cavendish simultaneously, and Ernest Walton, a nimble experimentalist who could repair watches—were greasing up a homemade "accelerator" in the former library of the laboratory to do something even more spectacular. In 1928, the twenty-five-year-old Gamow had realized that the proton, because of its wave nature, might tunnel into the nucleus of an atom, splitting it apart—even though, viewed as a particle, it had no chance of doing so.

Rutherford stormed good-naturedly in and out, giving orders, encouragement, or distraction, occasionally shocking himself by hanging his wet coat on a live terminal or lighting overdry tobacco and sending it up "like a volcano with a great cloud of smoke, flames, and piles of ash," as the experimentalist next door to Cockcroft and Walton noticed with amusement. And on either April 13 or 14—Cockcroft and Walton in their notebooks wrote down different dates—he observed the scintillations as pieces of the atom hit the fluorescent screen. *We've split the atom!* This jubilant cry was soon heard at the newspapers—the atom, the unsplittable, had been split. *Things fall apart; the center cannot hold.*

When the time came for Nobel nominations in September of 1932, after a year of meditation on the mysterious interconnections indicated by the wave equation for his *gedanken* particle and its *gedanken* box, Einstein had ceased to waffle and was confidently nominating only Schrödinger: "Our understanding of the quantum phenomena has been furthered most by his work in connection with the work of de Broglie." Ehrenfest, in a footnote to what would be his last paper, that ominous autumn of 1932, reminded everyone what that meant: "If we recall what an *uncanny theory of action-at-a-distance* is represented by Schrödinger's wave mechanics, we shall preserve a healthy nostalgia for a four-dimensional *theory of contact-action!* . . . Certain thought-experiments, designed by Einstein, but never published, are particularly suited for this purpose."

Ehrenfest had titled his paper "Some Exploratory Questions Regarding Quantum Mechanics." He was groping for solid ground, and wondering if it even existed or if the quantum theory was "senseless." Many of the same questions had also occurred to Pauli, for whom the paper was "a source of sheer pleasure." He wrote a reply, under the same title, attempting to put the quantum theory in terms of action-by-contact, without great success. But Ehrenfest was slightly reassured that he was not alone. "For fear of Bohr and you," Ehrenfest told him, "I have struggled more than a year with the decision to print the few lines, until, at last, despair drove me to it."

Schrödinger had spent the summer of 1932 in Berlin with the constant companionship of Itha Junger—once his laughing fourteen-year-old tutee, now a beautiful twenty-one-year-old and his faithful mistress of the previous four years. But as the days began to get colder, it became clear that she was pregnant. It was also clear that it was the end of the affair, and Schrödinger was hoping to move on to his assistant's wife. Still, yearning for a son, he fondly thought that Itha would give the child to him and Anny.

Instead, Itha had an abortion and left Berlin. Far from the cities she had grown up in and shared with Schrödinger, she married an Englishman. But that was not the end of the story. Schrödinger, with a series of mistresses, would have three daughters—the first with his assistant's wife in less than two years. But the scars from the end of the affair would leave Itha childless after a series of miscarriages.

Meanwhile, Schrödinger's unpublished notes of 1932 and 1933 show him beginning to investigate mathematically the action-at-a-distance in his equation that so agonized Ehrenfest and Einstein. Once in contact, two particles were described only by a shared wavefunction, ψ, long after they lost all physical contact. As before with wave mechanics—where he had foreseen the fuzzy outlines of de Broglie's idea, but did not pursue it until de Broglie independently made it clear and definite—he did not follow up this thought until someone else articulated it more clearly.

In this case, that would be Einstein, Podolsky, and Rosen, three years later.

In the little tan institute at 15 Blegdamsvej in Copenhagen, everyone has gathered in the main lecture room in late 1932 for what Gamow called "the customary stunt." Gamow himself is not present, since Stalin—already "dizzy with success" while Gamow's native Ukraine is purged and starved—refused him a passport to "fraternize with scientists of the capitalistic countries."

Bohr and Ehrenfest sit smiling and chatting in the third row, behind two empty front-row benches, elbows leaning on the little lecture-room desks. There's Lise Meitner, who would soon be the first to understand the portentous fission of uranium; there are Dirac and Heisenberg. The lights dim, and onto the stage walks Max Delbrück. (Inspired by a speech Bohr gave earlier that year exhorting *biologists* to search for complementarity, he will—though failing to vindicate Bohr's philosophy—become one of the founding fathers of the field of molecular biology.) He looks extremely young and fair and impish under his top hat. "I present to you, most honored guests, *The Blegdamsvej Faust.*" Everyone claps, and out walk three archangels, wearing carefully hand-drawn masks of the faces of famous astrophysicists.

Everyone laughs as the angels debate in rhymed German that carefully parallels Goethe's version of the play, which most of the audience read in school, when suddenly out pops Leon Rosenfeld, wearing a mask of Pauli's face—he is Mephistopheles. He leaps up to sit at the foot of the shrouded figure of God on a stool on top of the lab table.

As Mephistopheles addresses the Lord, the sheet is swept off to reveal Felix Bloch wearing a mask of Niels Bohr. Bloch climbs off his stool and intones, to laughter all around (Bohr, in the audience, is shaking his head with a huge grin):

> But must you interrupt these revels
> Just to complain, you Prince of Devils?
> Does Modern Physics never strike you right?

Mephisto-Pauli replies,

> No, Lord! I pity Physics only for its plight
> and in my doleful days it pains and sorely grieves me.
> No wonder I complain—but who believes me?

The Lord puts his finger to his chin. "You know this *Ehrenfest*?"
Mephisto throws up his hands: "The Critic!?" Someone runs in hold-

ing a cartoon of Ehrenfest, with light glowing around him and his hair three times as vertical and wild as usual. Under this cartoon is written "A Vision of Faust." As the Lord claims Ehrenfest-Faust as his knight, Mephisto brashly states that he can lead him astray. Ehrenfest's face—in the dark in the audience—has a wistful smile, his hand on his chin. *I am Faust,* he thinks. *How strange.*

The Lord sighs. Bloch, behind the Bohr mask, imitates one of the master's characteristic semi-German, semi-Danish phrases:

> *Oh, this is really dreadful! Must I say . . .*
> Ja, muss Ich sagen . . . *There is an essential*
> *Failure of classical concepts—a morass.*
> *One side remark—but keep it confidential—*
> *Now what do you propose to do with* Mass?

Mephisto laughs. "With Mass?" says Rosenfeld behind his Mephisto mask. "Why, just forget it!"

The Lord stumbles. "But . . . but this . . . is very *in*-ter-est-ing. Yet to try it . . ."

Mephisto cuts in: "Oh, *Quatsch* [shut up]! What rot you talk today! Be quiet!" At this point the whole room is roaring with laughter. They have all seen this scene between Pauli and Bohr so many times, and told stories of it even more often.

The Lord says, slowly shaking his head, as one would say to a small child: "But *Pauli,* Pauli, Pauli, we practically agree" (more laughter from the audience, who know very well that this is Bohr's way of saying that the other person is totally wrong—as von Weizsäcker described it, "wrapped in the gingerbread of his helplessly amiable way of talking").

"There's no misunderstanding—that I guarantee," continues the Lord, ". . . but if Mass and Charge go packing, what have you, by and large?"

Rosenfeld wobbles on the edge of the lab table in his Pauli-esque excitement:

> *Dear man, it's elementary!*
> *You ask me what remains?*
> *Why bless me, the* Neutrino!
> *Wake up and use your brains!*

Both the Lord and Mephisto are silent as they pace back and forth on the table. Then the Lord stops, and looks out into the audience. "I say this not to criticize"—Bohr is nodding and smiling in his chair, recognizing another of his favorite phrases—"but rather just to learn. . . . But now I have to leave you. Farewell! I shall return!" He jumps down from the lab table.

Mephisto says cheerfully,

> *From time to time it's pleasant to see the dear Old Man,*
> *I like to treat him nicely—as nicely as I can.*
> *He's charming and he's lordly, a shame to treat him foully:*
> *And, fancy!—he's so human he even speaks to Pauli!*

With that, he leaps off the lab table and skips off.

Delbrück returns, intoning: "*First Part. Scene: Faust's study,*" while waving a hand in the direction of the lab table. "Ehrenfest" walks in, carrying a huge pile of books, which he stacks before him. Sitting on the stool behind all this collected knowledge, he begins with a sigh to recite Faust's introductory speech (altered for the occasion)—

> *I have—alas—learned Valence Chemistry,*
> *Theory of Groups, of the Electric Field,*
> *And Transformation Theory as revealed*
> *By Sophus Lie in eighteen-ninety-three.*
> *Yet here I stand, for all my lore,*
> *No wiser than I was before.*
> *M.A. I'm called, and Doctor. Up and down,*
> *Round and about, the pupils have been guided*
> *By this poor errin' Faust . . .*

(the real Ehrenfest grins under his mustache)

> *. . . and witless clown;*
> *they break their heads on Physics, just as I did. . . .*
> *All doubts assail me. . . .*
> *And Pauli as the Devil himself I fear.*

Mephisto then bursts in, dressed as a traveling salesman, here to sell Ehrenfest on Pauli's neutrino, decked out as the radiant Gretchen, who sings her famous spinning-wheel song, with altered words:

> *My mass is zero*
> *My charge is the same*

You are my hero
Neutrino's my name.

The play goes on to have both a classical and a quantum-mechanical Walpurgis Night, recasts Mephistopheles' king-and-flea story as a tale of Einstein infested with unified-field theories, cavorts through Dirac's theory of the positron, visits "Oppie" (Oppenheimer), Kramers, Sommerfeld, and Tolman in "Mrs. Ann Arbor's Speakeasy" (the University of Michigan's Ann Arbor Summer School "near the Canadian border") where they are charmed by Mephisto's Gretchen-neutrino. Covering with its off-kilter sweep most of the physics (and in-jokes) of that year, the play ends with the semicomical death of Faust.

Mephisto: "The changing forms he wooed have never pleased him. . . . All's over now. How did his knowledge aid him?" Ehrenfest's face was drawn amid the laughter and though he laughed, too, his eyes were dark.

Finally, "Chadwick" arrives with a cardboard black ball balanced on his finger, intoning:

The Neutron *has come to be.*
Loaded with Mass is he.
Of Charge, forever free.
Pauli, do you agree?

Mephisto-Pauli admits:

That which experiment has found—
though theory had no part in—
Is always reckoned more than sound
to put your mind and heart in.

He nods to Chadwick's black ball, saying,

Good luck, you heavyweight Ersatz—
we welcome you with pleasure
But passion ever spins our plots,
and Gretchen is my treasure!

The "Mystical Chorus" (in the program, credited as "Everybody who can sing") breaks into a closing *ultimo* as clapping and cheering fill the room.

Von Weizsäcker later wrote about this play, "Our laughter about Bohr was the escape route which enabled us to say that, although often we could not understand him, we admired him almost without reservation and loved him without limit." Looking around at the smiling faces, the masks coming off, everyone laughing, one can see that this is goodness. The excitement that Rutherford's Cavendish Lab radiated that year like an unstable isotope was goodness. The discoveries and the hope of more were goodness. It is the most exciting year for physics since relativity. This is knowledge, this is greatness, this is progress, this is civilization.

On January 30 of the next year, 1933, Hitler took power in Germany.

Two months later, as spring began to come to the trees around Caputh, Hitler's brown-shirted *Sturm* troopers burst into Einstein's home. Seeing the abandoned *Tümmler* rocking in its mooring on the peaceful Havel outside, they confiscated what they reported as "the speed motor boat of Professor Einstein." The little sailboat was eventually sold (and would vanish from local record and local memory before Einstein could find it again, though he tried), with orders that it should not be "once again bought by public enemies." Einstein, in Pasadena, wrote to Berlin to officially resign from his position at the university. He would never again set foot on German soil.

Back in Berlin, Max Planck took it upon himself, as the representative of physics in Germany, to convince Hitler of the importance to Germany of the Jewish professors who were being dismissed from their appointments all over the country. Planck had barely gotten a word out of his mouth when the hard little dictator began raging and shouting incoherently at the distinguished seventy-five-year-old gentleman, as Einstein heard it, threatening Planck with imprisonment in a concentration camp. Planck may not have been in danger, but in a decade his son, implicated in a plot on the Führer's life, was gruesomely murdered by the state.

Meanwhile, Heisenberg found himself, like Planck and the rest of the "Aryan" professors who remained, a civil servant of the Third Reich, ordered to submit for scrutiny his parents' birth and marriage certificates, attend indoctrination camps, and preface all his lectures with "*Heil Hitler.*" In April, Pascual Jordan (Heisenberg and Born's stuttering co-creator of matrix mechanics) joined the Nazi Party. That May, outside the university walls, in the squares of the cities they had loved—Göttingen, Munich, Berlin—smoke rose from the burning of hundreds of books.

All around, things were becoming more and more temporary, people were on the move, no one knew where the solid ground would be. *Things fall apart; the center cannot hold.* The telephone woke Max Born up in the middle of the night: at the other end of the line was a harsh voice shouting menacing slogans. Born, still groggy with sleep, then was treated to a rendition of the Nazi fight song, the "Horst Wessel Lied." He and Hedi and their son, Gustav, left Germany in early May of 1933 (looking out the train window on books burning in a small town square) for Selva, just over the Italian border. The little town felt secret and protected, reachable only by hours of driving on vertiginous mountain roads, past snow-capped roadside shrines of bleeding Christs and onion-domed churches clinging to cliff and sky. It was hidden in a uniquely beautiful crease of the Tyrolean Dolomites, their huge—and shockingly near—rough peaks encircling the valley like the ramparts of heaven.

It was still deep winter when the Borns arrived in May. They rented rooms from a farmer. "So we settled down to our lonely life," wrote Born.

But soon spring came to the Dolomites, in an ecstasy of little yellow troll flowers, so many that the alpine meadows were a mass of waving yellow. The mountain peaks were jagged against perfect spring skies. Born remembered being overwhelmed by the beauty. "Hedi said that she longed to send a telegram to Hitler or to his minister of education, Rust, thanking him for giving us this spring in the Alps."

As the spring turned to summer and the edelweiss began to bloom, Weyl arrived with Born's daughters, who had finished their school term staying with Weyl and his wife. With the girls came the Borns' Airedale, Trixi. Devoted to Weyl, Anny Schrödinger next appeared in Selva (she and Erwin had also fled the insanities of Berlin into the Tyrol, and were not far away). Pauli came with his sister, who had been an actress on the stage of Max Reinhardt's theater in Berlin until Nazi rules left her jobless. And one, and then another, of Born's students found where he was and came out to join him. "Thus," wrote Born in his autobiography, "the University of Selva was founded, consisting of a bench in the forest, one professor, and two undergraduates. I cannot remember what I tried to teach them; but they were a welcome addition to our circle and accompanied the girls on walks and climbing expeditions."

Pauli had tried to persuade Heisenberg to join them "for a little mountain climbing," but instead Heisenberg, trying to save physics in Germany, was begging Born to return. He did not think someone as prestigious as his old professor could "be affected" by the anti-Jewish laws; if Born returned, Heisenberg believed that "the political transformation could take place by itself without any sort of damage to Göttingen physics." He thought they could live with the dualities, just as in quantum physics, and "certainly in the course of time the ugly will separate itself from the beautiful" and politics oppress physics no more. "Therefore, I would like to ask you not to make any decisions yet, but to wait and see how our country looks in the fall." Born, reading this, hardly knew whether to laugh or cry. Heisenberg's "duck and cover" strategy would barely save his own skin, and precious few of his Jewish friends or students, and certainly not physics in Germany. Born typed up what Heisenberg had written him and sent it to Ehrenfest: here is how the world is seen by "our well-meaning German colleagues."

Switzerland remained almost eerily unaffected by all the turmoil in Germany. Pauli had been made a professor at the E.T.H., and had a circle of terrific students who spent all the time they could with him, studying physics by day and the Zurich nightlife after dark. Many of them, including Pauli himself, would flee for safety to the United States before long.

But in early 1933, their thoughts were on more tame terrors, like being stuck in a car with the recently licensed Pauli in the driver's seat. Ehrenfest's student Hendrik Casimir tells of Pauli's "slightly disconcerting habit of saying from time to time, 'I'm driving rather well'—a statement he underlined by turning around to his passengers and releasing his hold on the wheel." If he was tipsy, the effect was even more disconcerting. Once, when Pauli was ebulliently driving home drunk from a conference, he took a nonexistent shortcut, his car overfull with five alarmed young physicists (sitting on seats and on the floor).

Fifteen years later, Casimir happened to meet again the one who had been in the front passenger seat that night. "When I asked him, 'Dave, do you remember that drive from Lucerne to Zurich?' his answer came promptly: 'Will I ever forget it?' "

Bohr's acolyte Rosenfeld had just given a lecture on a new paper he had written with Bohr, in which they demonstrated that yet another extension of quantum mechanics was consistent with the uncertainty principle. Rosenfeld was pleased to see the attentiveness with which the famous face with the black mustache and white mane of hair followed his words. But when Einstein stood up for the discussion, he talked of "uneasiness." Then he asked, "What would you say of the following situation?

"Suppose two particles are set in motion toward each other with the same very large momentum, and that they interact with each other for a very short time when they pass at known positions.

"Consider now an observer who gets hold of one of the particles, far away from the region of interaction, and measures its momentum; then, from the conditions of the experiment, he will obviously be able to deduce the momentum of the other particle"—since their momenta are exactly opposite.

"If, however, he chooses to measure the position of the first particle, he will be able to tell where the other particle is"—by a quick calculation based on the wave equation. But sharp momentum and sharp position, of course, are two incompatible quantum mechanical states.

"This is a perfectly correct and straightforward deduction from the principles of quantum mechanics; but is it not very paradoxical?" Einstein asked with an expression of innocence. "How can the final state of the second particle be influenced by a measurement performed on the first, after all physical interaction has ceased between them?"

The particles were out of the box—Einstein was even closer to the version of this argument that would in two years finally catch the attention of Bohr (and, by extension, Rosenfeld). But at the time, Rosenfeld thought that Einstein was just making "an illustration of the unfamiliar features of quantum phenomena."

"What are you doing, O child with white hair, about the pain of one who is searching for answers?" Einstein's best friend, Besso, had written from Zurich the September before, begging him to pay attention to his son— perhaps he could take Eduard on "one of your long journeys." Einstein had written back that he would bring him along "next year," in 1933. But next year was too late. By then, psychiatry, electroshock treatment, insulin shock treatment, and institutionalization had cured the once vivid boy of his extraordinary mind.

In May 1933, on his way to Oxford, Einstein stopped in Zurich to see his son for what was to be the last time.

There is a photograph of this moment: father and son sitting side by side on a carved floral couch in some formal room. King or ceremony rarely saw Einstein so well dressed. The handsome Tetel, also in crisp gray suit, waistcoat, and tie, looks older than twenty-three. He scowls at a large book open on his knee while his father stares sadly straight ahead, his violin loosely held under his elbow, bow in hand.

Max Born, in the Dolomites, wrote to Einstein, care of Ehrenfest, telling him of all that had happened. On May 30, 1933, Einstein wrote from Oxford, "Ehrenfest sent me your letter. I am glad that you have resigned your positions (you and Franck). Thank God there is no risk involved for either of you. But my heart aches at the thought of the young ones. . . ." He did not mean Born's children, but Born's students. He wrote of his and Niels Bohr's attempts to find money or positions for these exiles.

He finished with a wry postscript: "I've been promoted to an 'evil monster' in Germany, and all my money has been taken away from me. But I console myself with the thought that it would soon be gone anyway."

In Oxford, Einstein gave a speech about Born's statistical explanation of the Schrödinger equation: "I cannot but confess that I attach only a transitory importance to this interpretation. I still believe in the possibility of a model of reality—that is to say, of a theory which represents things themselves and not merely the probability of their occurrence.

"On the other hand it seems to me certain that we must give up the idea of a complete localization of the particles in a theoretical model. This seems to me to be the permanent upshot of Heisenberg's Uncertainty Principle." This did not leave him hopeless: "But an atomic theory in the true sense of the word (not merely on the basis of an interpretation) without localization of particles in a mathematical model, is perfectly thinkable." He explained how a field theory could accommodate this, finishing, "Not until the atomic structure had been successfully represented in such a manner would I consider the quantum-riddle solved."

"Many thanks for your kind letter," Born wrote back by return mail. "I wish I could help you to look after the young exiled physicists and others like them, but"—he lifted his eyes from the little rough desk out at the protecting Alps—"I am in the same position myself. . . ."

As for physics, "I have not given up, by any means, but I share Ehrenfest's opinion that those who are younger have a better chance of achieving something."

Early in September 1933, another Copenhagen conference was wrapping up. Dirac met Ehrenfest, leaving, on the doorstep of Bohr's house.

His clipped, high British tones earnest and sincere, Dirac said, "I hope you know how extremely useful you are at a conference."

Ehrenfest's eyes widened and he turned and rushed back into the house, leaving the tall, thin Dirac standing on the steps.

The door reopened. Ehrenfest walked back down to the step where Dirac was standing, and grasped his arm. He was crying, silent tears running down his unshaven cheeks. Dirac, who never knew what to do in emotional circumstances, saw frightening things swirling behind Ehrenfest's eyes.

"Dirac," said Ehrenfest, "what you have just said"—he took a breath—"coming from a young man like you. . . ." His eyes were so intense, and red with tears, set back so deeply under those dark brows. For Ehrenfest,

Dirac stood for all that shining tangle of new physics he thought he could no longer follow. "What you have said means very much to me, because, maybe"—another breath; this was agony—"a man such as I feels he has no longer the force to live."

He remained on the step, holding Dirac's arm. Horrified, Dirac looked into his eyes, unable to speak. Ehrenfest seemed to grope for something more to say, but turned and left without another word.

A few weeks later, Ehrenfest walked into the waiting room of the Professor Watering Instituut in Amsterdam, where his fifteen-year-old son Wassilji, who had Down syndrome, was cared for. Hitler had just passed a law "for the prevention of genetically impaired progeny," starting the organized sterilization of people who were not likely to produce the "master race." Soon the Hitler-ordered "mercy deaths" of disabled children would begin, the job done by doctors, in their offices.

Ehrenfest walked up to the desk and said in Dutch, "My name is Paul Ehrenfest and I have come to see my son Wassik," calling his son by the nickname he always used. The receptionist placed a call, as Ehrenfest sat quietly in one of a group of identical chairs.

A nurse led Wassik to the waiting room. His face shone at the sight of his father. They walked out of the Professor Watering Instituut and into the late September day, to a park nearby.

Then Ehrenfest, the beloved "conscience of the physicists," pulled out a handgun and shot first his son and then himself.

Afterward an unmailed letter was found in Ehrenfest's desk. It was dated August 14, 1933, a little over a month before, and addressed to "My dear friends," including Bohr, Einstein, James Franck, and Richard Chace Tolman:

> I absolutely do not know any more how to carry further during the next few months the burden of my life which has become unbearable . . . it is as good as certain that I shall kill myself. And if that will happen some time then I should like to know that I have written, calmly and without rush, to you whose friendship has played such a great role in my life. . . .
>
> In recent years it has become ever more difficult for me to follow developments [in physics] with understanding. After trying, ever more enervated and torn, I have finally given up in DESPERATION. . . . This made me completely "weary of life." . . . I did feel "condemned to live on" mainly because of economic cares for the children. . . . I tried other things . . . but that helps only briefly. . . .

Therefore I concentrated more and more on ever more precise details of suicide. . . . I have no other "practical" possibility than suicide, and that after having first killed Wassik. . . .

Forgive me. . . .

May you and those dear to you stay well.

In a railway car sits Max Born, looking out at the wintry 3 a.m. stars as the train rushes far away from their Italian refuge in Selva. His son, Gustav, is asleep curled up in one of the seats. Trixi's furry black-and-tan head lies in Born's lap. He feels like the only soul awake for miles, as the terrible year of 1933 draws to its shuddering end. A vision of yellow troll flowers blooms in his mind. "We knew by now all the delightful walks and little climbs around Selva and we loved them very much." And now they are leaving—Hedi and the girls have already gone on ahead—for "a foreign country and an uncertain future." The foreign country is Scotland, where Born will live for the next thirty years. The dog whimpers in her sleep, and Born lays a hand on her wiry back.

In December, Heisenberg, Schrödinger, and Dirac—with Anny Schrödinger, and Dirac's and Heisenberg's mothers—disembarked at Stockholm Central Station. Heisenberg was coming from Leipzig, Dirac from Cambridge, and Schrödinger had just settled in Oxford. Heisenberg was going to receive the deferred 1932 Nobel Prize, and Schrödinger and Dirac the 1933 one.

The previous month, Heisenberg had desperately tried to keep his old rival in Germany, "since he was neither Jewish nor otherwise endangered." He was angry at Schrödinger's defection and refused to understand that there could be moral reasons to leave. Meanwhile, Schrödinger's old friend Weyl, whose wife was Jewish, resigned from his post in Germany at the same time as Schrödinger did, and left for Princeton.

At the ceremonial banquet, Schrödinger ended his toast by saying, "I hope that I may come again soon . . . not to a celebration hall bedecked with flags and not with so many formal clothes in my luggage, but with two long skis over my shoulder and a knapsack on my back." Dirac gave an incoherent toast about how "anything to do with numbers should be capable of a theoretical solution," even economic depression, and religion would only get in the way (reminding Heisenberg of the Solvay conference six years earlier when Pauli had declared, "There is no God and

Dirac is his prophet"). Heisenberg simply thanked everyone for their hospitality.

No prize for peace was given that year.

Heisenberg wrote to Bohr a month later from Zurich, where the mail was not owned by the Nazis, "Concerning the Nobel Prize, I have a bad conscience regarding Schrödinger, Dirac, and Born. Schrödinger and Dirac both deserved an entire prize at least as much as I do, and I would have gladly shared with Born."

Thus went the final celebration of quantum physics before the whole world—and with it the tiny world of physics—changed irreversibly with a second world war and the explosion of two bombs.

Hitler's power was increasing by the day. Europe was tensing herself for what atrocities she knew not. Refugees from Germany were arriving in Copenhagen: the gentle experimentalist James Franck; the Hungarian biologist Georg von Hevesy, who would become a beloved fixture there for the next decade, terrorizing the institute with radioactive cats and other non-*gedanken,* nonphysics, radiobiology experiments. Meanwhile, still buffeted by the death of Ehrenfest, Bohr was to receive another awful blow in the summer of 1934. Before his eyes, while sailing on their little boat, his seventeen-year-old son, Christian, the beautiful, his eldest, was swept into the choppy water where he immediately drowned.

Things fell apart ever faster as Hitler took over Austria and a pivotal piece of Czechoslovakia in 1938, with the assent of a Britain and France desperate to appease him. Early in 1938, Hitler's laws broke apart the team of erstwhile close friends, physicist Lise Meitner and chemist Otto Hahn. From the chilly safety of Stockholm, Meitner guided Hahn, back in Berlin, in splitting the uranium atom late that year—the first step to a nuclear bomb.

As a newly naturalized American, von Neumann visited Bohr and his brother Harald at Bohr's Copenhagen mansion that September. "I had numerous conversations with the Bohrs and Mrs. Bohr, of course mostly political—but we even managed to talk an hour and a half on 'the interpretation of quantum mechanics,' " von Nuemann wrote his fiancée. "I'm sure we were showing off, the both of us, giving an exhibition, that we can worry about physics in September 1938. . . .

"It's all like a dream, a dream of a peculiarly mad quality: This enormous house, with all the gadgets Mr. Jacobsen (who built it) put in: a great hothouse winter garden, a doric collonade, all covered with [neoclassical sculpture]—the Bohrs quarreling whether Czechoslovakia ought to give in—and whether there is any hope for causality in quantum theory. . . ."

16

The Quantum-Mechanical Description of Reality

1934–1935

1: GRETE HERMANN AND CARL JUNG

In a small basement room of the physics institute at Leipzig, Heisenberg and Carl Friedrich von Weizsäcker are playing Ping-Pong beside a blackboard covered with equations. Von Weizsäcker, once a fifteen-year-old sitting in the back of a taxi while his hero told him he had just refuted the law of causality ("In that moment," Carl Friedrich remembers, "I decided to study physics to understand this"), is now a philosophically astute twenty-one-year-old. With dark, well-coiffed hair, he bears a diplomatic expression befitting the son of an ambassador.

Ambassador von Weizsäcker was so diplomatic, in fact, that he managed to keep his position while his government convulsed from republic to dictatorship. Living in Leipzig for these last four fateful years, Carl Friedrich has become a central figure in Heisenberg's group, and probably Heisenberg's closest friend. But the experience of learning quantum physics from the source, thrilling as it has been, has contained a certain disillusionment at its core. Early on, von Weizsäcker found that his mentor "did not concern himself with the philosophical problems in physics," the very things "for the sake of which I had been tempted by him to study the subject."

Heisenberg is always saying reasonably, "Physics is an honest trade; only after you have learned it have you the right to philosophize about it." Bluff, pragmatic statements like these do not hide the fact that Heisenberg relies completely on Carl Friedrich, his junior by a decade, in matters of philosophy, not just in the realm of physics but also in that of politics.

Outside it is a nervous, edgy spring. Here in this basement room, though, with its equations and table tennis, there are no Nazis, no moral crises, no outside world. The young men's faces are intense, but only with the heat of the game; their jackets are thrown up on pegs near the door.

The only sound is the scuff of their shoes and the hollow *ping!* of the plastic ball on wood, followed by an occasional exclamation.

Above, a door opens. Down the stairs comes the unfamiliar tread of a woman's shoes. Striding into their sight, the woman is thin, and has the monochrome look of a greyhound, with a narrow, withdrawn face and boyish, slicked-back, dun-colored hair. Von Weizsäcker's father would certainly not approve of her radical political views, but within the walls of the institute such subjects do not carry quite such importance.

"Are you Herr Professor Heisenberg?" she asks. "I am here to prove you wrong."

Heisenberg, still holding his Ping-Pong paddle, blinks.

"I believe you have declared causality 'empty of content' as a result of your famous principle," she continues. Not too long before, Heisenberg had committed himself to a few intentionally inflammatory words in the newsprint of the *Berliner Tageblatt*. "You said, 'Now it is the task of philosophy to come to terms with this new situation.' " She raises her eyebrows.

Her name is Grete Hermann, she is exactly Heisenberg's age, and she does not want to "come to terms." One of the few female members of a group usually referred to as "Noether's boys," she studied under the great mathematician Emmy Noether in Göttingen. In her thesis, she had done original work on the foundations of what would become known as computer algebra—one of the fields (along with game theory) which von Neumann, her complete opposite in almost every way, so fruitfully established and dominated. After her Ph.D., she stayed in Göttingen to become the private assistant of the philosopher Leonard Nelson, an electric personality who took it upon himself to unite philosophy, mathematics, and ethics. Nelson had gathered around him a circle dedicated to the work of the neo-Kantian Jakob Fries, founding, as an outgrowth of this circle, his own political party, the International Socialist Combat League.

Now both mentors are gone. An insomniac, Nelson worked himself to a feverish early death in 1927—a sacrifice to pure reason and "international socialism." Noether was banished by the coarser calculus of National Socialism, and now teaches in America after losing her post for being insufficiently Aryan (the next year she would have lost her post anyway, for being female). Nelson died just before his superabundant energy would have been most needed, and Grete remains to marshal the Combat League into active resistance against the Nazis. She edits a Socialist daily called *The Spark* and organizes underground anti-Nazi colloquia, finding philosophy even more exhilarating when mixed with danger.

If Heisenberg and von Weizsäcker, standing there with their Ping-Pong paddles, knew all this, they would be astonished that a bit of intellectual provocation in the newspaper could bring her to their doorstep in such troubled times. But Heisenberg's declaration of the dethroning of Kant was more than academic for one whose whole system of ethics and political action—her whole life—is based on that source. It was for Kant's philosophy that her hero had essentially crucified himself.

She will not give in easily.

"What prevents us," she asks, "from assuming that, as our physical knowledge increases, new formulas and rules might be added to quantum mechanics? What prevents us from believing that these might render exact predictions possible again?" Her steady eyes hold Heisenberg's. "Everything depends upon the answer to this question."

Von Weizsäcker's face lights up at the prospect of a philosophical argument, the sort of thing for which he usually does not find any serious takers. He offers her one of the two chairs in the room.

Half sitting on the Ping-Pong table, Heisenberg gamely starts to explain. "Nature in fact tells us that no new determinants exist, that our knowledge is complete without them." Grete raises her eyebrows and crosses her legs, leaning back in the chair and folding her arms across her chest.

Heisenberg begins to walk, Bohr-like, in an elliptical curve around the Ping-Pong table. In an experiment on a quantum object, he explains, two worlds are meeting—a quantum world and a classical world, which behave in totally different ways. Something crucial is lost in that meeting. Bohr had just that year published a collection of his essays, titled *Atomic Theory and the Description of Nature.* Over and over in its pages he emphasizes that measurement disturbs what is being measured, and this disturbance "is of such a nature that it deprives us of the foundation underlying the causal mode of description." This concept is what he had used with such success in his Solvay conference triumphs over Einstein. It creates an intuitive understanding of Heisenberg's uncertainty principle and the mysteries of the collapse of the wavefunction, but Einstein (with the assistance of Podolsky and Rosen) is about to alert Bohr that it is wrong, or at least far less intuitive than they had thought.

"Is a deterministic completion of quantum mechanics possible?" asks Heisenberg. He turns to face Grete Hermann the short way across the table—as if they are both referees debating whether or not a ball touched the net—and answers his own question. "This is why not.

"Both on the quantum-mechanical side, with the object, and on the classical side, with the apparatus"—he puts down his hands, leaning forward over the net toward his two-person audience—"the laws of that side hold precisely. The statistics come in here"—he grabs the net with a smooth and sudden gesture—"at the *Schnitt,* the cut. You can't measure a particle without disturbing its causal course.

"Now, if you were going to hope for, as you said, 'new formulas and rules' to be added to the quantum theory and bring causality back to it, they have to enter along the *Schnitt.*" He shakes the net. "But the *Schnitt* can always be moved—you can always describe something quantum-mechanically which you were describing classically before; you can always include a little more of your apparatus in the quantum-mechanical system, as long as some part of the apparatus remains classical." He walks back around the table to sit back down on it, legs swinging, facing Grete and Carl Friedrich. "But when the *Schnitt* moves, a contradiction between the lawlike consequences of the new hidden properties and the more fluid *relationships* of quantum theory will be unavoidable.

"Hence a deterministic completion of quantum mechanics is not possible. Von Neumann has just written a book, and one of the chapters shows this with much greater rigor."

"I've read that book," says Grete.

"Wonderful," says Heisenberg; "well, then, there you have it."

"I think if you look Neumann's crucial section over again," she says distinctly (no "von" for her!), "you'll notice that a necessary step to prove the nonexistence of such distinguishing characteristics involves the implicit assumption of their nonexistence."

Heisenberg and von Weizsäcker stare at her. "What did you say?"

"The possibility of other characteristics, upon which the course of the motion actually depends"—that is, the possibility of hidden variables—"is not excluded by Neumann's *results* so much as by his *assumptions,*" Grete says.

Heisenberg gets up silently and hands her an eraser and a piece of chalk from the dozen lying end to end along the blackboard's chalk tray.

Turning around in her chair, she clears a little space on the board and writes a short equation just behind her. $<P + Q> = <P> + <Q>$ sits innocently in its chalky clearing: the expected value of the position *and* momentum of a particle both measured at the same time is equal to the sum of the expected value for a measurement of position and the expected value for a measurement of momentum.

"With this assumption" (which is true enough for quantum mechanics) "stands and falls Neumann's proof," she says. "He's trying to read off the impossibility of such a deepening of our knowledge from your uncertainty relations." She looks over at Heisenberg. "If the position and momentum of a particle cannot both be measured with arbitrary accuracy, then how could one gain secure knowledge of the future trajectory, which is just determined by the present position and momentum of the body?" Von Weizsäcker frowns meditatively. Heisenberg's brow is furrowed.

"But this argument is based on a *subjective* view of the uncertainty relations. It's looking at the electron merely as a *particle,* and saying, since we can't ever know what the position and momentum are exactly, and its future trajectory is decided by these things, then the cause will forever be concealed from observation.

"But this is ignoring the fact that the electron is not at all just a classical particle, but it is also a wave. Because of this fact, the uncertainty relations are not about our limited *knowledge* but about the nature of the world." This prompts a nod from Heisenberg.

"If according to this reasoning the electron does *not* simultaneously have an exact position and an exact momentum, then its exact position and its exact momentum *cannot* be decisive for its further motion," Grete continues. "Having dropped this assumption, it becomes an open question whether one could not find *other* characteristics upon which the course of the motion would actually depend causally. Just because the formalism of quantum mechanics does not acknowledge such characteristics, it does not follow that one is justified to declare them impossible."

The two men are staring at her as if she has dropped from the sky.

"As far as I can tell," Grete says, smiling slightly and standing up, "it's just another example of the well-known fact that there can only be a single sufficient reason to renounce as fundamentally futile any further search for the causes of an observed process: *that one already knows these causes.* I am here to find them."

Heisenberg raises his eyebrows but he is nonetheless impressed. "We are going to have some good discussions this term," he says to von Weizsäcker. "Welcome to Leipzig, Fräulein Grete Hermann. You'll join in with us for the whole term, then?"

"Yes, I will," she says.

As Grete walks back up the stairs, she hears the Ping-Pong game recommencing.

Around the same time, in 1934, Pauli's life was beginning to make sense again. For the previous three years, the combined effect of his mother's suicide, his father's speedy remarriage, and his own divorce (his bride having run off with a "mediocre chemist") had been to force Pauli ever deeper into drink and depression. Recovery had come through "becoming acquainted with psychic matters" and "the proper activity of the soul," as he told his friend and assistant Ralph Kronig in October 1934, signing the letter: "Your old and new W. Pauli."

He had been introduced to "psychic matters" two years before by his new colleague at the E.T.H., Carl Jung. Emerging on the psychiatric scene in Zurich as Sigmund Freud's great rival, Jung theorized a "universal unconscious" full of primal "archetypes" that has since intrigued intellectuals with its supposedly uncanny echoes of the quantum theory. When Pauli met Jung, he saw the psychiatrist as offering a possible way out of his alcohol-soaked depression. But Jung had a more opportunistic reaction to Pauli when the formerly buoyant physicist came for his advice. As he described Pauli (without naming him) in his famous Tavistock lecture of 1935, "In that interview, I got a very definite impression of him: I saw that he was chock-full of archaic material, and I said to myself: 'Now I am going to make an interesting experiment to get that material absolutely pure, without any influence from myself. . . .' So I sent him to a woman doctor who was then just a beginner and who did not know much about archetypical material." (He did not mention that the "beginner" was one of his disciples, soaking up the dogma on archetypical material right from the source.)

Even in the fog of depression, Pauli was capable of fighting back to some extent, writing to the "woman doctor" in February of 1932, "I consulted Herrn Jung because of certain neurotic phenomena which make it easier for me to achieve academic successes than successes with women. Since with Herrn Jung rather the *contrary* is the case, he appeared to me to be quite the appropriate man to treat me."

And treat him he did, after half a year of the "experiment." In two years of Jung's personal analysis and friendship Pauli shed his depression. In 1934 he met and married Franca Bertram, who would be his companion for the rest of his life. Whether it was Jung or the sensible Franca who had really brought about Pauli's recovery, the psychoanalyst had left an indelible imprint on the *Geissel Gottes* (God's Whip). Pauli tapped into a fuzzy-thinking and credulous side of himself that Bohr and Heisenberg would barely recognize, thrilled when Jung would use his "material" in lectures, absorbed by the baroque richness of Jung's symbolism, and above all the cloudy world of dreams.

But in the autumn of 1934, half a year after his marriage to Franca, Pauli wrote to Jung, "I feel a certain need to get away from dream interpretation and dream analysis, and I would like to see what life has to bring me from the *outside*." He enclosed with the letter a token that suggested he was not taking his leave for good: a recent paper by Jordan, not on physics but on telepathy. Pauli felt that poor Jordan—driven inward by the oppression of an almost insurmountable stutter—had come close to Jung's idea of the "collective unconscious."

Jung, meanwhile, was seduced by a series of experiments on telepathy at Duke University performed by a botanist named J. B. Rhine. Rhine's monograph *Extra-Sensory Perception* (a term of his own coinage, usually shortened to "ESP") had just been published in 1934. Inspired to study the paranormal by a Sir Arthur Conan Doyle lecture, Rhine had displayed the extent of his scientific rigor quite early, in 1927—the year of Heisenberg's uncertainty principle, the Pauli matrices, and the Bohr-Einstein debate—by earnestly declaring a horse named Lady Wonder telepathic. A magician who investigated her the same year discovered that in fact she was reading subtle cues from her trainer's stance and expression.

Undeterred by the equine debacle, Rhine moved on to telepathic humans. In the most famous of his experiments, just finished in 1934, Rhine's graduate student turned over a deck of twenty-five "ESP cards," one per minute, in their parapsychology room on the top floor of the psychology laboratory. Across the Duke quadrangle in a cubicle in the library stacks, a clairvoyant divinity student attempted to guess which of five simple symbols was on each card. When they rejoined, they compared results. After seventy-four runs of twenty-five cards each, Rhine declared that the findings were statistically significant, with 10 percent more cards called correctly than mere chance would predict. Jung thought it was wonderful; Pauli, initially, would have none of it.

To the end of his life, Jung believed that the Rhine experiments had provided "scientific proof" of ESP, despite the increasing awareness that Rhine, though sincere, had only a tenuous grasp on the details of scientific method. "These experiments prove that the psyche at times functions outside of the spatio-temporal law of causality," Jung explained. "This indicates that our conceptions of space and time, and therefore of causality also, are incomplete."

While Jung's credulity is not a surprise, his power of persuasion over Pauli is. From his early twenties, Pauli could tell Bohr to "shut up" and Einstein that his ideas were "actually not so stupid." Feared by men with the stature of Ehrenfest and Max Born, Pauli was at his happiest with a letter from Heisenberg in his fist, "walking around like a caged lion in our apartment," as his first wife remembered, "formulating his answers in the most biting and witty manner possible." But, in the words of Franca

Pauli, "the extremely rational thinker subjected himself to total dependence on Jung's magical personality."

In 1950, Pauli wrote to Jung, "We had basically agreed in the past on the possibility and usefulness, and also (in view of the Rhine experiments) on the necessity, of a further principle for the interpretation of nature, other than the causal principle." They discussed a complementarity between causality and "synchronicity" (Jung's word to describe significant coincidences). Three years later, the subject came up again, Pauli speaking of the "observer" and "measurement" of quantum mechanics: "Today I do in fact believe that it is possible for the same archetype to be in evidence both in the selection of an experimental setup by an observer as well as in the result of the measurement," Pauli wrote to Jung, "—similar to the dice in Rhine's experiments."

A dream he had had in the waning months of 1934, soon after he had stopped seeing Jung, was still haunting him: "A man resembling Einstein was drawing a graph on the blackboard." In the dream, Einstein drew a simple upward-slanting line, labeled QUANTUM MECHANICS, bisecting a hatched surface representing DEEPER REALITY. "I saw quantum mechanics—and so-called official physics in general—as a one-dimensional *section* of a two-dimensional, more meaningful world," Pauli continued, "the second dimension of which could only be the unconscious and the archetypes."

Grete Hermann returned to Göttingen armed with all she had learned in a semester with Heisenberg and a philosophical friendship with von Weizsäcker. She had waded into deep complementarian waters, and emerged clasping her own personal harmonization of Kant and the quantum theory—far more important to her than any demonstration of the faultiness of von Neumann's no-hidden-variables proof. From the shaking pillars of quantum physics still upheld in Leipzig—the correspondence principle and the idea that "measurement disturbs the system"—she constructed an elaborate defense of the completeness of the Copenhagen interpretation (as any combination of the quantum-mechanical ideas of Bohr, Heisenberg, and Pauli—plus those of the often forgotten Born—came to be called) that impressed Heisenberg and von Weizsäcker.

She even included von Weizsäcker's version of Einstein's light-*blitz* box, which allowed her to make the important conclusion (soon to be emphasized by Bohr) that "the quantum-mechanical characterization is not, like the classical one, attributed to the physical system, as it were, 'in itself.' " That is, she explained, the state of a particle is not independent "of the observations through which one acquires knowledge of it": measurement doesn't just disturb, but *creates,* the measured characteristics. But like Heisenberg and von Weizsäcker, she didn't seem to notice that if it does so, it does so at a distance. The Copenhagen interpretation does not provide the language that allows a physicist to notice that something looks like spooky action-at-a-distance, let alone to recognize entanglement as a specific, quantifiable phenomenon. Grete, like so many others, passed over it in silence.

The resulting paper, in its most widely read form, ended up in the pages of *Die Naturwissenschaften,* a journal founded to foster communication between the many branches of the natural sciences. This was a dialogue that mathematical formulae were deemed to inhibit, and Grete's incisive passages on "Neumann" were left out. Nelson's neo-Kantian journal also printed the paper, in its entirety, but this was hardly a place where the devotees of von Neumann's defective proof were likely to ever discover it.

The year after her paper was published, 1936, Grete was forced to lay aside any further meditations on the foundations of quantum mechanics. Her anti-Nazi stance seriously endangering her life, she fled to Denmark and then England, throwing herself into the London branch of the International Socialist Combat League.

Thus it was that no one beyond Heisenberg, von Weizsäcker, and Grete's embattled circle knew of her dismissal of von Neumann's no-hidden-variables argument. And none of them, for mysterious reasons of their own, said anything more on the subject.

2: EINSTEIN, PODOLSKY, AND ROSEN

In 1934, Nathan Rosen was working at Princeton, having received his doctorate under Slater and married his high school sweetheart, Anna, a musicologist, fine art critic, and pianist. One day he knocked tentatively on the door of Einstein's Room 209 in Fine Hall. The Institute for Advanced Study was essentially being built around the world-famous émigré, who had arrived the previous October. For now, it was housed in the spare rooms of the Princeton math department, and at the time, von Neumann and Einstein made up half the faculty. After moving heaven and earth to bring his assistant, Walter Mayer, along with him and install him at the institute out of harm's way, Einstein found Mayer becoming restive at the thought of another assault on the Everest of unified field theory. Now here was Rosen—a gentle, boyish, eager-looking twenty-five-year-old from Brooklyn, who wanted to talk about Einstein's first unified field theory, the subject of Rosen's master's thesis under Slater.

The next day, walking across the green, Rosen got a jolt when Einstein approached him, saying with his heavy German accent, "Young man, what about working together with me?"

When Einstein and Rosen would meet in Room 209, it was not to discuss what the implications would be if two codependent atoms, forming one of the hydrogen molecules Rosen studied, were to separate like Einstein's *gedanken* photon from its *gedanken* box. Instead, they were peering into the giddy depths and vast expanses of Einstein's field equations. Emerging from these equations was a fascinating phenomenon. A black hole—a star, perhaps, that has collapsed so far under its own gravity that even light cannot escape from its pull—would create a tiny rip in space-time at its center, the eye of the hurricane. If two of these rips were to line up, two distant portions of space-time might be connected in a kind of mysterious shortcut—what was soon being called an Einstein-Rosen bridge (and would later be referred to as a "wormhole").

Also at the Institute for Advanced Study early in 1935 was Boris Podolsky. Already acquainted with Einstein and his series of light-*blitz* boxes, he learned of Rosen's analysis of the hydrogen molecule. It seems that it was Podolsky who put two and two together and realized that Rosen's intertwined twin hydrogen atoms formed a pre-existing case capable of demonstrating what Einstein had been talking about but never published.

Einstein, Podolsky, and Rosen's discussions resulted in a paper titled "Can Quantum-Mechanical Description of Physical Reality Be Considered Complete?" Einstein reported to Schrödinger that "for reasons of language"—Einstein at the time was limited to about five hundred English words—"it was written by Podolsky after much discussion." (The absent "the" in the title was an artifact of the English of someone whose first language was Russian.) Less certain is how the collaboration between the three authors proceeded. Rosen's memories on the subject were vague. Podolsky told his son that he and Rosen would talk with Einstein "when they thought they were onto something." Late in his life, Podolsky impishly, and unforgettably, told John Hart, his colleague in the physics department, "We added Einstein's name without asking."

The EPR argument, in any case, traveled along the same lines that Einstein's series of light-*blitz*-box *gedanken* experiments had done, though presented more thoroughly and with more complicated logical and quantum-mechanical analysis. Two "systems" (be they particles or boxes) interact and then separate. From a measurement of momentum on one system, the experimenter can learn the momentum of the faraway, untouched system. But if the experimenter decided to measure the position, instead, the position of the faraway system could be computed from the quantum-mechanical wavefunction of the nearby one.

So at this stage in the argument, there is an alternative: either measure the momentum here to learn the momentum there, or measure the position here to learn the position there.

But the paper famously (and significantly) defined an "element of reality": "*If, without in any way disturbing a system, we can predict with certainty the value of a physical quantity, then there exists an element of physical reality corresponding to this physical quantity.*" In that case, should not both characteristics of the faraway "system"—position and momentum—be deemed elements of reality? And if so, was not quantum mechanics incomplete for stating otherwise?

The last two paragraphs of the paper were similarly significant. "One would not arrive at our conclusion if one insisted that two or more physical quantities can be regarded as simultaneous elements of reality only *when they can be* simultaneously *measured or predicted,*" EPR admitted. "On this point of view, since either one or the other—but not both simultaneously—of the quantities P [momentum] and Q [position] can be predicted, they are not simultaneously real." But they were skeptical. "This makes the reality of P and Q depend upon the process of measurement carried out on the first system, which does not disturb the second system in any way. No reasonable definition of reality could be expected to permit this.

"While we have thus shown that the wave function does not provide a complete description of the physical reality, we have left open the question of whether or not such a description exists. We believe, however, that such a theory is possible."

Podolsky had to leave for California at about the time of the paper's submission, and it is not clear that Einstein saw it before he was suddenly confronted, on May 4, 1935 (eleven days before the paper appeared in *Physical Review*), with an article on page 11 of the Saturday *New York Times* entitled "EINSTEIN ATTACKS QUANTUM THEORY," complete with a hundred-word exegesis that the newspaper attributed to Podolsky.

But whatever is buried in history and in the complications of working with someone whose every move is deemed newsworthy, Podolsky had done a great service to the history of physics in committing the EPR argument to print. (Einstein, with his eyes on the big picture, sometimes neglected to do so. Von Neumann's proof of the impossibility of hidden variables is a case in point. Sometime around 1938, Einstein was sitting in his office in the Institute for Advanced Study with his assistants, Peter Bergmann and Valentin Bargmann, when the subject of von Neumann's proof came up in conversation. Einstein proceeded to open von Neumann's tome and point to the same assumption that, unbeknownst to him, Grete Hermann had just criticized a few years earlier. "Why should we believe in that?" he asked, and the conversation moved on. Einstein's distractedness in not alerting anyone, beyond the equally distracted Bergmann and Bargmann, marks the second time that von Neumann's proof managed to escape wider public knowledge of its error.)

When Podolsky applied for a position at the University of Cincinnati, where he would teach for a quarter of a century, Einstein wrote his reference: "Podolsky always goes directly to the heart of the problem."

3: BOHR AND PAULI

Einstein's judgment on Bohr's physics was like the old Sufi story of the seven blind men and the elephant. "An elephant," says the first, who has felt the flapping ears on a hot day, "is like a fan." "No, no, no, that's not right at all," says the second, who once got in the way of the tail swatting a fly. "An elephant is like a rope." The third, who has stumbled against a strong leg, says, "She is like a tree." The fourth, who has felt the warm, smooth tusk, declares her like a spear. The fifth, who walked by while she was bathing and was doused by her trunk, finds the elephant to be more like a hose. The sixth blind man muses that we must always speak of this ineffable creature in "classical" terms, like hose and tree, and these terms must be used in "complementary" ways, depending on what we are measuring: the elephant will not simultaneously feel like a fan *and* like a rope, but both are needed for the complete description. "We must, in general, be prepared to accept the fact," he says, "that complete elucidation of one and the same object may require diverse points of view which defy a unique description."

The seventh blind man is the mahout, and he walks away laughing.

Einstein was not the mahout, but he could hear the laugh: "Subtle is the Lord," he said, "but He is not malicious."

A thousand identical *Physical Review*s, volume 47, rolled off the presses in mid-May 1935, and began their journey out into the world. A thousand page 777s fatefully asked if the quantum-mechanical description of physical reality could be considered complete.

Almost immediately there were three radically different responses from the three people—Bohr, Pauli, and Schrödinger—who mattered the most.

At Bohr's institute in Copenhagen, "this onslaught came down on us as a bolt from the blue," remembered Bohr's assistant, Rosenfeld. Finally Einstein had really caught Bohr's attention. "Its effect on Bohr was remarkable. A new worry could not come at a less propitious time. Yet, as soon as Bohr had heard my report of Einstein's argument, everything else was abandoned: we had to clear up such a misunderstanding at once."

Bohr started with great confidence to take the same *gedanken* experiment and show "the right way to speak about it." But soon he hesitated, fumbling, his brow furrowing as his eyebrows overshadowed his eyes. "No . . . this won't do. . . . We must make it quite clear." He tried again and again, "with growing wonder at the unexpected subtlety of the argument."

Silence reigned, punctuated by Bohr suddenly turning to his assistant: "What *can* they mean? Do *you* understand it?"

Bohr later told an interviewer that when Dirac heard of EPR, he had a similar reaction: "Now we have to start all over again, because Einstein proved that it does not work."

The evening had drawn late, and Bohr was bewildered. "Well," he said, "I must sleep on it."

Pauli, meanwhile, was pacing around his apartment in Zurich, composing a letter to Heisenberg in as devastating terms as possible. "Einstein has once again expressed himself publicly on quantum mechanics, namely in the issue of *Physical Review* of 15 May (with Podolsky and Rosen—no good company, by the way). As is well known, this is a catastrophe every time it happens."

Pauli ended in a flourish:

> *Weil, so schließt er messerscharf*
> *Nicht sein kann, was nicht sein darf.*

These are the last two lines from the beloved nonsense poet Christian Morgenstern's *The Impossible Fact,* in which (in English translation) "Old Palmström, an aimless rover / walking in the wrong direction / at a busy intersection / is run over." Palmström ignores his death, for, in the final verse of the poem, it became:

> *as clear as air:*
> *Cars were not permitted there!*
> *And he comes to the conclusion:*

His mishap was an illusion,
And so he reasons pointedly:
That cannot be which should not be.

"I'll grant him," Pauli continued snidely, "that if a student in the early semesters had made such objections to me, I would have regarded him as very intelligent and hopeful. Because a certain danger exists of a confusion of public opinion—namely, in America—it might be good to send a comment to *Phys. Rev.*, and I'd like to convince YOU to do it." Pauli went into great depth for pages discussing this supposedly trivial argument and trying to prepare Heisenberg to think about it in the right way. It was all, of course, a tempest in a teapot, he wanted Heisenberg to understand: "It is probably only because I recently got an invitation to Princeton for the next winter semester that I have gone to such trouble on these things, which for us are just trivialities. It will be a lot of fun to go there: in any case, I want to make the Morgenstern motto popular. . . .

"Elderly gentlemen like Laue and Einstein"—both fifty-six to Pauli's and Heisenberg's thirty-five—"are haunted by the idea that quantum mechanics is correct but incomplete. They think that it can be completed by statements which are not part of quantum mechanics, without changing the statements which are part of quantum mechanics. . . . Maybe you could—in the reply to Einstein—clarify with authority that such a completion of quantum mechanics is impossible without changing its content."

In fact, just such an "impossible" hidden-variables completion already existed, presented by de Broglie to deafening silence at the 1927 Solvay conference. It would take the independent resurrection of the idea in 1952 (with even less positive response) to lead John Bell to his discovery, no thanks to Pauli.

"Quite independently of Einstein," Pauli continued, "it appears to me that, in providing a systematic foundation for quantum mechanics, one should *start* more from the composition and *separation* of systems than has until now (e.g., with Dirac) been the case. This is indeed—as Einstein has *correctly* felt—a very fundamental point."

Physical Review reached Schrödinger in Oxford, where he was cloistering himself from the Nazi world (but finding the all-male professorial dinners of the university a little stifling). The EPR paper struck him, as it had Bohr, like a bolt from the blue. But for Schrödinger, the lightning strike was one of inspiration. He wrote to Einstein, "I was very happy that in the paper just published in *P.R.* you have evidently caught dogmatic q.m. by the coattails with those things that we used to discuss so much in

Berlin." He analyzed the situation mathematically, and unlike Bohr, was already focusing on what, in two months, he would christen "entanglement."

In Copenhagen the next morning, Bohr walks through the door looking ecstatic. "Podolski!" he says with a swashbuckling motion of his hand. "Opodolski, Iopodolski, Siopodolski, Asiopodolski, Basiopodolski!"

Rosenfeld is startled, as well he might be.

Bohr's whole face lights up in a big grin. "Just a few appropriate lines from Holberg" (Ludvig Holberg was the poet, thinker, and prolific playwright of early eighteenth-century Copenhagen who single-handedly legitimized Danish as a literary language), "where the servant comes in and starts talking nonsense in *Ulysses von Ithaca.*"

Rosenfeld is still bewildered, but Bohr is exuberant: "Well . . . let's write that paper."

". . . that paper?" echoes Rosenfeld.

"Our response to Einstein and Podolsky and Rosen."

Rosenfeld begins nodding. "Oh, of course, right, right."

"The trend of their argumentation," says Bohr, "does not seem to me to adequately meet the actual situation with which we are faced in atomic physics."

Rosenfeld is nodding faster, eagerly awaiting the solution; he had thought as much.

"We will show that their criterion of physical reality contains an essential ambiguity when it is applied to quantum phenomena," says Bohr. "I shall therefore be glad"—he really does look glad—"to use this opportunity to explain in somewhat greater detail a general viewpoint" (he smiles at Rosenfeld), "conveniently termed 'complementarity'— which I have indicated on various previous occasions . . . and from which quantum mechanics within its scope would appear as a *completely rational* description of physical phenomena."

"It's so strange that Einstein won't appreciate complementarity," muses Rosenfeld, "when it is so similar with his own approach to these problems."

"Exactly," says Bohr. "I would also like to emphasize that—perhaps near the end of the paper. Complementarity—this new feature of natural philosophy—means a radical revision of our attitude as regards physical reality, in striking analogy, as has often been noted"—nodding once to Rosenfeld with joking formality—"with the fundamental modification of ideas brought about by the general theory of relativity. Once one is capable of making that revision in one's attitude, everything else falls into place."

Rosenfeld smiles with relief. "You seem to take a milder view of the case this morning."

"That's a sign that we are beginning to understand the problem," says Bohr. "It all became very clear to me when I began to consider the simple case of a particle passing through a slit in a diaphragm." He begins to pace. "So let us begin there." Rosenfeld picks up his pencil and pad. "Even if the momentum of the particle is completely known—" Bohr stops walking to explain, "My main purpose in wanting to repeat these simple, and in substance well-known, considerations, is to emphasize that in the phenomena concerned we are not dealing with an incomplete description. . . ." He pauses, then expands his sentence. "We are not dealing with an incomplete description characterized by the arbitrary picking out of different elements of physical reality at the cost of sacrificing other such elements—but with a rational discrimination between essentially different experimental arrangements and procedures. Any comparison between quantum mechanics and ordinary statistical mechanics—however useful it may be for the formal presentation of the theory—is essentially irrelevant."

Rosenfeld looks up from his transcription. "It doesn't have to do with ignorance. It's actually *impossible* to know more."

"That's right," says Bohr. "Let's put that in: Indeed"—he begins to walk again—"we have in each experimental arrangement not merely to do with an ignorance of the value of certain physical quantities . . ." He trails off, then retraces his steps. "In each experimental arrangement *suited for the study of proper quantum phenomena,* we have not merely to do with an ignorance of the value of certain physical quantities, but with the impossibility of defining these quantities in an unambiguous way.

"As in the simple case, in the special problem treated by Einstein, Podolsky, and Rosen, we are just concerned with a *discrimination between different experimental procedures which allow of the use of complementary classical concepts*—that should be underlined, Rosenfeld. We are concerned with a discrimination between different experimental procedures which allow of the *unambiguous* use of complementary classical concepts." Rosenfeld nods, underlines; Bohr stops and turns to him.

"Oh, Rosenfeld, you can imagine the relief I felt when this all began to make sense to me. . . . Last night, I was almost in despair." A broad smile. "But now!" He resumes pacing and dictating.

"We now see that the wording of the above-mentioned criterion of physical reality proposed by Einstein, Podolsky, and Rosen contains an ambiguity as regards the meaning of the expression 'without in any way disturbing a system.' Of course there is in a case like that just considered no question of a *mechanical* disturbance of the system under investigation during the last critical stage of the measuring procedure." (Good-bye to detailed mechanical drawings showing how one measurement physically disturbs the other: Bohr has advanced, or retreated, to a higher level of abstraction.) "But even at this stage there is essentially the question of *an influence on the very conditions which define the possible types of predictions—*"

Rosenfeld is writing as fast as he can. Without looking up, he asks, "Could you repeat that last sentence?"

Bohr obliges. "There is essentially the question of—this should be underlined, too—*an influence on the very conditions which define the possible types of predictions regarding the future behavior of the system.*

"Since these conditions constitute an inherent element of the description of any phenomenon to which the term 'physical reality' can be properly attached, we see that the argumentation of the mentioned authors does not justify their conclusion that quantum-mechanical description is essentially incomplete." Bohr turns on his heel.

Einstein, Podolsky, and Rosen have, of course, already anticipated this reply ("One would not arrive at our conclusion if one insisted that two or more physical quantities can be regarded as simultaneous elements of reality only when they can be simultaneously measured or predicted") and reiterated, as Einstein has been saying since 1930, that this made the reality of one thing depend on measurements made on the other. But Bohr has bigger fish to fry. He wants to win Einstein over to complementarity.

"In *fact*," Bohr expands, "it is only the mutual exclusion of any two experimental procedures which provides room for new physical laws, the coexistence of which might at first appear irreconcilable with the basic principles of science. It is just this entirely new situation that the notion of *complementarity* aims at characterizing." Bohr falls silent for a bit, then continues.

"We see throughout, the necessity of discriminating, in each experimental arrangement, between those parts of the physical system which are to be treated as *measuring instruments* and those which constitute the *objects under investigation.*" He stops. "It is true that this choice is a matter of convenience. . . ." He starts pacing again. "But it is of fundamental importance, because we must use classical concepts to interpret all quantum-mechanical measurements."

"Necessarily involving what Heisenberg calls the *Schnitt*," says Rosenfeld, looking up. The *Schnitt* (the cut) was the movable (thirty years later, John Bell would call it "shifty") split between the quantum under investigation and the classical measuring apparatus, which is ultimately made of quantum-mechanical atoms.

Bohr nods. "It is impossible to make a closer analysis of the reactions between the particle and the measuring instrument. We have to do here with a feature of *individuality* completely foreign to classical physics." This would be about as close as Bohr would come to entanglement—a suggestion that further analysis of the complicated interactions of measurement be precluded.

"The very existence of the quantum," Bohr explains, "entails the necessity of a final renunciation of the classical ideal of *causality* and a radical revision of our attitude toward the problem of physical reality." He stops again, looking pleased. "That's what I've thought out so far, but I do believe I am on the right track."

Though how closely anyone can really follow this argument is an open question, Rosenfeld is awed. "Their whole argumentation falls to pieces—for all its false brilliance."

"They do it *smartly*," says Bohr, "but what counts is to do it *right*."

Rosenfeld muses. "If I understand you rightly," he says, "this is really a case in which the authors paid too much attention to their own preconceived notions of reality instead of taking guidance, humbly—as you are always exhorting us to do—in what we can learn from nature herself."

Bohr starts pacing again. "Mmm. Well, let's not get out of hand. We've got to make sure we've really cracked it open. Shall we go back and work over their argument again? I want to work out what role the idea of *time* plays in the description of this phenomenon. . . ."

4: SCHRÖDINGER AND EINSTEIN

Einstein had not yet received Schrödinger's letter when, on June 17, he wrote to him of Bohr's point of view: "I consider the renunciation of a spatio-temporal setting for real events to be idealistic, even spiritualistic. This epistemology-soaked orgy ought to burn itself out." He was not sure where Schrödinger stood in all of this: "No doubt, however, you smile at me and think that, after all, many a young whore turns into an old praying sister, and many a young revolutionary becomes an old reactionary."

The next day, Schrödinger's letter arrived, and Einstein thanked him for it, explaining that he had not written the paper himself and apologizing that it "did not come out as well as I had originally wanted; rather the essential thing was, so to speak, smothered by the formalism." For example, he explained, "I don't give a *sausage*" whether or not incompatible observables—Bohr's favorite subject—are involved.

It all came down to the relationship of Schrödinger's equation to reality. What is the connection between the mathematical description of events, and the events themselves? In what way does the Schrödinger wavefunction, ψ, reflect the actual state that a particle found itself in? Reality, or the particle's real situation, is represented in these discussions by the word *state* or the phrase *state of affairs*. The wavefunction, ψ, must represent this real state of affairs somehow. But it was hard to even articulate what was meant by such a connection to reality, or even what was meant by *reality* or *state*.

In his letter to Schrödinger, Einstein characteristically cut through this briar patch of linguistics with a parable. He wanted to illuminate the main point that had been obscured in the EPR paper. "In front of me stand two boxes, with lids that can be opened, and into which I can look when they are open. This looking is called 'making an observation.' In addition there is a ball, which can be found in one or the other of the two boxes where an observation is made. Now I describe a state of affairs as follows: *The probability is one-half that the ball is in the first box.*" (This is all the Schrödinger equation will tell you.) "Is this a complete description?" asks Einstein, and then gives two different answers.

"*NO:* A complete description is: the ball *is* (or is not) in the first box. . . .

"*YES:* Before I open the box the ball is not in *one* of the two boxes. Being in a definite box only comes about when I lift the covers. . . .

"Naturally, the second 'spiritualist' or *Schrödingerian* interpretation is absurd," Einstein continued tactfully, "and the man on the street would only take the first, *Bornian,* interpretation seriously." Born might not have recognized his interpretation, which Einstein seemed to be using in this description only as far as he wanted to, but presumably Bohr would have recognized himself, even without being named: "But the Talmudic philosopher whistles at 'Reality' as at a bugaboo of naïveté, and declares that the two conceptions differ only in their mode of expression. . . .

"One cannot get at the Talmudist if one does not make use of a supplementary principle: the *separation principle,*" Einstein explains. "The contents of the second box are independent of what happens to the first. If one holds fast to the separation principle, only the Born description is possible, but now it is incomplete."

The flood of letters on the subject of EPR—Einstein to Schrödinger, Schrödinger to Pauli, Pauli to Heisenberg, Heisenberg to Bohr—

continued unchecked throughout the summer, sometimes as many as three of them being written on the same day.

"I'd really like to know," Schrödinger wrote to Pauli about EPR, "whether you really think that the Einstein case—let's call it thus—doesn't give anything to think about, but is completely clear and simple and self-evident. (This is what everybody said with whom I talked about it for the first time, because they had well learned their Copenhagen *credo in unum sanctum*. Three days later usually there came the statement: 'What I said earlier was of course wrong, much too complicated.' . . . But I've not yet received a clear explanation of why everything is so clear and simple. . . .)

"So hearty greetings, dear friend, from your old Schrödinger."

Schrödinger complained to Pauli of the murky use of the word *state* ("a word," he wrote, "that everyone uses, even Saint P.A.M."—Dirac— "but that doesn't add to its content"), and Pauli responded to this instantly.

"In my opinion *there is simply no problem*," Pauli replied about EPR, "and we know this state of affairs even without the Einstein example." Pauli believed, as he later wrote, that "an observer, by his indeterminable effects, creates a new situation" (and this observer-created situation is a quantum "state"; the observer creates reality by observing it). It was necessary, according to Pauli, that the process of "measurement" be an ineffable, indescribable, lawless event, and its outcome "like an ultimate fact without any cause." But enthroning measurement to the status of an unanalyzable god, creating worlds out of nothing, was not a point of view that either Einstein or Schrödinger could find useful.

"You have made me extremely happy with your two lovely letters of June 17 and 19," Schrödinger wrote to Einstein, "and the very detailed discussion of very personal things in the one and very impersonal things in the other. I am very grateful. But I am happiest of all about the *Physical Review* piece itself, because it works as well as a pike in a goldfish pond and has stirred everyone up. . . .

"I am now having fun and taking your note to its source to provoke the most diverse, clever people: London, Teller, Born, Pauli, Szilard, Weyl. The best response so far is from Pauli who at least admits that the use of the word 'state' for the psi-function is quite disreputable." To uncritically say that the wavefunction represents the real state of affairs of a particle is to throw a veil over a multitude of mysteries, since it was not at all obvious what that meant.

"What I have so far seen by way of published reactions," Schrödinger continued to Einstein, "is less witty. . . . It is as if one person said" (the exiled Austrian in Oxford tried to think of the most strange and faraway places) " 'It is bitter cold in Chicago,' and another answered: 'that is a fallacy, it is very hot in Florida.' . . .

"My great difficulty in even understanding the orthodoxy over this matter has prompted me, in a lengthy piece, to make the attempt to analyze the current interpretation situation once and for all from scratch. I do not know yet what and whether I will publish on it, but this is always the best way for me to make matters really clear to myself. Besides, a few things in the present foundation strike me as very strange."

The whole idea of speaking only in classical terms made him feel that "only with difficulty can precisely the most important statements of the new theory be forced into the Spanish boot." (The Spanish boot, whether or not lined with spikes, is an instrument of torture in which boards are cranked tighter and tighter around the foot.) Einstein's illustration of the action-at-a-distance that happens when what was wavelike and spread out over miles suddenly becomes particulate upon measurement he also continued to find curious. The third thing that bothered him was a feeling that we have been "prescribed with wise, philosophical expressions that these *measurements* are the only real things, that whatever goes beyond is metaphysics. Then in fact it does not trouble us at all that our claims about the *model* are monstrous."

"You are the only person with whom I am actually willing to come to terms," Einstein replied. "Almost all the other fellows do not look from the facts to the theory but from the theory to the facts; they cannot get out of the network of already accepted concepts; instead, comically, they only wriggle about inside."

He proceeded to describe his solution to what he called "the paradox": that Schrödinger's wavefunction ψ does not describe individuals at all, but only groups in a statistical way. "*You,* however, see something quite different as the reason for the inner difficulties. You see in ψ the representation of reality and would like to change its connection with the concepts of ordinary mechanics"—i.e., concepts like position and momentum that do not have much meaning for a wave—"or do away with them altogether. Only in this way could the theory be made to stand on its own two legs. This point of view is certainly coherent, but I do not believe that it is capable of avoiding the felt difficulties. I would like to show this by means of a crude macroscopic example."

Einstein then described a charge of gunpowder "that, by means of intrinsic forces, can spontaneously combust," and will do so, on average, in a year. "In principle, this can quite easily be represented quantum-mechanically. . . . But, according to your equation, after the course of the year . . . the ψ-function [wavefunction] then describes a sort of blend of not-yet and of already-exploded systems."

This blend is known to those who study waves as a superposition. Examples of classical superposition involving sound, water, or light waves abound: for example, the four individual voices of a barbershop quartet superposing into one harmonic wave. The superposition of two waves is just another wave (or the absence of a wave, if the two waves are exact opposites and cancel each other out).

The concept is stranger when it describes not waves but particles, as it does in the quantum world, where, for example, electrons routinely are in a superposition of two different locations, acting as if they are in two places at once. And the concept becomes positively ludicrous when it is applied to something like gunpowder. "Through no art of interpretation can this ψ-function [of gunpowder simultaneously dormant and exploded] be turned into an adequate description of a real state of affairs," wrote Einstein. "In reality there is just no intermediary between exploded and not-exploded."

Schrödinger responded first to Einstein's own interpretation of the wavefunction. He explained, diffidently but correctly—as John Bell would later prove—that "it doesn't work" to try to solve "the antinomy or paradox" by claiming that the wave equation merely describes a group of atoms, on average. With his tongue good-naturedly in his cheek, he quoted Einstein's words back at him: interpreting the wavefunction in this way would "change the connection with the concepts of ordinary mechanics"—this time because of the concept Schrödinger had just defined five days before, in a paper submitted to the Cambridge Philosophical Society on August 14, 1935.

In what he called a "Discussion of Probability Relations Between Separated Systems," Schrödinger, writing in English, described the EPR situation of two atoms interacting and separating again. After all the discussion that had gone before, he could barely bring himself to use the words *state* or *wavefunction*—instead he talked of the "representatives" of these atoms in the quantum-mechanical formalism. *As far as this formalism is concerned, these two atoms, no matter how far they separated, ceased to be individuals after their interaction.*

"I would not call that *one* but rather *the* characteristic trait of quantum mechanics," wrote Schrödinger, "the one that reinforces its entire departure from classical lines of thought. By the interaction, the two representatives have become entangled." Thus the word and concept of entanglement entered physics.

In August of 1935, the Nazi curse arrived at the door of Arnold Berliner, the founder and editor of *Die Naturwissenschaften,* who had asked Schrödinger to write something discussing the EPR paradox. Schrödinger's "general confession" on the subject—a long and wonderful exploration of the strangeness of quantum mechanics—was still lying on Berliner's desk, but (as Schrödinger told Einstein on August 19), "as of twenty-four hours ago, he was no longer editor." The random injustice of this deposition was all the more striking given Berliner's habitual kindness and wisdom, well known to the whole community of physicists (in particular, he had been instrumental in encouraging a young and insecure Max Born). Schrödinger wanted to withdraw the paper in support of Berliner and in protest of his treatment at the hands of the Nazis.

Caring more for his journal than himself, Berliner asked Schrödinger to publish the paper in *Naturwissenschaften* anyway, and it came out in three parts in the last three months of 1935, while Einstein tried to find a way to rescue the old man from Germany. Instead, Berliner would spend the war in his apartment, leaving as little as possible to avoid wearing the Nazi-mandated Star of David. The only bright spots were the weekly visits of von Laue, when both men were able for a few hours to retreat into a better, more cultured world, sitting and talking beside a bust of Berliner's friend Mahler, sculpted by Auguste Rodin. But in March of 1942, with the Nazis telling him he must leave his apartment by the end of the month, Berliner gave up hope and made his fatal decision. Von Laue received a gently worded letter, which obliquely described how Berliner sat down in his armchair and "went to sleep." Berliner's beloved *Naturwissenschaften* did not even recognize his death. The Nazis ordered that there be no funeral for the elderly Jewish suicide, and no one was to honor Berliner as he was laid in the ground. Von Laue ignored the orders and came to stand beside the grave as his friend's coffin was lowered into the earth.

Schrödinger's paper that Berliner had insisted on publishing was (after his wave equation of 1926) probably the most important paper of his life, certainly the most entertaining, and ultimately the most famous. In it, among many other things, he introduced the concept of "entanglement" in German: *Verschränkung.* (This is actually a somewhat different word than the English word he had introduced half a year earlier, and Bohr might say they have complementary meanings—where the English word colloquially conveys mess, the German one suggests order: a German speaker will define it by folding his arms across his chest, to illustrate *cross-linking.*)

Leading up to the most famous thought-experiment of all of physics, echoing his correspondence with Einstein, Schrödinger discussed the

Born interpretation of the wavefunction as a list of probabilities: "Does one not get the feeling that the essential content of what is being said can only with some difficulty be forced into the Spanish boot of probabilistic predictions for finding this or that classical measurement result?" The wavefunction does not show itself to be a list of anything: instead, all the so-called options are added together as if they are simultaneous. This superposition is the characteristic of waves. But when certain kinds of measurement are made, the superposed quantum-mechanical wavefunction ceases to be literally accurate, and it is then that (with great success, but no explanation) it is interpreted as the tote board of the gambling God.

But without the help of this interpretation—and its magical "collapse of the wavefunction"—Schrödinger's equation loses all connection to the outside world. He was thinking of Einstein's exploding gunpowder and his ball-in-a-box.

"One can even construct quite burlesque cases," wrote Schrödinger. He described "a cat shut up in a steel chamber with a diabolical apparatus (which one must keep out of the direct clutches of the cat)." This apparatus involves a vial of poison which will be smashed by a hammer; the hammer is triggered by the decay of a single radioactive atom. If the atom decays, the cat breathes the poison; if not, the cat remains safe. There is so little radioactive substance "that in the course of an hour *perhaps* one atom of it disintegrates, but also with equal probability not even one. . . .

"If one has left this entire system to itself for an hour, then one will say to himself that the cat is still living, if in that time no atom has disintegrated. The first atomic decay would have poisoned it. The ψ-function of the entire system would express this situation by having the living and the dead cat mixed or smeared out (pardon the expression)." With this reductio ad absurdum—a cat in a superposition of simultaneous death and life—Schrödinger demonstrated the desperate state of a theory that required measurement to make it work.

Schrödinger was getting bolder and more confident. The quantum theory really was more fascinating than any of its inventors had realized. In October, in the midst of writing the paper, he sent off a letter to Bohr teasing him about Bohr's "avoidance of the Einstein paradox. . . . There must be quite definite and clear grounds," Schrödinger hoped, "why you repeatedly declare that one *must* interpret observations classically. . . . It must belong to your deepest conviction—and I cannot understand on what you base it." Schrödinger wrote that "I should like very much to see and talk with you again, but the times are now little suited for pleasure trips."

Heisenberg was meanwhile withdrawing into his shell, writing to his mother around the same time, "I must be satisfied to oversee in the small field of science the values that must become important for the future. That is in the general chaos the only clear thing that is left for me to do. The world out there is really ugly, but the work is beautiful."

The final section of what Schrödinger called his "general confession" (but which now is universally known as "the cat-paradox paper") was published before Christmas of 1935. In early 1936 he got the chance to discuss all these things with Bohr, as he wrote to Einstein. "Recently in London spent a few hours with Niels Bohr, who in his kind, courteous way, repeatedly said that he found it 'appalling,' even found it 'high treason' that people like Laue and I, but in particular someone like you, should want to strike a blow against quantum mechanics with the known paradoxical situation, which is so necessarily contained in the way of things, so supported by experiment. It is as if we are trying to force nature to accept our preconceived conception of 'reality.' He speaks with the deep inner conviction of an extraordinarily intelligent man, so that it is difficult for one to remain unmoved in one's position.

"I found it good that they strive in such a friendly way to bring one over to the Bohr-Heisenberg point of view. . . . I told Bohr that I'd be happy if he could convince me that everything is in order, and I'd be much more peaceful."

Nineteen thirty-five was both the climax of the fight for the meaning of quantum theory and the year of its armistice. There were to be no more public skirmishes over the soul of the quantum theory, and in subsequent years, Einstein, Schrödinger, and von Laue were mostly allowed to drift along undisturbed in their skepticism.

"It seems hard to look in God's cards," said Einstein in 1942. "But I cannot for a moment believe that he plays dice and makes use of 'telepathic' means (as the current quantum theory alleges he does)."

A few years later, Schrödinger wrote to him, "God knows I am no friend of the probability theory, I have hated it from the first moment when our dear friend Max Born gave it birth. For it could be seen how easy and simple it made everything, in principle—everything ironed out and the true problems concealed. Everybody must jump on this bandwagon. And not a year passed before *Probability* became an official credo, and it still is."

"You are the only contemporary physicist, besides Laue, who sees that one cannot get around the assumption of reality—if only one is honest,"

Einstein wrote Schrödinger after World War II. "Most of them simply do not see what sort of risky game they are playing with reality—reality as something independent of what is experimentally established. Their interpretation is, however, refuted most elegantly by your system of radioactive atom + amplifier + charge of gun powder + cat in a box, in which the ψ-function of the system contains both the cat alive and blown to bits. Nobody," he said, "really doubts that the presence or absence of the cat is something independent of the act of observation."

Schrödinger agreed. "No reasonable person would express a conjecture as to whether Caesar rolled a five with his dice at the Rubicon." (As he crossed the river Rubicon, which separated his province Gaul from Italy—knowing that by doing so he was provoking war—Caesar famously said, "*Iacta alea est!*": the die is cast.) "But the quantum mechanics people sometimes act as if probabilistic statements were to be applied *only* to events whose reality is vague."

Bohr, too, remained concerned. One day in 1948, Abraham Pais walked into the guest office at the Institute for Advanced Study in Princeton. Einstein, who did not like the big room, was glad to offer it to Bohr, and he himself was using the little assistant's room off to the side. Pais found Bohr sitting at the desk with his head in his hands. "Pais," he said, "oh Pais, I am sick of myself." He put his head in his hands again. "I am sick of myself." It turned out that he had been talking with Einstein about quantum mechanics. "I don't understand why I can't convince him."

Pais didn't understand it either. As he later wrote, "There was Einstein, so wise, yet so incomprehensibly unyielding. There was Bohr, and he could just as well have said: Einstein is a great man, I am fond of him, but in regard to quantum physics he is out to lunch. Let him."

It was Einstein whom Pais had come in to discuss. Bohr was writing a tribute in honor of Einstein's seventieth birthday in which he planned to review their famous decade-long debate, and Pais was his scribe for this project.

Bohr took a deep breath and stood up. "Sit down," he said, and then grinned. "I always need an origin for the coordinate system."

Pais sat down and took out a pen and paper.

Bohr began to dictate: "The discussions with Einstein which form the theme of this article have extended over many years—years which have witnessed great progress in the field of atomic physics. Sometimes there has not been much time— Some meetings— Our meetings . . ." His voice dropped as he sank into thought, pacing around the desk in what Pais (at the center of the coordinate system) described as "an eccentric ellipse," occasionally repeating, "Our meetings—"

Bohr stopped and turned. The sentence had come to him. "Whether our actual meetings have been of short or long duration, they have always left a lasting impression—a deep and lasting impression on my mind."

Pais was scribbling away.

". . . And even when entering on topics apparently far removed from the problems under debate at our meetings, I have, so to speak, been arguing with Einstein all the time."

Moved, Pais looked up at Bohr, whose voice was dropping as he circled. "I have been arguing with Einstein all the time. . . . I have been arguing with Einstein. . . ." He paced, slower and slower, hands behind his back, muttering, "Einstein . . . Einstein . . ." and finally stopped, gazing out the window, without seeing anything.

The assistant's door soundlessly swung open behind him and Einstein himself tiptoed into the room. There was "an urchin smile on his face," and he signaled Pais to be silent. "I was at a loss what to do," remembered Pais, "especially because I had at that moment not the faintest idea of what Einstein was up to." Very quietly Einstein lifted the lid of Bohr's tobacco pot, and began to fill his pipe.

At that moment, Bohr got his thought and spun around. "Einstein is—" and he stopped in complete shock. "There they were, face to face, as if Bohr had summoned him forth," wrote Pais. "It is an understatement to say that for a moment Bohr was speechless."

"I am sorry, Bohr," said Einstein, as Bohr burst out laughing. "But you know my doctor forbade me to buy tobacco. . . ."

In October 1960 (the year before Schrödinger's death, and five years after the death of Einstein), Schrödinger, back in Vienna, wrote to Max Born, back in Germany. "Maxel, you know I love you and nothing can change that.

"But I do need to give you for once a thorough head washing. So stand still. The impudence with which you assert time and again that the Copenhagen interpretation is practically universally accepted, assert it without reservations, even before an audience of the laity—who are completely at your mercy—it's at the limit of the estimable. . . .

"Have you no anxiety about the verdict of history?" he demanded. "Are you so convinced that the human race will succumb before long to your own folly?"

After Schrödinger died, Born, who was his foil in so many aspects of their long and full lives, wrote his most touching eulogy: "His private life seemed strange to bourgeois people like ourselves. But all this does not matter. He was a most loveable person, independent, amusing, temperamental, kind, and generous, and he had a most perfect and efficient brain."

Niels Bohr died the next year. On his blackboard were left two drawings, a record of what he had been thinking about the night before he died. The first looked like a spiral staircase—a Riemann surface—which was Bohr's favorite metaphor for the ambiguity of language, the way you can arrive back at the same word in your thought and it can have a whole new layer of meaning than when you first thought it. But how, he used to ask, can you communicate this to another person?

The second drawing, almost vibrating on its chalky spring, was Einstein's light-filled box.

The Search and
the Indictment

1940–1952

David Bohm

17

Princeton

April–June 10, 1949

NIGHT IS FALLING. It is late April in Princeton, New Jersey. Two young men are walking about the grounds of the university, talking in low voices.

"Einstein told me I should refuse to play along with their games," says the older one, a thirty-two-year-old assistant professor at Princeton named David Bohm. His face betrays his pride to speak of the great man, whom he has met only briefly. "He thinks that to appear before the committee would validate the hearings. But then," he relates with a queasy look, "he said, 'You might have to sit for a while.' "

A subpoena had arrived on Bohm's desk a few days ago, calling him to testify about his time as Oppenheimer's student in the prewar Berkeley campus full of brilliant physicists and at least one leak to the bomb program of the Soviet Union. World War II is over, the Cold War has begun, and in the last decade the world of the physicists has changed again and again.

Bohm is known for not letting his friends get a word in edgewise, but tonight his companion—Eugene Gross, his twenty-three-year-old former student—is grateful for this. He wouldn't know how to help. Gross is now a graduate student at Harvard, but Bohm and he remain close friends, still writing papers together on plasmas, the form of matter that makes up the stars. On the side, Bohm's other writing project is a book on the quantum theory, a subject that he teaches at an evening graduate class. On a similar walk, he would tell another young professor, Murray Gell-Mann, that "as a Marxist, he had difficulty believing in quantum mechanics." But he is diligently reading Bohr, hoping for enlightenment.

Bohm's words are coming faster and faster: "And then Oppenheimer—did I tell you Rossi was here and we saw Oppenheimer?" Giovanni Rossi Lomanitz, Bohm's roommate and fellow Oppenheimer disciple back in Berkeley, was a brilliant, outspoken twenty-one-year-old Ph.D. student who had come to California from Oklahoma in a burst of national publicity over an uncle who was on trial for some form of extreme unionist activity.

"We were on Nassau Street. And out of the barbershop came Oppenheimer. We told him what had happened, and he said, 'Oh my God, all is lost.' "

This last statement Bohm reports with a bit of bemused flair, in the Oppenheimeresque cadences that he, like all Oppie's students, had tried so hard to acquire back at Berkeley. Now, in Princeton after the war, Bohm is in the shadow of both Oppenheimer and Einstein: two living legends. Einstein, in fact, had once received a letter saying, "If you and Oppenheimer will run for President and Vice President of the United States in 1952 I shall vote for you." But while eight-year-old girls try to bribe Einstein with fudge for help with their math homework, Oppenheimer is famous for quoting the *Bhagavad Gita* upon the first test of the atom bomb: "I am become Death, the Shatterer of Worlds." Just returned from Los Alamos with little over a hundred pounds on his tall frame, he wafts through the Institute for Advanced Study like a disembodied spirit, staring through people with haunting blue eyes.

But the redbrick institute with its white cupola and white-haired geniuses, whose work the young professors and graduate students at Princeton study and emulate and try to supersede, is a few winding country miles from the university. Here, rising up behind Bohm, Gothic Princeton looks like a herd of dark cathedrals grazing on the close-clipped lawns. Bohm is in his element while walking; his favorite way to think about physics is on a ramble like this, walking around the campus as an idea turns around in his mind. He feeds his ideas on fresh air, coffee, and nighttime conversation. When he goes back later to the chalkboard or the notebook, the math falls into place as if it were wanting to please him.

But tonight they are talking of problems that are harder to solve.

"According to him they're taking it all pretty seriously, and there's an FBI man on the committee. Oppie looked at us and said, 'Promise me you'll tell the truth.' Oppie is a paranoiac, that's what Rossi says," Bohm finishes in a rush of words.

Late that night, after the physics department has all gone home, Bohm sits in his office before an empty desk, tossing coins in the air and catching them again and again.

"He is totally free of guile and competitiveness, and it would be easy to take advantage of him," Gross later wrote of Bohm. "Indeed, his students and friends, mostly younger than he is, felt a powerful urge to protect such a precious being."

"Mr. Bohm," said the senior investigator for the House Un-American Activities Committee, "have you ever been a member of the Young Communist League?"

It was May 25, 1949. Bohm, sitting in the cold room in the Old House Office Building in Washington, facing the six Representatives present, repeated the phrases his lawyer had given him, sounding nervous and ill at ease. "I can't answer that question, on the ground that it might tend to incriminate and degrade me, and also, I think it infringes on my rights as guaranteed by the First Amendment."

The chairman said, a little incredulous, "Would you mind repeating your answer?"

Bohm repeated his answer.

The chairman said, "I am a little curious to know how his rights under the First Amendment to the Constitution would be infringed by his answer to that question." But the hearing went on.

"Mr. Bohm, are you now or have you ever been a member of the Communist Party?"

"I decline to answer the question for the same reasons as just stated."

"While employed by the Manhattan Engineering District, did you ever attend Communist Party meetings?"

"I decline to answer that question for the same reasons as already stated."

Around the quiet room, faces were hardening, little muscles in hands or shoulders tensing, breaths taken. Digging in for an unexpected siege.

The questions—pages of them—were implacably asked, and nervously, politely, blankly answered with the same statement.

There was a moment of relief when they came to the question "Are you a member of or affiliated with any political party or association?"

Whereupon Bohm said, "Yes, I am. I would say 'Yes' to that question."

He was prompted: "What party or association is that?"

All ears perked to hear the confession.

Bohm leaned over to confer with his lawyer.

"I would say definitely that I voted the Democratic ticket."

The representative from Missouri who asked the question must have been truly annoyed as he said, "That is not responsive to my question. I asked if you were a member of any political party or association."

Bohm said, with grating innocence, "How does one become a member of the Democratic Party?"

Princeton University quickly declared that Bohm was regarded "by his

Princeton colleagues as a thorough American and at no time has there been any reason for questioning his loyalty."

His loyalty, nonetheless, was challenged at another round of questioning on June 10. Bohm was told to help "protect our country," which statement he answered, characteristically, with an analogy: "I believe that in some cases many people feel that security has been so—" He knows he is treading on thin ice here, regroups, and tries to put it in the best possible light. "People have concentrated so much on security that they are not able to do the job on hand." He looks around at the skeptical stares. "In other words, I mean, as an analogy, take the individual who was so afraid to cross the street he would never be able to do anything. You would have to take a certain medium attitude in that."

Two different committee members spoke at once after this satisfyingly suspicious comment, with one finally prevailing. "Do you not think that persons who have to classify information had better err on the side of overcaution than the other side?"

Bohm, now deep into the argument, said, "To a certain extent, but there is always a limit. You have got to draw a dividing line somewhere."

The committee members got huffier and started moving in for the kill. Finally one of the committee said, "I say it is better to err on the right side."

To which Bohm replied drily, "I say it is better not to err at all."

The meeting was adjourned.

"He used to make jokes about it," remembered his student Ken Ford, "real gallows humor. There was a faculty lecture on unresolvable paradoxes"—If the barber is the man who shaves all men who do not shave themselves, who shaves the barber?—"and Dave said that 'Congress should appoint a committee to investigate all committees that do not investigate themselves.' "

18

Berkeley

1941–1945

It all had started with Oppenheimer.

"I loved Oppenheimer," said Bohm, and in this he was not alone.

Oppenheimer taught every spring term at Caltech (after the Berkeley term was over), a long-limbed stork swooping down the California coastal cliffs, and one term, in 1941, finding Caltech cynical and uninspiring, Bohm followed him back up the coast.

J. Robert Oppenheimer was raised in an eleventh-floor apartment overlooking the Hudson River, among his gregarious father Julius's van Goghs and Fauves, by an elegant, gentle mother named Ella who had studied art in Europe and always wore a glove to cover her artificial hand. Science arrived in Robert's life in the form of a mineral collection from his impecunious self-educated grandfather back in Germany—by twelve, Robert was reading papers to the New York Mineralogical Club. But a post-high-school trip to visit Bohemian mines resulted in dysentery; as a rest cure, his parents sent him with a favorite teacher to New Mexico, where he fell in love with horseback riding, high piney plateaus, and open spaces. After Harvard, he sought the centers of quantum theory. He went to Rutherford's Cambridge (where he became almost suicidal over his experimental and social clumsiness), Born's Göttingen (where his arrogance, quick mind, and what his students would later call "the blue glare treatment" reduced the sensitive Born to a puddle of nerves); and two places where he felt at home: Ehrenfest's Leiden, where he befriended Dirac and was christened "Opje" (later Americanized to "Oppie"), and, best of all, Pauli's Zurich, where anything Oppenheimer could dish out Pauli could handle effortlessly.

He arrived as a twenty-five-year-old professor in Berkeley a month before the stock market crash of 1929. "In addition to a superb literary style, he brought to his lectures a degree of sophistication in physics previously unknown in the United States," remembered Hans Bethe, one of the greatest theorists of the thirties and forties. "Here was a man who obviously understood all the deep secrets of quantum mechanics and who yet made it clear that the most important questions were unanswered." And he knew so much. "He became an almost mythical figure, especially to experimenters," remembered his friend the great experimentalist I. I. Rabi. "He could display his great knowledge in their own fields but then could take off into the blue of abstract theory where they could not follow."

"I didn't start out to make a school," Oppenheimer recalled. "I started really as the propagator of the theory (quantum theory) which I loved, about which I continued to learn more, and which was not well understood but which was very rich." His influence on the students for whom he claimed not to be looking was phenomenal.

Oppenheimer's students watched and learned and tried to walk and talk as much like him as possible, even down to the sound he made when thinking: *nim-nim-nim*. He took them to restaurants, concerts, read them Plato in Greek; he taught them to eat his incendiary chili, drink his fine wine, and light other people's cigarettes. Rabi claimed he could recognize an Oppenheimer student across a crowded room.

In a 1934 letter to his younger brother (in which he sent greetings from "many physicisti, all the californiacs"), Oppenheimer wrote, "I take it that . . . physics has gotten now very much under your skin, physics and the obvious excellences of the life it brings." It was this that Oppenheimer imparted to his students. He had created, purposefully or not, the theoretical department at Berkeley along the lines of Rutherford's Cavendish Lab at Cambridge or Bohr's institute at Copenhagen: a combination of cutting-edge physics institution and personality cult. When told, in the thirties, that there was no good American physics, Pauli responded, "Oh? Haven't you heard of Oppie and his nim-nim-nim boys?"

Oppenheimer was to impart two theories to David Bohm in the short time that Bohm was his student. One theory was Oppenheimer's entire intellectual life, and the other almost took that life away from him.

The first was the quantum theory, as put forth by Bohr and his students. Bohm had left Caltech in 1941 "a convinced classicist," deeply skeptical of the quantum theory. He and a friend and fellow grad student at Berkeley, Joe Weinberg, would argue late into the night, Weinberg secure in the quantum theory and Bohm calling Weinberg's emphasis on mathematics "Pythagorean mysticism." "Physics," said Bohm, "has changed from its earlier form, when it tried to explain things and give some physical picture. Now the essence is regarded as mathematical. It's felt that the truth is in the formulas." Bohm the lover of analogy felt he could never be at home with a theory like that.

But Oppenheimer was fascinating. Bohm's slow process of acceptance and then rejection again of the quantum theory was to be the defining struggle of his life, the ramifications of which would spill out into the history of physics and unwittingly lay the foundation for John Bell.

Bohm absorbed much more quickly the other theory that the hitherto apolitical Oppenheimer had started to impart to his students, naïve and tired of cynicism and Depression, in 1936. Bohm joined the Communist Party a year after his arrival in Berkeley, in November 1942, encouraged by his new friends in the physics department. Though he found its meetings dull and drifted away in a few months, he only grew more excited about the original ideas. Oppenheimer himself admitted to having been a "fellow traveler"—the euphemism for a Communist sympathizer—but was heard to report jauntily, "I would say I traveled much less fellow after 1939," the year of the Nazi-Soviet pact. If so, this change was not obvious to his worshipful students.

Then he disappeared. He was doing something classified for the government, "the Manhattan Project," and his students were left without him in wartime Berkeley. Grad students started disappearing, too, as if they had died and gone to the bosom of Abraham. Instead, they were surely happier: they were in the bosom of Oppenheimer doing secret and important things among the otherworldly red New Mexico plateaus. Those left behind knew only that it was classified work, and Oppie was in charge (though Bohm later claimed, "We knew that people were working on uranium so we could guess it might be a bomb"). Most of Bohm's friends—the Left-most wing of Oppie's Left-leaning students—were not called.

In March of 1943, when Oppenheimer asked for the transfer of Bohm to Los Alamos, General Leslie Groves, who was in charge of the Manhattan Project, told him that Bohm could not come ("we had a little quadratic letter code," Oppenheimer remembered, for this sort of information), with the limp excuse that Bohm had relatives in Nazi Germany. But there was much that neither Oppie nor his students knew.

Oppenheimer and Ernest O. Lawrence, the heads of Berkeley's theoretical and experimental physics departments, had joined the nascent bomb project in early 1942. At the same time, the Army made a security investigation of the campus. A year later, while Oppenheimer was asking for Bohm at Los Alamos, the Army's surveillance paid off when some unidentified man ("Scientist X") came to Steve Nelson, a local Communist leader (and friend of Oppenheimer's wife, Kitty), at his house. He read to Nelson a formula, and was apparently handed money.

Colonel Boris Pash, who led counterintelligence in San Francisco's beautiful Presidio military base, recalled, "We had very little information. The only thing we had definite was that the man's name was Joe, and the fact that he had sisters living in New York." They began to scrutinize the Berkeley Radiation Lab. The dashing Giovanni Rossi Lomanitz, he with the famous unionist uncle from Oklahoma, was their first suspect for "Scientist X."

Rossi was rarely seen without Bohm, Joe Weinberg, and Max Friedman, counterintelligence soon noted. Bohm and his friends were soon being followed everywhere, to classes and Communist-front organizations.

By June of 1943, counterintelligence reported that its investigations had revealed "Scientist X" to be Joe Weinberg of the Radiation Lab at Berkeley. By July, Rossi Lomanitz was drafted and switched from uranium separation work to boot camp, over Ernest Lawrence's howls of protest.

That September, Oppenheimer casually mentioned to the security guards at Los Alamos that "it was known among the physicists" that people at Berkeley were being approached for espionage. A week later he found himself summoned to Washington. His interviewer was the man who had led the Berkeley investigation, Colonel John Lansdale. He was now the head of security for the whole atomic-bomb project, though his pleasant, wholesome-looking face made him appear more like a father of small children than an Army bloodhound. Lansdale discovered immediately that Oppenheimer had no intention of revealing who had been "approached" or who the approacher was, on the grounds that "I would regard it as a low trick to involve someone where I would bet dollars to donuts [that he had done nothing wrong]."

"Here we are," said Lansdale in the middle of a long session of fruitless questioning, "we know that information is streaming out from this place every day. . . . Now, what shall we do? Shall we sit back and say, well, my God, maybe the guy recanted . . . ?"

Oppenheimer nodded with wrinkled brow, intellectually interested in the problem. "Hard for me to say," he said, "because of my own personal trends"; then he gave Lansdale an "Of course you understand, old boy" glance from the famous blue eyes.

"Well," said Lansdale, "is there anything else that you believe you can tell me that could give us any assistance?"

"Let me walk around the room and think," said Oppenheimer.

He got up and paced, then said abruptly, "I can tell you that I doubt very seriously whether—well, I don't know Bohm very well, but I doubt very seriously whether Weinberg would do anything along the lines of what we were talking about." He went on to say something about an older German student of his, Bernard Peters, which the hidden tape recorder did not catch. Oppenheimer had found Peters working as a longshoreman on Fisherman's Wharf in San Francisco and brought him into the physics department. He was later to state that it was during this time period that he began to think, because of "the way he talked about things," that Peters was "a dangerous man to be on a secret war project." Beyond these insights, the conversation wound on, without result.

"I wish, Colonel," said Oppenheimer, "that I could do what you want. I can't deny that I could give you that information. I wish I could do it."

"Well, I want to say that personally I like you very much and"— Lansdale grinned sheepishly—"I wish you'd stop being so formal and calling me Colonel, 'cause I haven't had it long enough to get used to it."

"I remember at first you were a captain, I think." Oppenheimer lifted his head and raised his pipe to his half-opened mouth.

"And it hasn't been so long," said Lansdale, "since I was a first lieutenant, and I wish I could get out of the Army and back to practicing law, where I don't have these troubles."

Oppenheimer was nodding in businesslike sympathy. "You've got a very mean job."

More than fifty years later, the controversy still rages on, more muddied than ever, about just how much Oppenheimer had to do with the Soviets. In 2002, the husband-and-wife team of Jerrold and Leona Schecter published a book titled *Sacred Secrets: How Soviet Intelligence Operations Changed American History.* The book caused a scandal among physicists and many people instantly denounced it (to the point where its authors felt their book had been hurt by "a conspiracy of silence," citing, perhaps oversensitively, the fact that the *New York Times* review, in recommending the book, mentioned "errors"). Their information about Oppenheimer comes from an ex-KGB agent named Sudoplatov who has not always turned out to be telling the truth in his other confessions.

Amid all the confusion and conflicting stories, the Schecters print in their appendix a (quite possibly inflated) report to the head of the KGB, dated October 4, 1944, which reads, "In 1942 one of the leaders of scientific work on uranium in the USA, Professor R. Oppenheimer, while being an unlisted member of the apparatus of Comrade Browder, informed us about the beginning of work. . . . He provided cooperation in access to research for several of our tested sources."

In January 1944, Oppenheimer was on a train to Santa Fe. In his compartment was a security man from Los Alamos, Major Peer de Silva, who began prodding him about his former students, asking who, among

Bohm and his friends, Oppenheimer thought was "truly dangerous." De Silva reported:

"He named David Joseph Bohm and Bernard Peters as being so. Oppenheimer stated, however, that somehow he did not believe that Bohm's temperament and personality were those of a dangerous person and implied that his dangerousness lay in the possibility of being influenced by others. Peters, on the other hand, he described as being a 'crazy person' and one whose actions would be unpredictable. He described Peters as being 'quite a Red' and stated that his background was filled with incidents"—he had fought street battles with Nazis in Germany and then escaped from Dachau—"which indicated his tendency toward direct action."

In March of 1944, Oppenheimer came back to Berkeley for a visit, and Bohm came to see him. He and Joe Weinberg were having a predictably tough time teaching Oppie's legendary course in his absence. Bohm wondered if things had changed and there might be a possibility of being transferred to Project Y (the Manhattan Project), since in his present situation he had "a strange feeling of insecurity." The word choice is unintentionally ironic but characteristic of Bohm, standing earnestly before Oppenheimer.

Oppenheimer said he would let him know. Later, he asked Major de Silva—the same man who claimed that Oppenheimer had described Bohm to him two months before as "truly dangerous"—if he would have objections to Bohm coming to Los Alamos.

"The undersigned," wrote de Silva, duly reporting the incident, "answered yes."

Still, Bohm participated in war work in a way. The Army wanted to learn about plasmas—the stuff of the stars, northern lights, lightning, St. Elmo's fire, and even the alien glow of a neon pizzeria sign. Paralleling the ancient Greek universe (composed of earth, water, air, and fire), plasma is the fourth state of matter, after solid, liquid, and gas. Hot gas has become plasma when most of its atoms are dissociated into positive ions and electrons freely floating around each other. Bohm discovered that the electrons in a metal—which float among its nuclei, belonging to the whole metal but not to any part—form a plasma, too.

He was fascinated by the collective behavior of electrons in a plasma. Plasmas have long lent themselves to allegory: to Bohm, they symbolized the perfect Marxist state. He became the leading plasma theorist in the country, an expert on arcane topics such as "plasma oscillation" and what is still known as Bohm diffusion.

After the war, Bohm became a Princeton professor, on the recommendation of Oppenheimer, at the Institute for Advanced Study, a few small woods and open fields away. Oppenheimer's big secret had now exploded over Hiroshima and Nagasaki. Bohm still did not know Oppenheimer's other secret, which, justifiably or not, made both of them, and their friends, marked men.

Bohm asked the department if he could teach a small graduate class on his longtime nemesis, quantum mechanics. In preparation, he returned to the Oppenheimer class notes taken years ago by his friend Bernard Peters, the concentration-camp escapee and longshoreman whom Oppenheimer had inducted into Berkeley's ivory tower.

19

Quantum Theory at Princeton

1946–1948

WHEN I WAS A BOY a certain prayer we said every day in Hebrew contained the words to love God with all your heart, all your soul, all your mind," remembered Bohm in a 1987 interview. "My understanding of . . . this notion of wholeness—not necessarily directed toward God but as a way of living—had a tremendous impact on me."

Poring over Peters's notes on Oppenheimer, Bohm finally rediscovered in the quantum theory that old atmosphere of wholeness. During rereading after rereading, a deep physical sense began to slowly emerge from the fog of Bohr's prewar semiphilosophical papers, and Bohm started to write a textbook called *Quantum Theory.*

The theory is usually taught by inching ever closer to the quantum from the classical point of view, but Bohm found understanding through the most foreign and elusive—the least classical—aspects of the theory. "The notion of spin particularly fascinated me," he remembered. "The idea that when something is spinning in a certain direction, it could also spin in the other direction but that somehow the two directions together would be a spin in a third direction. I felt that somehow that described experience with the processes of the mind." He was able to "produce in myself an analogy, in my state of being," of spin. "I can't really articulate it. It had to do with a sense of tensions in the body. . . ."

In quantum mechanics—which Bohm analyzed in three parts, as quantized motion, statistical causality, and indivisible wholeness—"I came closer to my intuitive sense of nature."

He remembered that "even as a child I was fascinated by the puzzle, indeed the mystery, of what is the nature of movement." He got his students thinking about the way perception really works: "If we think of an object in a given position, we simply cannot think of its velocity at the same time. . . . A blurred photograph of a speeding car suggests to us that the car is moving. . . . On the other hand, a sharp picture of a moving car, taken with a very fast camera, does not." The uncertainty principle, so counterintuitive and strange, "is actually in close accord with our simple picture," Bohm told his students.

Our planets and our cannons have brought us differential equations, Bohm told them. The idea of continuous motion that we now think is so natural arrived with eighteenth-century ballistics and astronomy. "Many of the ancient Greeks were unable to grasp the idea of continuous motion," he reminded the class, "as those who have studied Zeno's paradoxes will know. One of the most famous of these paradoxes concerns an arrow in flight. Since at each instant of time the arrow is occupying a definite position, it cannot at the same time be moving."

In the case of causality, just as with the continuous trajectory, "once again, quantum theory was a step in the direction of the less sophisticated ideas that arrive in ordinary experience," declared Bohm—ordinary experience "where one seldom encounters an exact relation between cause and effect, but instead usually thinks of a cause as producing a tendency in a given direction." He reassured the students (or shocked them) by saying that, "contrary to general opinion, quantum theory is less mathematical in its philosophical basis than is classical theory, for, as we have seen, it does not assume that the world is constructed according to a precisely defined mathematical plan."

With quantized motion and statistical causality effortlessly embraced, Bohm reached his most deeply felt concept—indivisible wholeness. "Even in the classical limit," Bohm pointed out, "we recognize that the separation between object and environment is an abstraction." For example, a bacterium: "In a few hours most of the matter that was originally in the bacterium may have been expelled and replaced by matter from the surrounding medium. In the meantime, the bacterium may also have changed into a spore.

"How are we justified in thinking of this as a continuation of the same living system that we saw originally?" Causality and continuity allow us to do so, despite the microscopic blurring of object and environment. "In a system whose behavior depends critically on the transfer of a few quanta, however," Bohm explained, "the separation of the world into parts is a non-permissible abstraction because the very nature of the parts (for instance, wave or particle) depends on factors that cannot be ascribed uniquely to either part, and are not even subject to complete control or prediction."

"He was a wonderful teacher," remembered Ken Ford, who took Bohm's evening quantum-theory seminar in the spring of 1949 (timed,

Ford figured, not to conflict with Bohm's three-night-a-week moviegoing schedule). "He didn't just teach us standard problem solving, he tried to get us to understand the meanings behind the equations.

"He never said, 'go away,' when he was in his office, so everyone liked him," said Ford, laughing. "We spent a tremendous amount of time together," wrote Bohm's student Eugene Gross. "There are advantages to having a bachelor for a mentor." "He had no concern for personal comfort or appearance," said Ford. "No social life. His whole being was physics."

Looking back years later, Bohm felt that "understanding reality as a coherent whole" had been the goal of all his work, from the mathematical to the philosophical, and he wanted to understand the mind as a part of that whole. Three dense pages of his *Quantum Theory* cover what he enthusiastically called "the remarkable point-by-point analogy between the thought processes and quantum processes." Some of these points are unexpectedly witty, like his correspondence of Logic to Thought as classical physics is to quantum mechanics—though it's probably for moments like this that Dirac's brother-in-law Eugene Wigner condemned the textbook as "too much schmooze." But vivid imagery, Bohm explained, can be "helpful in giving us a better 'feeling' for quantum theory."

"If a person tries to observe what he is thinking about at the very moment that he is reflecting on a particular subject," Bohm began his favorite analogy, "it is generally agreed that he introduces unpredictable and uncontrollable changes in the way his thoughts proceed thereafter." His students could not have expected that this discussion was headed in an unusual direction. "If we compare (1) the instantaneous state of a thought with the position of a particle, and (2) the general direction of change of that thought with the particle's momentum," continued Bohm, "we have a strong analogy."

He went beyond mere analogy, however. Stressing the speculative nature of such a possibility, he brought up Bohr's idea that quantum-mechanical limitations might play a role in human thought.

"Even if this hypothesis should be wrong," continued Bohm to his students, "and even if we could describe the brain's functions in terms of classical theory alone, the analogy between thought and quantum processes would still have important consequences: we would have what amounts to a classical system that provides a good analogy to quantum theory. At the least this would be very instructive. It might, for example, give us a means for describing effects like those of the quantum theory in terms of hidden variables."

Hidden variables? What other teacher and what other textbook would even mention such a thing? Despite the carefully added warning "(It would not, however, prove that such hidden variables exist)," this was a big departure from standard operations.

At the Solvay conference in 1927, de Broglie had sought to explain the skein of probabilities that made up quantum mechanics with an underlying hidden causal structure—to give a muscled skeleton to its mysteriously moving but apparently vacant skin. De Broglie had been disregarded then, his ideas insufficiently in tune with *der Kopenhagener Geist*. After six years of ignoring the theory, justification for doing so was reassuringly supplied with von Neumann's proof that an underlying causal structure could not exist, keeping the innards of the quantum theoretical beast numinous and ineffable. Von Neumann was one of the greatest minds of the twentieth century, and few aside from the distracted Einstein and the ignored Grete Hermann wondered if it could perhaps be von Neumann who was mistaken. De Broglie himself did not wait for von Neumann's proof; by then he had already recanted.

Quite early in his textbook, Bohm told his students that "every indication" pointed to the continued success of quantum mechanics, and to its ultimate conquering of the nucleus and relativity. "Until we find some real evidence for a breakdown . . . it seems, therefore," he concluded, "almost certainly of no use to search for hidden variables."

Yet he returned again and again to the idea throughout his book, always with the caveat that he would later explain in detail why hidden variables are a lost cause. He made this promised explanation not a dismissal or a footnote but the climax of *Quantum Theory*. Already this is unusual enough, but the surprised reader finds that Bohm's argument against hidden variables resurrects the ten-years-forgotten problem of Einstein, Podolsky, and Rosen.

Importantly for all future discussions of EPR (or ERP, as Bohm calls it), he reformulated the *gedanken* experiment in terms of spin, definitively sharpening it to such an extent that the original treatment is relegated to historical curiosum (two options in each direction—spin up or spin down—keep the argument simpler than when it dealt with a near infinity of possible positions and momenta).

Bohm started his discussion of the EPR *gedanken* experiment with two atoms, once bound together in a molecule (hence with opposite spin, "insofar as the spin may be said to have any definite direction at all"). The molecule is delicately "disintegrated" and the atoms are now moving apart from one another. We are free to measure the spin of either atom in any direction (x, y, or z—"but not more than one of these") by passing it through a Stern-Gerlach magnet.

The argument then follows as usual: without disturbing the second atom, we could arbitrarily know its spin in any single direction from a measurement of the first atom. Nothing about the way the atoms were separated from each other could prepare them for what the angle of the Stern-Gerlach magnet will be. Presumably it's all the same to the second atom whatever we choose to do to the first, and any measurement on the first merely happens to reveal part of the other unmeasured atom's preexisting state. Since the first atom might be measured in any direction, the unmeasured atom must really have spin in all three dimensions. "Because the wave function can specify, at most, only one of these components at a time with complete precision, we are then led to the conclusion," explained Bohm, "that the wave function does not provide a complete description of all the elements of reality existing in the second atom."

Though Bohm found this "a serious criticism of the generally accepted interpretation of quantum mechanics," he believed that it was not a fatal attack, because of an implicit assumption that the EPR paper did not even question, but that Einstein himself had worried about ever since 1909: the world can correctly be analyzed in terms of distinct and separately existing "elements of reality."

And this separability was the one thing above all that Bohm did not believe. This is why hidden variables are inconsistent with the quantum theory, he explained. A system composed of causal, separable elements is exactly the kind of classical construct that vanishes in the face of what he described elsewhere in his book as quantum theory's "interwoven potentialities."

"We conclude then," Bohm said, "that no theory of mechanically* determined hidden variables can lead to all of the results of the quantum theory."

---

*"Mechanical" meaning "separable parts obeying causal laws." A characteristic footnote elsewhere in the book reads, "The term 'quantum mechanics' is very much a misnomer. It should, perhaps, be called 'quantum non-mechanics.' "

20

Princeton

June 15–December 1949

Bohm was wandering the halls jangling a bunch of coins in his pocket when one of his grad students, intent on a question about plasmas, startled him. It was five days after Bohm's testimony to the HUAC. "Dave, I was just—." He took in Bohm's preoccupied, miserable face. "Dave?"

Bohm did not stop jingling the coins. "Yes?" he said after a couple of seconds.

"Dave, are you O.K.?" He felt weird asking his professor this question in the midday halls of the physics building, with undergrads pouring into their various classrooms all around.

"Did you ever know Bernard Peters?" Bohm asked.

The student thought about the stories he had heard. "Was he the guy that Oppenheimer discovered working as a longshoreman?"

Bohm nodded. "He had escaped Dachau, amazingly enough, and was putting his wife through medical school in San Francisco. We were friends back at Berkeley." Bohm was talking faster and faster. "He was so smart, and he was older than the rest of us, and had been through so much, and he and Oppenheimer were friends and now Oppenheimer—" Bohm's voice collapsed. "Have you heard about Oppenheimer and the House Un-American Activities Committee?"

The hallway suddenly went empty as the hour struck and classes began.

The grad student shook his head.

"I don't really understand," said Bohm, rattling away. "But apparently it's all over the newspapers in Rochester, where Bernard's a professor now.... They asked Oppie about Joe Weinberg and Rossi—friends of mine back in Berkeley," Bohm explained. "And he was—protective, like I would expect him to be." *Jangle jangle.* Everyone was used to Bohm's coin-rattling, but it seemed louder and more insistent today, in the quiet hallway. "But apparently long ago he had told them that Bernard Peters was a crazy Red that people should watch out for." Bohm pulled his hand out of his pocket, holding the coins all in one loose fist and shaking it, squinting his eyes. "This was all over the papers."

"Jeez—" said the student. "Did he take any of it back?"

"No, no, no, no, no—that's it—he repeated himself, he said it was all true. Poor Bernard. All these years Bernard has thought of Oppie as the best teacher he's ever had, as his *friend,* and all this time Oppie was ready to hand him over to the—to the Feds."

"I wonder if he said anything about me," said Bohm in a small voice.

On September 23, 1949, four months after Bohm was brought before the House Un-American Activities Committee, President Harry S. Truman announced that the U.S.S.R. had tested "an atomic device." The decryption of thousands of cables between the KGB and its agents in San Francisco, New York, and Washington had revealed Berkeley and Los Alamos as primary targets of Soviet espionage, and now the quality of the job these spies had done could be tested by airmen flying over Siberia, who were the first to relay back to the United States that the bomb had been built and successfully detonated.

On a night in early December, Bohm was arrested for "contempt of Congress" in his refusal to answer questions, and the marshal took him to Trenton so that he could obtain bail. In the dark, the scene flashing past was the grim naked woods of the end of the year. Bohm looked wistfully at one neon light and thought about plasma, millions of overheated individuals behaving as one perfect society, telling him to DRINK COCA-COLA.

The marshal asked him conversationally about physics. He told Bohm that he had come from Hungary and was now a loyal American.

"I hope that you were not disloyal," he said to Bohm.

The morning Bohm returned from Trenton, he found that he was suspended from teaching and forbidden to set foot on campus. A few of the graduate students approached the president of Princeton to try to appeal this decision. "After a brief exchange," remembered Silvan Schweber, one of these students, "we were reprimanded, reminded that 'Gentlemen, there is a war on!' and were invited to leave."

Thus Bohm, courtesy of the House Un-American Activities Committee, ended up with a paid year and a half with nothing to do but go to the library, think about plasmas (during the time he was officially absent from Princeton he wrote four papers on plasmas with Princeton friends and students), and ruminate on the quantum theory.

21

Quantum Theory

1951

IT IS ANOTHER EVENING, a year and a half later, and there is Bohm, still pacing about Princeton. This time his companion is physics' new wonder boy: Murray Gell-Mann. He is a twenty-two-year-old postdoctoral student at the nearby Institute for Advanced Study, having just received his Ph.D. from M.I.T. after an early undergraduate sprint through Yale.

They stop in a small coffee shop. Things are good and bad for Bohm. He was finally acquitted (the Supreme Court decided in December 1950 that the Fifth Amendment protected those, like Bohm, who had refused to incriminate themselves), but he was not reappointed to his Princeton post. For the moment, however, they are not discussing that. Bohm is ecstatic. He has finished his *Quantum Theory* textbook, and has been sending it around for opinions. "Einstein called me!" Bohm said. "He told me he wanted to discuss the book with me. And Bohr hasn't written me back, but Pauli has, and he was actually enthusiastic!" His smile of triumph splits his face.

They head back out into the dusk, under the orange street lamps. Gell-Mann, too, feels infected by Bohm's excitement. "You think you can convince Einstein?"

Bohm grins broadly. "We'll see."

Two days later, they are back at the coffee shop. Bohm seems sapped of energy. He doesn't want to walk; he doesn't know where he is going anymore.

Gell-Mann breaks the silence: "How was your meeting with Einstein?"

"He talked me out of it." Bohm places his coffee cup on the table, punctuating the statement. This was slightly worse than Gell-Mann was expecting.

Bohm is looking at his cup. "I'm back where I was before I wrote the book."

"What did he say?" asks Gell-Mann.

"He told me that I had explained Bohr's point of view as well as could probably be done, but he still wasn't convinced." Bohm takes a sip of his coffee, and remembers the old man standing at the window in his office at the institute, smoking and talking quietly. He remembers the effortless persuasiveness of genius lulling him like the tobacco smoke swirling in the light. "What came out was that he felt that the theory was incomplete, not in the sense that it failed to be the final truth about the universe as a whole, but rather in the sense that a watch is incomplete if an essential part is missing."

Bohm finishes his coffee with a wry expression on his face. They stand up, leaving their money on the table.

Then Bohm says, "What Einstein said was so close to my intuitive sense—before—of what was wrong with quantum theory. It's a theory that can't go beyond appearances." He shrugs his shoulders, then smiles weakly. "Remember," he says, "I'm a Marxist: we like our theories deterministic." When he looks up, the hazy orange glow of the street lamps obscures the stars.

Albert Einstein to P. M. S. Blackett at the University of Manchester, England:

April 17, 1951

Dear Professor Blackett:

Dr. David Bohm of Princeton University, whom I know personally and from his scientific work, has told me that he has applied for a fellowship connected with your University.

Dr. Bohm has a clear mind, is very energetic in his scientific work and of rare independence in his scientific judgment. He will be an asset to any circle of scientists he joins. There is also a second reason to give Dr. Bohm an opportunity for work, preferably outside the United States. Mr. Bohm himself has not been in any way politically active. He has refused, however, to answer official questions concerning colleagues. This admirable attitude was the cause of an official indictment and subsequent termination—or rather non-renewal—of his appointment at Princeton University.

I would be very grateful to you indeed if you would consider favorably Dr. Bohm's application.

Yours very sincerely,
Albert Einstein

.   .   .

"I had taught a course on the quantum theory for three years," wrote Bohm, "and written the book primarily in order to try to obtain a better understanding of the whole subject, and especially of Bohr's very deep and subtle treatment of it. However, after the work was finished, I looked back over what I had done and still felt somewhat dissatisfied."

The problem that weighed on his mind most heavily, as he explained later, "was that the wave function could only be discussed in terms of the results of an experiment or an observation, which has to be treated as a set of phenomena that are ultimately not further analyzable or explainable in any terms at all. So the theory could not go beyond the phenomena."

Bohm then set out to single-handedly create a theory that could.

22

Hidden Variables and Hiding Out

1951–1952

"THE USUAL INTERPRETATION of the quantum theory is self-consistent, but it involves an assumption that cannot be tested experimentally," Bohm wrote to start his infamous paper—"viz., that the most complete possible specification of an individual system is in terms of a wave function that determines only probable results of actual measurement processes." It was 1951. *Quantum Theory* would soon appear in bookstores. But as his book was being printed, the author himself was refuting its main point in a long two-part article in *Physical Review*, the most authoritative journal in the field.

Bohm continued: "The only way of investigating the truth of this assumption is by trying to find some other interpretation of the quantum theory in terms of (at present) 'hidden' variables, which in principle determine the precise behavior of an individual system, but which are in practice averaged over in measurements of the types that can now be carried out." He sat back from his desk for a second and looked at the almost empty paper, its few lines scrawled in his boyish handwriting. He could barely believe what he was about to write.

"In this paper and in a subsequent paper, an interpretation of the quantum theory in terms of just such 'hidden' variables is suggested." It is not a new theory, he explained, just a reinterpretation: "It is shown that as long as the mathematical theory retains its present general form, this suggested interpretation leads to precisely the same results for all physical processes as does the usual interpretation. Nevertheless, the suggested interpretation provides a broader conceptual framework than the usual interpretation, because it makes possible precise and continuous description of all processes, even at the quantum level."

After subsuming all the ideas of quantum mechanics into his worldview, this was a drastic reorientation. But Bohm was buoyant. "This broader conceptual framework allows more general mathematical formulations of the theory than are allowed by the usual interpretation." He went on to explain his hope that quantum mechanics might, as seemed possible at the time, break down at distances shorter than the diameter of a small nucleus.

"In any case, the mere possibility of such an interpretation proves that it is not necessary for us to give up a precise, rational, and objective description of individual systems at a quantum level of accuracy."

He sent the paper to *Physical Review* just before Independence Day, 1951. So commenced David Bohm's long, lonely fight against Copenhagen, which would lead in the end to John Bell's pivotal breakthrough.

Only after four dense and unrushed *Physical Review* pages, mostly on the uncertainty principle, does Bohm start to discuss his own idea, set out in a subtitle: "A New Physical Interpretation of Schrödinger's Equation." Never a great reader of other people's work, Bohm did not know until after he had finished his article that this "new physical interpretation" of the wavefunction was actually a resurrection, in a more fully realized form, of de Broglie's pilot wave that guides the quantum particle.

The pilot-wave interpretation dealt with the Schrödinger equation as something corporeal, like a stream—albeit a weirdly woven, invisible stream—rather than a catalogue of possibilities that mysteriously "collapsed" into a single choice during the act of measurement. The quantum particles, in Bohm's and de Broglie's imagination, are like twigs in this invisible stream—here riding the broad current, there eddying among trapped leaves, over there driven under the foam on the sheltered side of an underwater rock. The forces that an ordinary stream exerts on an ordinary twig are mostly reducible to gravity or electromagnetism, and so it is for quantum particles in the stream of the Schrödinger equation. But not all of the effects of this equation can be ascribed to these forces, and Bohm tentatively called the remainder "a 'quantum-mechanical' potential."

Associated with this potential was "an objectively real field" that Bohm dubbed the "ψ-field." This field, as he boldly wrote in the second half of his paper, gives "a simple explanation of the origin of quantum-mechanical correlations of distant objects," as in the EPR scenario. "The 'quantum-mechanical' forces may be said to transmit uncontrollable disturbances instantaneously from one particle to another through the medium of the ψ-field." Mediated by a field or not, an instantaneous effect is still "spooky" action-at-a-distance, exactly the explanation Einstein could not accept.

"There is, of course," wrote Bohm, "no reason why a particle should not be acted on by a ψ-field, as well as by an electromagnetic field, a gravitational field . . . and perhaps by still other fields that have not yet been discovered." For those with the stomach for it, all the strangeness of the quantum theory—minus its pernicious and queasy subjectivity—could spring from this unknown and eerie force in the invisible stream.

The first half of Bohm's paper (not the half with the "simple explanation" of EPR in terms of an instantaneously acting field) comes reassuringly bookended by references to Einstein, who had turned Bohm's thinking so dramatically around. On page one: "Einstein . . . has always believed that, even at the quantum level, there must exist precisely definable elements"—such as those that Bohm is postulating—"determining the actual behavior of each individual system, and not merely its probable behavior." Likewise, "Einstein has always regarded the present form of the quantum theory as incomplete." And Bohm wraps up the first half of the paper with thanks to "Dr. Einstein for several interesting and stimulating discussions."

The paper ends with a comment that he or Einstein might well have made during those discussions:

> For the purpose of a theory is not only to correlate the results of observations that we already know how to make, but also to suggest the need for new kinds of observations and to predict their results. In fact, the better a theory is able to suggest the need for new kinds of observations and to predict their results correctly, the more confidence we have that this theory is likely to be a good representation of the actual properties of matter.

This is a strange way to close a paper that begins with a reassurance that it "leads to precisely the same results for all physical processes as does the usual interpretation."

"I can't believe I should have been the one to see this." Bohm's hands fluttered and his face looked almost prayerful as he finished describing his new theory. He was sitting in his old friend Mort Weiss's Long Island living room, his back to the window. Evening was falling; a yellow convertible driving by glowed in the dusk. "To break through—after thirty years of everyone believing there's only one way!"

Weiss grinned. Bohm's excitement reminded him of their days at Penn State—the late nights reading and arguing with each other, teaching themselves modern physics over the sweet day-old pies Weiss would bring back from the diner where he worked. Bohm was fulfilling his wildest undergraduate dreams.

"Let me guess," Weiss said. "You worked it all out on a walk."

Bohm grinned. "I don't understand how anyone can think inside a building."

In college, Bohm had been famous for his reliance on the "floods of oxygen" obtained by walking (he took a two-hour post-lunch walk through the hills every day, along with his evening campus strolls), and for his sweet tooth, which Weiss saw had not changed either as Bohm devoured the celebratory ice cream Weiss brought out.

They reminisced even further back, to their childhood in Wilkes-Barre, Pennsylvania, about trips with the tent Bohm's father bought them so they could go camping on their own because the Boy Scouts didn't serve kosher food. One of their childhood friends, Sam Savitt, was now an illustrator of horse books; another, with whom they used to study physics after school, dropped out to work in the mines. Even in that coal town, there had never been any danger for Bohm and Weiss, with their middle-class Jewish parents, of such a fate.

They also talked of Bohm's book: "Oppenheimer told people that the best thing I could do when I was done with the book was to dig a hole and bury it." But Oppenheimer had also written a recommendation for Bohm, and with another recommendation from Einstein, Bohm had finally gotten an appointment, in São Paulo, Brazil.

"Have you seen a yellow convertible driving by the house?" asked Bohm suddenly.

"Yes, I have," said Weiss, surprised. "Why?"

"I'm being followed," said Bohm.

That night Bohm decided to leave under the cover of darkness for Florida, and Weiss came with him on the commuter train taking Bohm to Penn Station. Holding the bar above his head, his jacket hunched up around his shoulder, Bohm talked about his theory, letting go to gesture often with his fluttering hands.

A man opposite them was reading the paper. On the page facing Weiss was a picture of Bohm, that thin elfish face with wistful eyes, with a caption: "All they ever got from him was his name." Weiss wanted to laugh from the surreality of it all. "Dave, look: you don't need to wait for the *Phys. Rev.* to get your name in print!"

Bohm looked over his shoulder. His eyes dilated. He quickly looked away.

"I think you did a brave thing," said Weiss, quietly.

"Your dad always thought I would come to a bad end, politically," said Bohm.

Weiss said, "I know. I used to go upstairs to bed while you two argued communism versus capitalism late into the night, and then"—he shook his head at the memory—"you always refused a ride home. Even back then, you were always walking at night." His brow suddenly furrowed as he looked at Bohm. "Dave . . . your nose . . . it really isn't the same, is it?"

Bohm looked at him quizzically. Then he remembered: heavy late-night snow falling past bright Penn State windows, the two boys' breath coming out in clouds as they returned from a walk, Weiss tossing snow-balls into an unaccountably open window in their dorm. Bohm then had opened the front door and walked into the fist of an angry and snowy vic-tim lying in wait.

"No," said Bohm, reaching up to feel where his nose had broken, "it's not."

Weiss said sincerely, "Dave, I'm really sorry. I hope the crucial floods of oxygen haven't been affected."

Bohm laughed. "Don't worry."

"You never held a grudge, Dave," Weiss said.

They got off at Penn Station, and Bohm boarded the Florida-bound train, away from actual or imagined watching eyes.

In October, Bohm set out for São Paulo. As he sat inside the plane, the rain lashed the tarmac, and the wind banged the drops against the double-paned oval windows. Bohm sat in the blue-and-brown checked seat, watching the rain and willing the plane to leave soon.

He had agonized as his passport was being processed: *Isn't it taking longer than it should?* After consulting with Oppie, Einstein had assured Bohm that he thought no one would try to keep him from leaving: "It may be that Oppenheimer takes the case too lightly but I too am inclined to think nothing serious is behind the matter." This, from his two greatest mentors, was heartening, and his passport arrived with no apparent problems.

The plane began to move. Bohm stared out the window, watching the dreary landscape streak by. Then the plane pulled to a halt. The pilot's voice announced: "The plane will return to the terminal due to a passport irregularity of one of the passengers."

Bohm felt as if there were a sudden volt of electricity running through his lungs, colliding with his stomach. His knuckles, for some reason grasping his seat belt, were white. He tried to swallow.

The stewardess was walking toward him, looking as if she were about to ask him if he would like ice cubes in his orange juice. His heart was beating so fast it scared him. The seat belt hurt the insides of his hands, but he couldn't let go. *It may be that Oppenheimer takes the case too lightly . . .*

She walked by.

It may be— What just happened?

She hadn't even glanced in his direction. A few rows past Bohm's, she stopped and leaned over one of the passengers. A small Indian man who didn't seem to know much English followed her off the plane.

The plane began to taxi again down the runway. Bohm collapsed, his head on his knees. The plane lifted off the runway, like a miracle, into the rainy skies.

"I know, liftoff is just so terrible," said a sympathetic female voice beside him, "but my sister says air travel is now safer than driving a car. Can you believe it?"

Bohm sat up again, nodding weakly to his neighbor. He felt as if his stomach had still not received word of his deliverance. He swallowed, took a breath, tried to get his heart to behave normally.

"I wonder why they took that foreign gentleman off the plane," she mused.

23

Brazil

1952

Brazil had seemed awful from the moment Bohm realized, panicking yet again, that there was no one at the airport to pick him up. He gesticulated his way to a hotel, where he sat on the bed in his room trying to learn Portuguese from a little book.

The next day, stumbling over his new Portuguese words, he got in contact with Jayme Tiomno, a Princeton graduate student, who took him to his new home on Avenida Angelica, and then through the big white pillars of the university on Rua Maria Antônia, introducing him to a student who would show him around. On a wall was the seal of the college, and written beneath: VENCERÁS PELA CIÊNCIA. The student explained with pride how "sophisticated" the building was.

Bohm asked, "What does this mean? *Ciência*—science, right?"

"Yes, yes, true science, how do you say?—knowledge—the words say, 'You shall—conquer,' yes? 'You shall conquer through—knowledge,' yes?" The student moved on. "Let me show you the physics building."

Bohm stayed a second longer, looking at VENCERÁS PELA CIÊNCIA, thinking of his paper, which would come out in *Physical Review* in only a few months.

When he started teaching, he was enthused by the amount of good he could accomplish there. He felt needed. He worked with Tiomno on spin in his hidden variables. He had never before read the physics journals too assiduously, but now they formed one of his few links with other physicists when they trickled across the hemispheres to him. He walked from the university to his home on Avenida Angelica, feeling sick from the smell of decaying food, which haunted his sleep and all his eating. He threw up often and lost weight, looking more gaunt and ethereal. His face grew more lined.

He lay in bed. One humid night, he thought, *I have discovered the sixth law of thermodynamics, valid only in Brazil: everything that is sup-*

posed to move is stationary, and everything that is supposed to be station-ary moves.

He lay in the ocean water. "I like to float and watch the waves break," he wrote his friend Hanna Loewy. "One gets a feeling of unity with this warm sea, and sometimes I wish I could dissolve in it and spread out to its furthermost shores."

He lay in the hospital with various stomach complaints, feverish and fretful.

His student and collaborator Tod Staver left Brazil for Massachusetts, and a little while later Bohm was to learn that he had died in a skiing acci-dent. Jayme Tiomno moved back to Rio de Janeiro. Bohm had not made new friends, had not learned to cook, and felt "somewhat shaken in my convictions about the value of my work."

In December, three months after Bohm had arrived in Brazil, at a meeting of the Brazilian Academy of Sciences in Belo Horizonte, he saw Feynman.

Richard Feynman was taking a sabbatical year teaching in Rio, a little way up the Atlantic coastline from São Paulo. It was amazing to see that familiar face and the good-old-boy act, hear all the stories and the Feyn-man laugh. Bohm felt as if his own personal universe had been restored to some kind of order: Feynman was here.

After the conference, Feynman said, "Come on, Dave, let's go some-where. Do you know where there's an interesting bar?"

Bohm didn't. "O.K., great!" said Feynman, without irony. "Come on!"

Feynman hailed a cab, then asked the driver a question in Portuguese. Bohm, sitting back on the fake-leather cab seat, understood a little of what Feynman said ("where something samba something interesting something"). The cabby was nodding, looking ahead, but attentive, and replied at length as he poked his meter.

"*Bem bom,*" said Feynman to the cabby, and "O.K., great!" as he leaned back in the seat. "He's going to take us to the Savassi district, that's the place to go."

Bohm felt sort of giddy and collegiate: here they were, Dick and he, two bachelors not much over thirty, and not that bad to look at, going out on the town—in Brazil. Fun, right? O.K., great! He wished that he wasn't so dressed up. Feynman always wore any old thing, could go effortlessly from bongo-drumming to physics conference to nightclub, and look out of place in a way that was better than looking *in* place. He was like Einstein, who wandered around Princeton wearing what Bohm thought of as some genuinely odd things (no socks), but it was O.K.—it was more than O.K., it was part of the whole mystique. Feynman was becoming an expert at generating mystique himself.

"So, Dave," said Feynman, "how's everything going for you in São Paulo?"

Bohm was nodding, eyebrows raised: "Ah . . . it's O.K. . . . there are several good students here. . . ."

"Tell me, Dave, do you find that it's all memorized? The students down in Rio have memorized everything, yet nothing has ever been translated for them into meaningful words. . . . There's so much for us to do out here. There's so much to teach."

"We're helping establish physics down here," said Bohm. "The department in São Paulo was founded less than twenty years ago."

"Yeah, it's exciting," said Feynman. "You know, I don't believe I can really do without teaching. I remember when I was at Princeton and I could see what happened to those great minds at the Institute for Advanced Study. They'd been specially selected for their tremendous brains and were now given this opportunity to sit in this lovely house by the woods there, with no classes to teach, with no obligations whatsoever. So these poor bastards can sit and think clearly all by themselves"—he laughed—"That would choke me up like *nothing* else."

Bohm was laughing now too. "The Princetitute: home of the greatest stagnating brains of the century."

"You need someone to bother you! They just don't have anything, any students, any interaction with experimentalists, anything to give them"—Feynman snapped his fingers—"the spark of an idea."

"But here . . ." said Bohm slowly, "I worry about stagnation myself. In explaining physics to people who do not understand English very well, the imagination is not stimulated nearly so much. It's a little frightening to realize that the language barrier will make really close contacts with most people rather difficult."

"You'll get it, Dave," said Feynman. "It won't take as long as you think."

The cabbie stopped, said something in Portuguese to Feynman. "O.K., great." Feynman paid him while Bohm was reaching for his wallet. "This bar is supposed to have good sambistas playing tonight. Oh, Dave, I've gotten so crazy about samba. I've joined a school now."

They walked out of the skyscrapered street into the close dimness of the bar.

"School—in the sense of fish," explained Feynman, "not in the sense of Princeton."

"Oh . . . so—you play the drums?"

"Well, to play with my sambista school, I'm learning this great instrument called the *frigideira;* it's like a toy frying pan that you beat with a little metal stick." Feynman sat on a bar stool and moved his hands to suggest the instrument. "It's great. We're getting new music ready for the Rio Carnival, it's really exciting. You have to dress up—I think I might be Mephistopheles."

Bohm was suddenly and completely overwhelmed by the jealousy that had been creeping up on him ever since he had run into Feynman. Feynman was actually happy in Brazil. He thought it was *O.K., great!* For Bohm, there were the stomach problems, the hospital stay, the fear of traffic, the fear of restaurants, the fear of people talking Portuguese, and a new fear about his passport. He was miserably, corrosively jealous of Feynman and his vibrancy, his fluency in any situation, how everything always worked out for him.

"Dick, they took my passport away."

"They what? Who?"

"The American consulate. They told me I had to go there for 'registration and inspection of passports,' and suddenly, there I am, they say I can come and get it if I want to go back to the States, otherwise they'll keep it."

"Jeeez . . ." Feynman scowled. "Any idea what the point is?"

"They want to make sure I stay here, and don't pass on any of their precious secrets to Russia, as if I know any."

Feynman laughed broadly. "Well, Dave, you heard them. No hobnobbing with Reds."

"The evening after they took my passport away, two days ago, my roommate saw a car circling the house for over an hour."

"Jeeez . . ." said Feynman again. And then: "You know, it was probably just somebody who was lost. These South American cities are awfully confusing, especially at night."

Bohm shrugged and raised his eyebrows. "I'm being watched, Dick, even down here," he said simply, and Feynman could not tell if it was the truth or Bohm's paranoia speaking.

Feynman leaned back, his arms across his chest. "Well, I don't know, Dave . . ."

But Bohm was on a roll. "I realized I am not really out of the U.S., even here."

"You're not going back, are you, Dave? Yeah, I can see it. Look, this type of stuff makes me depressed. And, hey, you haven't told me, but word on the street is that you've submitted something interesting to *Phys. Rev.*"

Suddenly Bohm glowed, so obviously that Feynman noticed the change.

"Tell me about it," said Feynman, leaning, arms crossed, on the bar.

"I have reformulated quantum mechanics in terms of hidden variables."

Feynman's eyebrows shot up toward his wild hair.

"You see, I was looking at the WKB approximation to the Schrödinger equation"—a useful semiclassical tool—"and thinking, why couldn't the particles actually have paths?" Feynman's lower eyelids squinted up, his head tilted. "And then, if they did, if they moved along them in a totally deterministic way, then what would I have to do to transform the approximation into something that would reproduce the results of the quantum theory?"

Somewhere in there Feynman's generous suspension of disbelief had come crashing down. "Dave, you're totally crazy. You can't have determinism and quantum leaps. You're completely crazy."

"No, Dick, wait, the whole thing works; let me talk you through it. The key thing is this new kind of potential energy that fell quite naturally out of Schrödinger's equation. I called it the quantum potential."

Several drinks and lots of hand-waving and napkin-writing later, Bohm felt he had explained the fine points and convinced Feynman that the whole thing was logically consistent. He later wrote to his friend the mathematician Miriam Yevick that Feynman was "terrifically impressed with it."

Feeling the momentum of Feynman's approval, Bohm rushed on: "See, Dick—remember how exciting a really new theory can be? You used to be so interested in speculating, finding out new ideas. I think Bethe's trapped you back at Cornell in this depressing place where you spend all your time doing endless calculations. Bethe's a human calculator."

Feynman looked serious. "Dave," he said, "Hans Bethe is one of the greatest scientists and greatest men and"—he broke out in a grin—"greatest human calculators of my acquaintance. One of my favorite things is doing math with Hans. We compete. We see who's quicker, who has better tricks. It's lots of fun. Even though he always beats me."

Bohm had been lulled by Feynman's open-mindedness into thinking that he could actually woo him away from the orthodox quantum theory. As Feynman himself was fond of saying, why are smart men so stupid in bars?

"Look, Dave," Feynman continued, "I'm glad for you that you've pulled this theory together. It's something great. But it's not my thing. I don't work that way—you know me, I like the problems. For me, I don't see any problem here: quantum mechanics works, why mess with it? It's O.K., Dave," he said, as Bohm was about to answer the rhetorical question, "I believe that you have seen something, and that's great. You should do that, if that's what you want to do.

"Do you want a beer? I want a beer," said Feynman. "*Duas cervejas,*" he said to the bartender.

"No," said Bohm hurriedly, "*só uma cerveja.*" Then he asked, "So, Dick, have you heard from home?"

"I have to say, not that much. You?"

"No, no, I don't hear much at all. I got a letter from Einstein; that made my day. And Pauli wrote; he practically concedes my idea is logical. But it takes so long for letters to get here."

"No kidding," said Feynman. "For a while I was talking with Fermi, we were, you know, writing back and forth about mesons,"—mysterious new particles just found in 1947—"and it was so frustrating to float out a good idea and then have to wait and wait." He took a swig of his beer. "I have one instant connection though." He grinned. "I can hook up to Caltech by ham radio."

"You what?"

"I've got this friend, this blind guy, he's a ham-radio operator, he sets me up once a week."

"That's so amazing to be able to talk physics with people back home like that—I— I feel so cut off. Things are dated by the time they get *published,* let alone by the time they arrive down here."

Feynman nodded, dragging his finger around the edge of his beer mug. "You know, I get lonely and then sometimes I think I should get married again."

"Yes, I thought that too; I thought maybe I should get married before I came here . . ."

"But then I think, look at all the nifty girls down here!" said Feynman, determinedly waving an arm in the general direction of the rest of the bar.

"Yeah . . ." said Bohm. "But anyway, it's not just companionship, it's scientific companionship that's missing down here. . . . For a while at least Tiomno was around, but now you've got him."

"Yeah, yeah. I know what you're saying. Look, I feel lonely even with my once-a-week ham-radio link and Tiomno," Feynman said. He was laughing, but ruefully, elbows on the bar, looking at Bohm.

Bohm felt the jealousy lift away. Feynman, even Feynman, with all the physicists and all the girls who worshipped him, with all his vitality and ease, was lonely, too.

Bohm and Feynman stood on the curb outside the bar just before dawn. Feynman was joking around, hailing cars that weren't cabs, and Bohm was laughing as he had not laughed in three months.

"I now think he is my friend for life," Bohm wrote when he got home.

24

Letters from the World

1952

THE DAY HE ARRIVED back in São Paulo, Bohm received another letter from Pauli, sent on December 3, 1951.

> I do not see any longer the possibility of any logical contradiction as long as your results agree completely with those of the usual wave mechanics and as long as no means is given to measure the values of your hidden parameters both in the apparatus and in the observe system.

Bohm smiled at the final phrase—"observe system." In his own way, he was telling Bohm it was all right! And if Pauli couldn't find a problem, then no one could. Bohm read on:

"As far as the whole matter stands now, your 'extra wave-mechanical predictions' are still a check which cannot be cashed."

Bohm leaned back in his chair, reading it again. *The hidden variables are a check which cannot be cashed.* It was, he realized, the best he could hope for from Pauli, just as Feynman's reaction was the best he could hope for from Feynman.

Bohm was not shy to speak of the "instantaneous interactions between distant particles" (i.e., nonlocality) of his "quantum potential," which depends directly on the state of the whole, not on individual parts. "I want to emphasize," Bohm wrote, "how radically new are these implications. . . . They are hinted at only vaguely and indirectly in the subtle arguments of Bohr." Where Bohr had thrown a veil, Bohm had cast light "by putting quantum and classical theories in terms of the same intuitively understandable concepts," allowing "a clear and sharp perception of how the two theories differ. I felt that such an insight was important in itself." Bohm would have to wait for Bell before this was appreciated.

Louis de Broglie, on the other hand, did not answer when Bohm sent him a copy of his paper. Instead he published objections to it in a French

215

journal. News of this drifted down to the bewildered Bohm. He wrote to his old friend Miriam Yevick that de Broglie had not "really read my article, but simply reiterated Pauli's criticisms which led him to abandon the theory, but did not point out my conclusion that these objections are not valid. He's going to look a little silly. That's what he gets from rushing into print five months before my article came out."

He began to suspect the Brazilian postal system of sloppiness, while he waited for other responses and read and reread Pauli's letter. "What I am afraid of is that the big-shots will treat my article with a conspiracy of silence; perhaps implying privately to the smaller shots that while there is nothing demonstrably illogical about the article, it really is just a philosophical point, of no practical interest," he wrote to Miriam.

January came around, and he knew the 1952 *Physical Review* must have come out. "It is hard to predict the reception of my article, but I am happy that in the long run it will have a big effect," he told her expectantly.

As January wore on, Bohm ate at more and more expensive restaurants as his stomach troubles worsened, living every day for the moment the mail came, seeing American agents poisoning his life wherever he went. He wrote to Miriam of his "passionate desire to fight this stupefying spirit of formalism and pragmatism in physics," which was causing the physics community to turn a blind eye to his paper. "It cuts at one's insides like a hot knife being twisted inside your heart."

Meanwhile his old friends—Oppie's other grad students—were all in trouble. Rossi Lomanitz couldn't get work, and Joe Weinberg was on trial, accused of giving secrets to the Russians as "Scientist X." Bernard Peters, teaching at an institute in Bombay, found that his passport had expired and would not be renewed. He became a German citizen again.

The South American winter rolled around and a letter arrived, not from Copenhagen, but from Bohr's sidekick, Rosenfeld, now in Manchester, England. Rosenfeld, as Pauli described him only two years later (1954), had become "the square root of Bohr times Trotsky." Pais called him "the self-anointed defender of the complementary faith, *plus royaliste que le roi.*" Unlike Bohm, he apparently felt no conflict between his quantum theory and his strong views on dialectical materialism.

                                                    30st May 1952, Manchester.

[Bohm smiled at the "30st" of May, and read,]

Dear Bohm,
        I certainly shall not enter into any controversy with you or anybody else on the subject of complementarity, for the simple reason that

there is not the slightest controversial point about it. But I shall gladly reply to your kind letter in the spirit of a friendly conversation about some of the points you raise.

Obviously you are thinking that my assertive attitude in this matter is extravagant, inasmuch as I might be forgetting that I and even Bohr are also subject to error. I may reassure you by telling you something of my experience in working on this subject with Bohr. In our work on the problem of measurability of fields we have made probably all the errors that could conceivably be made before reaching at last solid ground. It is just because we have undergone this process of purification through error that we feel so sure of our results. To give only one example: we have not hesitated to throw doubt on the validity even of Maxwell's equations (in vacus) when a certain argument (later recognized as erroneous) suggested to us that they could be wrong. I tell you this to show you that there is nothing dogmatic about our attitude of mind, and that there is no truth in your suspicion that we may just be talking ourselves into complementarity by a kind of magical incantation. I am inclined to retort that it is just among your Parisian admirers that I notice some disquieting signs of primitive mentality.

The difficulty of access to complementarity which you mention is the result of the essentially metaphysical attitude which is inculcated to most people from their very childhood by the dominating influence of religion or idealistic philosophy on education. The remedy to this situation is surely not to avoid the issue but to shed off this metaphysics and learn to look at things dialectically.

. . . The main thing is not to accept any other guidance than that of Nature herself.

> With kind regards,
> yours sincerely,
> L. Rosenfeld

Bohm sat in his office above the Rua Maria Antônia, with the São Paulo traffic shrieking along the streets below him, irritated and tired. He had waited nearly half a year, and this self-parody ("even Bohr is subject to error") was the only response he would get from Copenhagen. *If he had seen de Broglie he wouldn't be talking about "your Parisian admirers" anymore,* he thought.

How can I be down here in this humid chill not able to fight for my theory? He thought wistfully of *Vencerás pela ciência.* It was so sure of itself—you will conquer through knowledge.

"We are both friends of Dave—we should know each other," wrote Miriam Yevick to Eugene Gross the month after Bohm departed for Brazil. Gross had been Dave's first graduate student, enraptured by his first lecture. "In his low-key fashion, Dave Bohm had opened up a vast panorama," remembered Gross decades later. Gross would write his first four papers with Bohm, mining the rich territory of the plasmas Bohm had talked about in that lecture.

Early in 1952, Miriam and her husband, George, a physicist who was also a friend of Bohm's, drove up to M.I.T., where Gross was now a post-doc, to meet him and his wife, Sonia. As Gross came walking toward them, Miriam was surprised by how familiar he looked. Studying in Princeton's grand old Fine Hall she used to see him pacing up and down, jingling his coins in his pocket, in a way so characteristic of Bohm that it would give her a start.

The two couples made friends quickly—"We spent three days talking," Miriam remembered later. The men were both more down-to-earth, pragmatic physicists than their friend Bohm, and Sonia, though studying chemistry, had much in common with the mathematician Miriam, who had spent part of her college days learning to blow glass and nights tending to liquid-nitrogen cooling tanks.

The conversation eventually turned to the contents of the new edition of *Physical Review.*

Miriam said, "When do you think people will start responding to Dave's theory? It's so frustrating for him to be so far away from all news."

Gross's kind, sardonic face formed a grimace. "I don't know if he's ever going to hear much of a reaction."

Miriam said, "You don't really think people will just ignore it? He's done what von Neumann said was impossible—that must be worth something."

"Physics isn't philosophy, though," said George in his quick voice. "The ordinary way of doing quantum physics works: it's too much effort to think about another interpretation."

"But that's ridiculous!" said Miriam. "What if he's right?"

"Physics isn't as clean and mental as math."

"It's a theory, Eugene! How much more clean and mental can you get? It's just a theory."

Bohm's hidden-variables theory made Gross feel tired. He wouldn't want to have to do all his calculations with it. Not only was it less mathematically elegant, but he was used to quantum mechanics the way it was. The only reason he would switch was if—

"He's got to get results, Miriam," he said. "He's got to get results!" His fist slammed the table for emphasis. "He's asking people to do what they never have wanted to do: switch from one worldview to another for a philosophical reason which doesn't affect their daily life. But if he gets results, then it won't be a question of taste or whether or not the math is cumbersome. Everyone will switch. Otherwise, right or not, he should just shelve it and move on."

"Gene's right," said George. "Can he work spin into his theory? Things like that he can't just talk his way around."

Gross laughed. "I remember once at a party, Dave, tongue in cheek, constructed an elaborate and convincing theory of the existence of ghosts and devils." He shook his head with amusement at the memory. "It's incredible how persuasive he can be, how good he is at constructing coherent intellectual structures."

"But you can only get so far with that," said George.

"So," said Miriam, "you're saying that Dave's new theory is an elaborate and convincing demonstration of the existence of ghosts and devils?"

Gross grinned. "I hadn't thought of it that way. . . . Funny, though—that was the terminology Einstein used when he was playing with a theory like Dave's. He called it *Gespensterfelder*: ghost waves guiding the particles."

"But he never published that," said George, "and he certainly didn't wait around to see if people would listen to him."

Miriam was meditative. "Dave's problem is that it matters to him what other people think—it matters too much."

Gross, suddenly serious again, said, "It matters to him because he's so completely, calmly, passionately absorbed in the search—he wants to share what he discovers. His lack of competitiveness or guile amazes me." His brow creased. "I've thought about this, and I can only use old-fashioned language to describe his impact on me. He's a secular saint."

That evening, Miriam wrote to Bohm about what Gross had said.

Bohm only saw the negative: "One of my best friends seemed to be turning away from me and running with the tide." *Results! Sometimes results take decades to arrive. Einstein and relativity! Copernicus, for goodness' sake!* No one these days was producing anything but trivialities. Feynman and Julian Schwinger may have produced "resultlets" with their quantum electrodynamics, but "these little mice are all that come out of twenty years of labor by the mountain of thousands of theoretical physicists. I alone am supposed in a year or two to produce a scientific revolution comparable to that of Newton, Einstein, Schrödinger, and Dirac all rolled into one. . . . As for Pais and the rest of the 'Princetitute,' what those little farts think is of no consequence to me. In the past six years, almost no work at all has come out of that place."

He had heard through the grapevine that Niels Bohr had called the theory "very foolish," after he had gotten over his initial surprise that the feat had been accomplished at all. Von Neumann, Bohm had been told, thought the idea was consistent and even "very elegant," higher praise than Pauli had given before the paper had come out. But now Bohm was in no mood to take scraps from the table. When recounting to Miriam this praise of von Neumann's, Bohm wrote, "(the unprincipled bum!)."

Mathematical abstraction, he explained, can deceive. "It grants an illusory sense that, as you say, nothing has been put over on you. . . . In his proof that there can be no causal interpretation of quantum theory, von Neumann implicitly assumed that there can be no 'hidden variables' in the measuring apparatus. Nobody could have seen from von Neumann's math that this was assumed. I was able to see it because I had a counterexample"—Bohm's own theory—"so that I knew, by God, that von Neumann had 'put something over' on everybody, including himself."

When another of his friends expressed misgivings, Bohm lost his usual dry wit and spluttered almost incoherently to Miriam that it was "comparable to a man who shoots you in the back, and then begs pardon saying that he was using a theorem from which he deduced that bullets come out from a gun at an angle of ninety degrees relative to the barrel and that he was therefore really shooting at someone else."

Schrödinger "did not deign to write me himself, but he deigned to let his secretary tell me that His Eminence feels that it is irrelevant that mechanical models can be found for the quantum theory, since these models cannot include the mathematical transformation theory"—Dirac and Jordan's generalization of quantum mechanics—"which everyone knows is the real heart of quantum theory. Of course, His Eminence did not find it necessary to read my papers, where it is explicitly pointed out that my model not only explains the results of this transformation theory, but also points out the limitations of this theory. . . . In Portuguese, I would call Schrödinger *un burro,* and I leave it for you to guess the translation."

And still, Bohm wrote to Miriam, "I am convinced that I am on the right track."

25

Standing Up to Oppenheimer

1952–1957

J UST BEFORE THE WAR, a young Max Dresden left Amsterdam for Ann Arbor to get his Ph.D. at the University of Michigan. Ann Arbor was swiftly becoming a destination for young physicists, under the guidance of Pauli and Heisenberg's old school friend Otto Laporte, and the Dutch discoverers of "spin," Sam Goudsmit and George Uhlenbeck (all in the United States because of Hitler). Dresden's American move was to be permanent, and he would become a beloved institution for a quarter of a century at the State University of New York at Stony Brook. There he wrote a masterly biography of Bohr's "cardinal," Kramers, that tells the larger story of the great era of quantum physics. It is an unusual book, since its hero is the one repeatedly disappointed, amid the victories of his closest friends and colleagues (Kramers came near to discovering matrix mechanics, the Dirac equation, and quantum electrodynamics, but each time failed to make the last imaginative leap).

In 1952, however, when Dresden was a new professor at the University of Kansas (already well appreciated for his jokes and his wide-ranging enthusiasm for physics, art, and literature), his students presented him with David Bohm's paper. At first he told them, "Oh, well, von Neumann has shown. . . ." But his students were so fascinated that he read it, too. He was surprised that he could not immediately find the fatal flaw. He went back to von Neumann, at which point it began to dawn on him that von Neumann's argument did not apply to Bohm's hidden variables.

Dresden asked Oppenheimer for his opinion. "We consider it juvenile deviationism," Oppenheimer said. He hadn't read the paper. "We don't waste our time." But when Dresden said the issue troubled him, Oppenheimer proposed that he give a seminar on Bohm's theory at the Institute for Advanced Study.

Presenting this seminar turned out to be a surreal experience. Pais, too, called Bohm's work "juvenile deviationism." "A public nuisance," said someone else. People complained about Bohm's fellow-traveling more than his physics—though Dresden relates that amid the disdain, some hard questions were asked about the physics of Bohm's interpretation, which he himself could not answer.

Oppenheimer summed up the situation: "If we cannot disprove Bohm, then we must agree to ignore him."

This edict of Oppenheimer's incurred little resistance among physicists, but a mathematician tried to stand up to him. This was the tragic John Nash (the subject of the movie *A Beautiful Mind*), who had had his greatest insight—a discovery about the theory of games, called the Nash equilibrium—at Princeton in late 1949, while Bohm was roaming about the same grounds struggling with the quantum theory. A decade later, at the Institute for Advanced Study, Nash got in a fight with Oppenheimer over the quantum theory, which the mathematician had learned from reading von Neumann's famous book (in German).

All that survives of this argument is a letter Nash wrote Oppenheimer in the summer of 1957. He apologized for how aggressive he had been, but voiced his frustration at "most physicists (also some mathematicians who have studied Quantum Theory)" whom he found "quite too dogmatic in their attitudes." They treat "anyone with any sort of questioning attitude or a belief in 'hidden parameters' . . . as stupid or at best a quite ignorant person." Reading Heisenberg's 1925 paper on matrix mechanics brought some clarity: "To me one of the best things about the Heisenberg paper is its restriction to the observable quantities. . . .

"I want to find a different and more satisfying under-picture of a non-observable reality."

Nash's biographer Sylvia Nasar notes that "it was this attempt that Nash would blame, decades later in a lecture to psychiatrists, for triggering his mental illness—calling his attempt to resolve the contradictions in quantum theory, on which he embarked in the summer of 1957, 'possibly overreaching and psychologically destabilizing.' "

By February 1958, Nash's always eccentric ways suddenly took a horrific turn for the worse. He embarked on a long journey of losing and, then, very slowly, reclaiming his mind, a struggle that culminated in the 1994 Nobel Prize. It was awarded to him for the Nash equilibrium, discovered almost half a century earlier.

The quantum theory seemed to have almost destroyed Bohm, too, but the story was not finished. His theory would not have such a grand comeback as Nash's, but, in the hands of an obscure physicist in a remote land, it would reveal the mysterious mathematical inequality of Bell's Theorem—an idea that would prove even more important than equilibrium.

26

Letters from Einstein

1952–1954

Early in May of 1952, Max Born wrote to Einstein about death. They were both in their early seventies, but younger friends had died recently, including Kramers. "So we old fellows become more and more lonely, and I am writing to you in order to keep intact the few remaining links with our contemporaries which still exist." He relayed the best wishes of his wife, and both of their greetings to Einstein's stepdaughter Margot.

Einstein's reply was quick:

Dear Born
. . . You are right. One feels as if one were an Ichthyosaurus, left behind by accident. Most of our dear friends, but thank God also some of the less dear, are already gone. . . .

Have you noticed that Bohm believes (as de Broglie did, by the way, 25 years ago) that he is able to interpret the quantum theory in deterministic terms? That way seems too cheap to me. But you, of course, can judge this better than I.

Kindest regards to you both
Yours, Einstein

At the end of the month, Hedi Born replied to this letter:

Dear friend Albert Einstein
. . . Old age is not so bad really, provided one does not have too many twinges. What have you got against being an Ichthyosaurus? They were, after all, rather vigorous little beasts, probably able to look back on the experiences of a very long lifetime.
In any case, we two old ones will go on thinking of you and Margot

with unchanging loyalty, even if we should never be able to meet
again.

With all my heart
I remain
Your old Hedi

In Born's notes on their collected letters, which he annotated before
his own death, in 1969, he commented, "Today one hardly ever hears
about this attempt of Bohm's, or similar ones by de Broglie."

Einstein mentioned Bohm again a year and a half later. The occasion
was Born's retirement at the age of seventy from the University of Edinburgh, and the volume of papers that was to be presented at the
celebration.

12 October, 1953

Dear Born,

. . . For the presentation volume to be dedicated to you, I have
written a little nursery song about physics, which has startled
Bohm and de Broglie a little. It is meant to demonstrate the indispensability of your statistical interpretation of quantum mechanics, which Schrödinger, too, has recently tried to avoid. Perhaps it
will give you some amusement. After all, it seems to be our lot to
be answerable for the soap bubbles we blow. This may well have
been so contrived by that same "non-dice-playing God" who has
caused so much bitter resentment against me, not only amongst
the quantum theoreticians but also among the faithful of the
Church of the Atheists.

Best regards, also to your wife
Yours,
A. Einstein

Einstein's "nursery song" had included, once again, his skepticism that
Born's interpretation could really be the last word.

26 November, 1953

Dear Einstein,

The presentation of the volume was made yesterday during a little
celebration at the university. It gives me tremendous pleasure that
so many of my old friends and colleagues have contributed to it.

For the time being I have read only a few of the articles—yours was the first, of course, and you are also the first to receive my heartfelt thanks. . . .

Incidentally, Pauli has come up with an idea (in the presentation volume for de Broglie's 50th birthday) which slays Bohm not only philosophically but physically as well. . . .

With sincerest thanks and kindest regards, also from Hedi,

Yours, Max Born

By 1954 Bohm—slain, as Born wrote so offhandedly, both philosophically and physically—was ever more desperate to leave Brazil; he wrote agonized epistles to all his friends describing the rotting food that kept him sick, the constant and relentless building projects that kept him sleepless, and the unintellectual atmosphere that kept him uninspired. The United States (symbolizing despotism) and Brazil (symbolizing chaos) loomed over his subconscious.

Pauli's supposedly devastating idea turned out to be that Bohm's work represented "artificial metaphysics." It is hardly a consolation, still, to find that your ideas are so disliked that weak claims against them are considered overpowering.

Eventually, even Einstein heard about Bohm's misery and wrote him a characteristically practical and perceptive letter in January of 1954, which began: "Dear Bohm, Lilli Loewy has shown me all your letters concerning your feeling of distress for being closed out and closed in at the same time. What impressed me most was the instability of your belly, a matter where I have myself extended experience."

This sympathy and fatherly tone from the great man loosed a flood of emotion from Bohm, who responded with a five-page enumeration of his miseries, in his boyish cursive hand. A brief digression describing the utter corruption of the Brazilian government ended with the tragicomical statement, "Now I want to return to my own problems." Nathan Rosen had given him a "not particularly good" offer to come to the Technion in Haifa, Israel—Rosen had just arrived there and would soon build what was a small technical college into a prominent center of physics— which Bohm had decided to accept, but did not know if either Brazil or the United States would let him go. Would Einstein (who had just been invited to assume the Israeli presidency) write a letter which might facilitate a visa for Israel?

Bohm finished on a slight upswing: "I am beginning to think in a new direction." He was still searching for a causal underlayer to the quantum theory.

Einstein wrote back in his unflappably kind, ironic way. His final paragraph is made even more poignant by his death a year later.

February 10, 1954

Dear Bohm:

I was very impressed by your letter. . . . I was really very happy to hear that Rosen is trying to call you there and I have already written to him. I am, of course, very glad to do everything to facilitate the realization of this plan; so do not hesitate to write me as soon as you see a possibility that I could help in the matter.

Your picture of your idyllic environment made a vivid impression on me. I believe your report was exhaustive with nothing left out.

I am glad that you are deeply immersed seeking an objective description of the phenomena and that you feel that the task is much more difficult than you felt hitherto. You should not be depressed by the enormity of the problem. If God has created the world his primary worry was certainly not to make its understanding easy for us. I feel it strongly since fifty years.

With kind regards and wishes,
yours, Albert Einstein.

Epilogue to the Story of Bohm

1954

IT TOOK A LONG TIME for Bohm to forgive Oppenheimer, and it certainly did not happen until Oppenheimer had more or less destroyed himself during his appearance before the House Un-American Activities Committee. By all accounts, he came out a shattered man, and was perhaps easier to forgive that way.

But in 1954, Oppenheimer's troubles were only beginning, and Bohm wrote to Miriam,

> I have just heard from a friend that J.O. may be called before a committee soon (of course he has ways of getting out of it perhaps) . . . [It's all going to happen to] the great man himself, whose face once appeared with infinite sadness on the cover of *Time Magazine*. Perhaps he will really have reason to be sad too. As I once said, he looked like Jesus Christ in the picture. I think a better image would be a linear combination of J.C. and Judas, or of Judas trying to look like J.C. An interesting case of mistaken identification, don't you think?

Bohm taught in Israel, and then, ever unsatisfied, in 1957 moved on to Britain, with a favorite, brilliant grad student from the Technion at Haifa, Yakir Aharonov, following close behind. Though his papers with Aharonov are well respected for the gains they represent in physical understanding, his interests (communism finally renounced) began to turn toward mystical philosophical musings, sitting at the feet of the guru Krishnamurti. His vision of the world became ever more abstract (and ever more popular with the public). He propounded, with his gently fluttering hands and earnest face, complicated ideas with broad strokes and beautiful (if somewhat vague) words.

"I would say," wrote Bohm in the introduction to his best seller, the philosophico-physical book *Wholeness and the Implicate Order,* "that in my scientific and physical work, my main concern has been with understanding the nature of reality in general and of consciousness in particular as a coherent whole, which is never static or complete, but which is an unending process of movement and unfoldment."

Feynman would become even more wildly popular.

In November 1964, he returned to his old snowy stomping grounds in Ithaca to give Cornell's annual "Messenger Lectures" for the general public on the subject "The Character of Physical Law," which were subsequently made into a book. With his clarity, his humor, his accent undiminished in spite of years of living far from Brooklyn, he charmed and inspired the crowds that packed into the lecture room to watch him cavort across the stage below them. The lectures displayed his love of Nature itself, of the way things really work, of the specific, the actual, the find-out-able, and they gave an overview of how one of the greatest physicists of his generation thought about his subject.

The sixth lecture was titled "Probability and Uncertainty." Now was the time to talk about quantum mechanics, where, he told his listeners, "our imagination is stretched to the utmost—not, as in fiction, to imagine things which are not really there, but just to comprehend those things which are there."

He warned his listeners that these things would be difficult to understand. "But the difficulty," he assured them, "really is psychological and exists in the perpetual torment that results from your saying to yourself, 'But how can it be like that?'—which is a reflection of an uncontrolled and utterly vain desire to see it in terms of something familiar. I will not describe it in terms of an analogy with something familiar; I will simply describe it.

"There was a time when the newspapers said that only twelve men understood the theory of relativity. I do not believe there ever was such a time. There might have been a time when only one man did, because he was the only guy who caught on, before he wrote his paper." (It was typical Feynman to dodge the mention of anyone but the first person singular.) "But after people read the paper a lot of people understood the theory of relativity in some way or other, certainly more than twelve. On the other hand, I think I can safely say that nobody understands quantum mechanics.

"So do not take the lecture too seriously, feeling that you really have to understand in terms of some model what I am going to describe, but just relax and enjoy it. I am going to tell you what nature behaves like. If you will simply admit that maybe she does behave like this, you will find she's a delightful, entrancing thing. Do not keep saying to yourself, 'But how can it be like that?' because you will get 'down the drain,' into a blind alley from which nobody has yet escaped. Nobody knows how it can be like that."

Feynman described the famous two-slit experiment, where a stream of electrons (for example) flows through a barrier with two holes. On the other side of the barrier an electrical system notes each electron's arrival. They arrive as particles, but they interfere with one another like waves. This is true even if only a single electron is traversing the barrier at a time. A series of very many such electrons, each passing alone through the double-slitted barrier, do not strike a detector on the other side in the pattern particles would leave (a broad clump just beyond each of the two slits). The pattern at the detector is instead striated and diffracted, made up of the parallel stripes that are the trademark of a wave. This means that the individual electrons mostly hit the detector in places no ordinary particle trajectory would land them—as if a single electron somehow passed through both slits or was "guided" by a wave. If you light their path to watch them, however, the interference disappears; they act just like the particles we thought they were.

"The question now is, how does it really work? What machinery is actually producing this thing? Nobody knows any machinery. Nobody can give you a deeper explanation of this phenomenon than I have given. Nobody can go beyond a description of it. . . . Physics has given up, if the original purpose was—and everybody thought it was—to know enough so that given the circumstances we can predict what will happen next."

There is a theory that tries anyway, said Feynman. The motion of the electron in a theory like this is determined by "some very complicated things back at the source: it has internal wheels, internal gears." (*You've missed the point,* Bohm, lover of "wholeness," would have said.)

"That," said Feynman, "is called the hidden variable theory." He looked out at his audience. "That theory cannot be true." If we were able to know in advance which hole the electron would go through, he continued, "then whether we have the light on or off has nothing to do with it." It is impossible, he continued, to explain the interference pattern as merely the sum of the contributions of electrons going through one hole or the other—a corollary to what John Bell (unmentioned by Feynman) had shown earlier that same year.

"It is not our ignorance of the internal gears, of the internal complications, that makes nature appear to have probability in it. It seems to be somehow intrinsic. Someone has said it this way—'Nature herself does not even know which way the electron is going to go.' "

Feynman did not mention what Bell in 1964 had made very clear. The

key difference was that Bohm's theory allowed the behavior of its particles to be influenced by faraway things. Just two years before Feynman's
speech and Bell's theorem, Bohm had written, "Even in its present incomplete form, the theory does answer the basic criticisms of those who
regarded such a theory as impossible.

"It seems that some consideration of theories involving hidden variables is at present needed to help us avoid dogmatic preconceptions," he
continued to insist. "Such preconceptions not only restrict our thinking
in an unjustifiable way but also similarly restrict the kinds of experiments
that we are likely to perform."

The Discovery

1952–1979

John F. Clauser with his machine inspired by John S. Bell, 1976

27

Things Change

1952

JOHN BELL, a red-haired, twenty-three-year-old college graduate from Northern Ireland, sat outside the English hostel in the Berkshire Downs where he was living, surrounded by the pieces of his motorcycle. Taking apart, fixing, and reassembling these bikes was the major hobby of the accelerator design group of Britain's Atomic Energy Research Establishment, young men who spent all day devising the theory of how to push particles to ever higher speeds in order to smash them to pieces. (The energy of destruction created new particles and new opportunities to understand the constitution of matter.) The motorcycles also sometimes smashed: Bell's new red beard hid a scar around his mouth that was a souvenir from a wreck.

It was 1952, and he was filled with an excitement that had nothing to do with high speeds. He had just read Bohm's paper. Bohm had accomplished what Bell had dreamed of—a theory with entities that were real regardless of the actions of experimentalists, which nonetheless produced the same results as quantum mechanics.

Bell saw that the great von Neumann "must have been just wrong" when he declared such a hidden-variable theory impossible. He would soon identify the same logical flaw that Grete Hermann had descried some seventeen years before, though her remarks were now buried by history.

"The kind of professions I had heard about in the family," Bell remembered, "were carpenters, blacksmiths, laborers, farm workers, and horse dealers. My father's first profession was horse dealer. He stopped going to school at the age of eight—his parents paid fines from time to time for that." His family had lived in Ireland for generations, "but we are from the Protestant tribe," explained Bell, "so the real Irish people regard us as colonists." This situation was not stressed by Bell's parents; Annie, his mother, had many Catholic friends, and her knitting skills adorned their daughters' First Communion dresses.

Bell spent much of his childhood in Depression and wartime Belfast at the library. At home he was called "Stewart"—his middle name—or "The Prof." At eleven, an age when money pressures made many Irish children drop out of school, including his older sister, Ruby, and the two younger Bell boys, Stewart told his mother he wanted to be a scientist. The Second World War had just begun. During the Battle of Britain, with his father sending money home from the army, the family somehow managed to pay the tuition for high school. After graduation at age sixteen, as the war raged on, Bell got a job as a lab technician in the physics department of Queen's University in Belfast, where kind professors gave him books to read and allowed him to attend their lectures. The next year, Bell entered Queen's as a student, and the war finally ended.

Bell was "a young man of high caliber," according to Bill Walkinshaw, head of the accelerator design group, remembering "great pleasure at the sharpness of John's mind and the challenge of keeping up with him." Two years before, he had snatched the newly hired twenty-one-year-old away from his nuclear reactor work (also for the Atomic Energy Research Establishment at Harwell), when its leader, Klaus Fuchs, was arrested for betraying Manhattan Project secrets to the Russians. Walkinshaw admired Bell's independence and, as a certain smart Scottish girl, Mary Ross, also working in his group, recalled, "did not mind John's Celtic temperament." Bell's special liking for particle dynamics was crucial in their work, calculating how particles would travel through various experimental schemes to accelerate them; perhaps it also heightened his appreciation of a quantum physics, like Bohm's, that allowed those particles to have definite paths, even when no observer was present.

Mary was to be Bell's best friend and lifelong coauthor, marrying him two years after they met in the Berkshire Downs. "He always liked the theory of anything to be well understood," she remembered. "As a boy he studied *Every Boy and Girl a Swimmer* and then followed the instructions." Ballroom dancing and skiing followed in a similar manner. At high school in Belfast "he had a theoretical course in bricklaying, but," she remarked, "as far as I know, no practical experience."

The director of Atomic Energy Research in Britain was the recently knighted Sir John Cockcroft, who split the atom with Ernest Walton in the library of the Cavendish when Bell was a child. (Cockcroft was also interested in bricklaying and on Sunday drives in Cambridge collected old bricks from farmers with which to refurbish the crumbling buildings of his college, spurning the garish factory-made ones that were used to renovate other colleges.) The Nobel Prize–winning Cockcroft-Walton machine was now merely the warm-up track to feed particles into machines many times larger and faster. Gone were the prewar days when cozy physics department labs like those at the Cavendish and Berkeley formed the cutting edge of high-energy physics. Now Brookhaven National Laboratory, on Long Island, a consortium of nine East Coast universities, with money from the federal government, boasted the fastest particle accelerator in the world, with the Stanford Linear Accelerator Center (SLAC) hot on its heels.

The European physicists, barely done burying the dead of the world war and eager for something hopeful and positive, were not immune to the building excitement over particle smashing. A united Europe should have a European center for particle physics, suggested de Broglie. As Bell was reading Bohm's papers for the first time, eleven countries were signing an agreement to set up a Conseil Européen pour la Recherche Nucléaire (CERN). Accelerator designers came to the CERN headquarters—a villa at the Geneva airport—to advise on a multinational accelerator bigger and faster than any yet built. Among these, in 1952, were Bill Walkinshaw and John Bell, experts on the just-discovered principle of "strong focusing," which would make this accelerator possible.

Bell had not read von Neumann's book. Written two decades before, it had never been translated into English, remaining a landmark rather than a resource. Bell knew of the no-hidden-variables theorem from "the beautiful book by Born, *Natural Philosophy of Cause and Chance,* which," he remembered, "was in fact one of the highlights of my physics education." Max Born stated von Neumann's result firmly, without going into detail.

"Having read this," Bell wrote thirty years later, "I relegated the question to the back of my mind and got on with more practical things." But now he realized he had to see von Neumann's proof for himself. In this he was lucky: Franz Mandl, his colleague in the accelerator design group, was fluent in German and interested in the subject. Mandl also considered himself lucky to have such an interesting person with whom to debate. For months they studied and shouted over von Neumann's proof and Bohm's papers in detail.

"I can remember how excited he was," wrote Mary Bell years later. "In his own words, 'the papers were for me a revelation.' When he had digested them, he gave a talk about them to the Theory Division. There were interruptions, of course, from Franz Mandl, with whom he had had many fierce arguments."

"Franz . . . told me something of what von Neumann was saying," Bell remembered. "I already felt that I saw what von Neumann's unreasonable axiom was."

It was becoming clear to Bell's supervisors that he was a theorist, and soon after this the Atomic Energy Research powers that be tapped him to do some postgraduate work under one of the best physicists in Britain, while still receiving his salary. In 1953, Bell went north to Birmingham, to the physics department of Rudolf (soon to be Sir Rudolf) Peierls.

Peierls had the great gift of being able to instantly understand and, with penetrating questions, to clarify any new subject in physics. He created an atmosphere full of intellectual investigations, social gaiety, and students living at his home (with his kind and bossy Russian wife, Genia, organizing every inch of their lives).

Soon after Bell arrived in Birmingham, Peierls asked him to give a talk. Fresh with the excitement of Bohm's papers and von Neumann's mistake, Bell offered two options: accelerators or the foundations of quantum mechanics. Peierls's time with Bohr and Pauli had been the keynote of his career; he certainly saw no need to reopen the book on quantum theory's foundations. Bell took the hint that talking about Bohm would be bad for his career and gave the speech on accelerators.

Ten years would pass before Bell picked up the subject again.

28

What Is Proved by Impossibility Proofs

1963–1964

THE PROTON SYNCHROTRON, planned at the Geneva airport in 1952, was built in nearby Meyrin, Switzerland, in 1959. It was a circular underground tunnel one-half mile in circumference in which to accelerate and smash protons, and analyze the results. Brookhaven in the United States followed with its own Proton Synchrotron the next year. For most of the sixties these two machines would be the world's preeminent accelerators.

CERN was by now a huge, bland-colored herd of box-shaped buildings. Big black numbers like brands on their tan flanks identified the otherwise anonymous interconnected lab and office structures. This numbering system seems quite as erratic as if the buildings were actually bovine and ambulatory, with, for example, a knot of buildings numbered in the low 500s, including the cafeteria, grazing intermingled with the low 60s, while hundreds of buildings distant, Building 65 stands protectively around Building 604. Prolonged map-gazing produces no glint of hidden order, though the tangled streets sport familiar names. Route Einstein parallels Route Democritus until an intersection with Route Pauli, when it becomes Route Yukawa (named for the Japanese physicist Hideki Yukawa who, in 1935, predicted the existence of the meson, the particle that carries the strong force holding the atomic nucleus together). Route Bohr skirts the big lawn in front of the cafeteria.

Inside the jumbled and linked buildings, CERN is labyrinthine. Endless tan halls funnel into other tan halls, then dead-end. Along these halls, with plastic venetian blinds protecting physicists from their industrial view, lie small cell-like offices.

Young scientists coming here to work for the first time often feel isolated and anonymous. They must join one of the little self-formed groups in order to work. Even the theorists work in small teams. For some, the right opening never occurs, and after a few months, they vanish. Even for those who stick it out, the experience is less than welcoming. Bell used to joke to Mary, in their early years at CERN, that after a mere six months he felt comfortable saying "Good morning" to everyone he passed on his hall.

But CERN finally becomes welcoming over food. The big cafeteria is a place of green plants, big glass windows, sun, and people willing and wont to talk. Round tables fill most of the room, each surrounded by a little flock of lavender or tan plastic chairs that perch on two bent metal legs. Here, unlike at a university, and particularly unlike a European university, even the really important people are known by their first names and any question is fair game.

John Bell leaned back in one of the birdlike CERN cafeteria chairs, his eyes squinting as he looked sidelong at the man across the round table from him. "I don't believe your impossibility proof can be right." There was a slightly amused look on his face.

Josef Maria Jauch was affronted. A revered professor visiting from the nearby University of Geneva, he had just given a seminar on his strengthening of von Neumann's famous no-hidden-variables theorem. Jauch was fifty at the time, Bell only thirty-five, and though Jauch recognized Bell as an up-and-coming new CERN physicist who had done good work in quantum field theory already, it was hardly the same thing as having written a much-praised and well-known book on the subject, as Jauch had done a decade before, with Fritz Rohrlich. And Bell's other subjects in nuclear and accelerator physics hardly gave him special knowledge on the subject of von Neumann's impossibility proof.

For Bell, the fact that Jauch had actually brought up this subject was liberating. None of his friends at CERN knew of his interest in the subject, or would have been interested had he told them. Before Jauch said anything in reply, Bell continued, from his perilous angle in the chair, "You see, though I notice that not too many people remark on this, Bohm ten years ago did the impossible. It's been done. A hidden-variables theory has been created. There simply must be something wrong with von Neumann's theorem. The existence of Bohm's pilot-wave picture shows this." (Bell referred to Bohm's hidden-variable theory by de Broglie's name for his similar version of 1927. In both cases real particles are "piloted" by the nonlocal quantum wave.) "So, before you can strengthen it, you'll have to find out what's going on there."

"Bohm's theory, and de Broglie's before him, is a very ingenious solution to the problems which one has to face when one wants to maintain a 'realistic' interpretation of these phenomena," said Jauch. He looked sternly at Bell. "It is an important issue to discuss, but I think we cannot get any further without at first admitting that the quest for hidden variables has its roots in an ideology that is past—in the determinism of nineteenth-century materialism. It is hardly a vision of a future theory. And that past is receding more and more quickly as it gives way to new forms of scientific thought, under the impact of new evidence."

"Well, yes," said Bell, "so they say."

"There's no mystery why," said Jauch. "The pilot-wave picture is a trivial reconciliation of quantum phenomena with the classical ideals of the past." He looked at Bell with the on-guard expression of a teacher faced with the one student who persists in passionately bringing up irrelevancies in the middle of class.

"Yes," said Bell, "but is that a bad thing?" He came forward in his chair with a small crash and leaned, elbows on the table, opposite Jauch. "Here we have a closed set of equations, whose solutions are to be taken seriously, and not mutilated when embarrassing. The subjectivity of 'orthodox' quantum physics, the necessary reference to the 'observer,' is eliminated." He shook his head a little: "The way de Broglie was laughed out of court (and Bohm ignored) I regard as disgraceful."

" 'Laughed out of court' may not be exactly the words you want to use to describe what happened," said Jauch. "It is the scientific process that less good theories get passed over in favor of better ones."

Bell's pale eyebrows shot upward on his forehead. "What happened was not science as usual. Their arguments were not refuted, they were simply trampled on."

Jauch's hand rose to his face to rub something out of one eye from behind his glasses. "It is possible, as de Broglie did, to speculate on what a causal infrastructure would have to look like, in a theory that merely avoids leading to predictions which disagree with the known facts. The argument cannot of course be decided unless one can construct a causal theory which permits a confrontation with experimental facts. And of course, no one has succeeded in doing so."

"That is no reason to ignore the theory!" Bell burst out, his brogue thickening. "Why do people never bring it up, even if only to point out what is wrong with it? Why did von Neumann not consider it himself, for that matter? More extraordinarily, why do people go on constructing 'impossibility proofs'? Why are you doing it? When even Pauli and Heisenberg could produce no more devastating criticism of Bohm's version than to brand it as 'metaphysical' and 'ideological'? Why is the pilot-wave picture ignored in textbooks? Should it not be taught, not as the only way, but as an antidote to the prevailing complacency? To show that vagueness, subjectivity, and indeterminism are not forced on us by experimental facts, but by deliberate theoretical choice?" Bell's forehead was furrowed. He was almost quivering in frustration.

Jauch's head tilted in a professorial way, his piercing eyes hidden by the sun glinting off his glasses. "There will almost always be more than one theory which agrees with all the facts known at a certain point," he said, his voice calm, with its Swiss-German accent. "External confirmation is not the only criterion for the truth of a theory. As Einstein himself said in his 'Autobiographical Notes,' there is a second criterion: the demand for inner perfection, including logical simplicity. The neglect of this second criterion leads to absurdity."

"Of course it is the same 'Autobiographical Notes' which suggests that the hidden-variable program has some interest," said Bell.

Undeterred, Jauch continued, "The situation really presents a remarkable similarity with the confrontation between the followers of Ptolemaeus and Copernicus. Then, as now, the question could not be decided on empirical grounds alone, since both systems were capable of correctly describing the observed phenomena." In order to keep the earth at the center of the universe, Ptolemy's system became ever more elaborate and ungainly. By putting the sun at the center, Copernicus swept away these contortions, explaining the world and the night skies with dramatic power and simplicity. For Jauch, Bohm's theory was murky and artificial, like earth-centered astronomy, and the Copenhagen interpretation was as clear as Copernicus's. "Then as now," he said, "the new view is often opposed by an appeal to reasons which are completely invalid."

Bell actually looked a little angry. He took a breath. "Among the books I would like to write—" he said, and then smiled with irony. "I would like to write half a dozen books, which means I won't write any." He leaned back in his chair again. "But one of these books would trace the history of the hidden-variable question and especially the psychology behind people's peculiar reactions to it. Why were people so intolerant of de Broglie's gropings and of Bohm?" Leaning forward, he said in a clear voice: "For twenty-five years, people were saying that hidden-variables theories were impossible. After Bohm did it, some of the same people said that now it was trivial. They did a fantastic somersault. First they convinced themselves, in all sorts of ways, that it couldn't be done. And then it becomes 'trivial.' " He raised his hands off the table with a bemused expression.

"Even Bohm did not cling to his theory," said Jauch calmly. "Even he found his 'quantum potential' rather artificial." The wavefunction in quantum mechanics can spread arbitrarily far. If it transmits influences instantaneously from wherever it spreads, there is no limit to what faraway factors might be implicated in an experimental outcome. "Why should we not also regard the influence of the phase of the moon," he went on, "or which constellation the sun is in, or the state of consciousness of the experimenter? If hidden variables are needed to render atomic events causal, why should we stop at that point and not admit all kinds of occult causal relationships? The door is left wide open."

Bell had a faraway look on his face and he was nodding slightly. "Terrible things happen in the Bohm theory," he said musingly. "The trajectories assigned to the elementary particles instantaneously change when anyone moves a magnet anywhere in the universe."

Jauch looked relieved at Bell's agreement. "Indeed."

"I wonder," Bell continued, "if this is a defect of his particular picture, or is it somehow intrinsic to the whole situation?"

"Well, I think we can safely say," said Jauch, "that it is a defect of Bohm's theory, and a rather egregious one at that."

"But remember the Einstein-Podolsky-Rosen paper," said Bell, becoming excited for no reason that Jauch could see.

"Well, really, Bohr explained what was wrong with that idea," said Jauch.

"Did he?" asked Bell, with a growing smile.

"Of course he did," said Jauch, "eliminating the false route that Einstein and the others were on, so that they—and we—are left with fewer possibilities among which we may find the one that will open our understanding for the basic complementarity which pervades all of our physical universe."

"Mmm," said Bell. He looked, as he often did, as if he were laughing deep down. "While imagining that I understand the position of Einstein as regards to the EPR correlations, I have very little understanding of the position of Bohr, despite earnest scrutiny."

"I have never hidden my profound humility when I speak about complementarity," said Jauch. He paused, and then elaborated in words that hardly made Bohr's famously opaque explanation any more satisfying. "Do we not see in the EPR correlations another example of that all-pervading principle of complementarity which excludes the simultaneous applicability of concepts to the real objects of the world? Rather than being frustrated by this limitation of our conceptual grasp of the reality, we can find in this unification of opposites the deepest and most satisfactory result of the dialectical process in our struggle for understanding."

"Yes, yes," said Bell, "I forget that rather than being disturbed by the ambiguity in principle—by the shiftiness of the division between 'quantum system' and 'classical apparatus'—Bohr seems to take satisfaction in it. He put forward his philosophy of 'complementarity' not to resolve these contradictions and ambiguities, but rather to reconcile us to them."

"It is, without a doubt, the sum and substance of our experience with the phenomena of microphysics," said Jauch. "Instead of being a principle which expresses the limitation of our ability to know"—he looked at Bell—"it expresses the very essence of the objective rendering of the physical phenomena, in the unambiguous language pertaining to factual evidence."

Bell looked at Jauch as if he wasn't quite certain the other man hadn't been making a joke. "I have a question about complementarity," he said, in the voice of one who is changing the topic slightly. "Because it seems to me that Bohr used the word with the reverse of its usual meaning." He grinned, tipping his head to the side. "Consider, for example, the elephant. From the front she is head, trunk, and two legs. From the back she is bottom, tail, and two legs. From the sides she is otherwise, and from the top and bottom different again. These various views are complementary in the usual sense of the word. They supplement one another, they are consistent with one another, and they are all entailed by the unifying concept 'elephant.' " Bell's hands gestured to suggest this. His eyebrows then lowered. "But Bohr, Bohr wouldn't—it is my impression that to suppose Bohr used the word in this ordinary way would have been regarded by him as missing his point and trivializing his thought. He seems to insist rather that we must use in our analysis elements which contradict one another, which do not add up to, or derive from, a whole. By 'complementarity' he meant, it seems to me, the reverse: contradictariness."

"I'm sure you have heard Bohr's statement that the opposite of a deep truth is also a deep truth," said Jauch.

"Yes," said Bell, "and that 'truth and clarity are complementary.' He certainly did like aphorisms like that. Perhaps he took a subtle satisfaction in the use of a familiar word with the reverse of its familiar meaning. It emphasizes the bizarre nature of the quantum world, the inadequacy of everyday notions and classical concepts, and lays stress on how far we have left behind naïve nineteenth-century materialism."

"The difference between the modern quantum physics and the classical physics of the past is less drastic than you think," said Jauch. "You will agree that chance is everywhere in nature and that we have no evidence in any branch of science that things happen with certainty. Nevertheless we agree that some things happen with such high probability that for all practical purposes it is reasonable to suppose that they happen with certainty. It seems then that the acid test of any science is to be able to make statements about the occurrence of events whose rightness is overwhelmingly probable. I am inclined to believe that once this point of view is admitted, the difference between classical and quantum physics becomes much attenuated. What seemed before an almost unreconcilable division of the world into two opposite camps now becomes more like complementary aspects of one and the same object."

"Well, yes, I think that is what Bohr might say, but he seems to have been extraordinarily insensitive to the fact that we have this beautiful mathematics, and we don't know which part of the world it should be applied to."

"We do know," said Jauch, with a bit of carefully controlled annoyance. "Bohr insisted that one had to regard the apparatus as classical."

"Yes, and there was no doubt that he was convinced that he had solved the problem and, in so doing, had not only contributed to atomic physics, but to epistemology, to philosophy, to humanity in general," said Bell with a smile. "There are astonishing passages—have you come across this?—in his writings in which he is sort of patronizing the ancient Far Eastern philosophers, almost saying that he had solved the problems that had defeated them. It's an extraordinary thing for me, the character of Bohr—absolutely puzzling. In my mind, there are two Bohrs: one is a very pragmatic fellow who insists that the apparatus is classical, and the other is a very arrogant, pontificating man who makes enormous claims for what he has done."

"Well, really, it is hard to overestimate what Bohr has done," said Jauch, more annoyed. "I think you're going too far. The principle of complementarity . . . is possibly one of the greatest discoveries in the scientific history of mankind, with ramifications on many other levels of science."

"I would never deny that Bohr's prestige is justly immense. But isn't it a little strange in Bohr that, at least as far as I can see, you don't find *any* discussion of *where* the division between his classical apparatus and quantum system occurs? For me it is the indispensability, and above all the shiftiness, of such a division that is the big surprise of quantum mechanics . . . and the hidden-variable approach is one way to get rid of that division. If you gave definite properties—'hidden variables'—to the elementary quantum particles, you don't have to be concerned that the classical apparatus has definite properties. Everything has definite properties. It is just that they are more under our control for big things than for little things."

"Now you're back to your hidden variables." Jauch leaned back in his chair, bending back a frond of the potted plant behind him. "You know, I think they resemble, as a theory, the French law courts, where you are suspected of being guilty unless you can prove your innocence to the satisfaction of the prosecution—which is, in a way, an easier system for the prosecution, just as hidden variables are in a way easier on the physicist. Like a hard-pressed prosecutor, Bohm finds randomness just as unacceptable as any crime, and he requires a culprit be found. But culprits are everywhere! When we have so many in the dock, the situation becomes bewildering." Jauch smiled tightly. "Perhaps we have uncovered only a small part of a profound, all-pervading conspiracy to annoy us with random sequences at some crucial points in our quest for understanding."

Bell was looking off into the distance. "Does every hidden-variable theory have to be so hideously nonlocal in order to work?" Would any completion of quantum mechanics in which particles always have a position and a real state have to rely on nonlocality—the spooky action-at-a-distance that Einstein had derided?

Bell leaned over the table, his finger tracing some design on its shiny surface: "The Einstein-Podolsky-Rosen setup"—finger tapping at his invisible design—"would be the critical one, wouldn't it, because it leads to distant correlations. They ended their paper by stating that if you somehow completed the quantum-mechanical description, the nonlocality would only be apparent. The underlying theory, unlike Bohm's, would be local."

Jauch looked at Bell in amazement. "You're so stubborn!" He was shaking his head. "I can't help but"—he had an incredulous and semi-exhausted smile on his face—"I can't help but admire you."

Through the following decade, Bell was not the only one who continued to think about the hidden-variables problem. Jauch remained intent on the parallels he saw between this debate and the long-ago one over the Copernican system versus the earth-centered Ptolemaic system with its preposterous epicycles—circles upon circles, where, if enough were added, the planetary orbits could be predicted. Feeling that the times were very similar to 1630 when Galileo wrote his "Dialogue on the Two Chief World Systems," Jauch wrote a book titled *Are Quanta Real?: A Galilean Dialogue,* which he sets in the "Fall of 1970, in a villa on the shores of Lake Geneva."

Jauch brings Galileo's three characters—Filippo Salviati, the wise one; Giovanfrancesco Sagredo, the seeker; and simple Simplicio—back onstage to "give us the benefit of their wisdom at a juncture of history which is, perhaps, comparable in importance to that of three hundred years ago." Salviati is now the sage of Bohr's "complementarity," and Simplicio is the one who doesn't see what's so bad about hidden-variables theories.

"Many of the passages used," writes Jauch in his introduction, "are more or less faithful reproductions of actual conversations or statements from correspondence and published material. The three interlocutors do not represent actual persons, however. They are composite characters, each representing a current tendency. I hope that living persons who find themselves thus 'quoted' are satisfied as to the accurateness with which their opinions are represented."

While Bohm could be recognized in Simplicio in one paragraph (quoting Lenin and talking about "wholeness"), there is never a trace of Bell—Simplicio is a straw man with clumsy arguments. This despite the fact that Jauch started writing this book not more than five years after meeting Bell, who had emerged during that time as the preeminent force in the opposition to the dual world of "complementarity" that Jauch loved so much.

What Simplicio really represents is just how little Jauch understood the force of Bell's one-world argument.

Jauch's own argument oscillates between cogent and insightful on one hand, and cloudy and complementarian on the other, in a manner palely reminiscent of his old teacher Pauli. It turns out that this is for the same reason. At the end of the first day—the book comprises four days of conversation—Salviati, who always has the second-to-last word in these discussions, encourages the reader to seek new scientific concepts in "ideologies and their systems of images," which are "the symbolic expression of the archetypal structure of the human soul." And Sagredo, who always has the last word (which tends to consist of reverent banalities about the depth of Salviati's wisdom), responds that this statement "needs to be contemplated by us all."

These injunctions might bewilder the reader, but the explanation comes in the notes, which begin, "In the psychology of C. G. Jung. . . ."

This, in turn, clarifies why the climax is given over to a long dream, which Simplicio recounts to the other two on the third day. Complete with a whispering Jungian anima and a discussion of the significance of "the passage from the number three to four," the notes explain that the dream deals with "the symbolic acceptance of the principle of complementarity, for which Simplicio is not ready."

The notes also explain that this dream conveys a moral: "What is the good of winning the whole world if one loses one's own soul? Evidently there are two ways of winning and losing. . . . Simplicio is not yet capable of distinguishing between the two."

The book ends grandiosely, with Salviati's declaration in the second-to-last paragraph: "Thus does microphysics lead to insights which . . . give reasons for hope of a better understanding of all our experiences, including the moral and social behavior of man," commenting that "I address these words particularly to Simplicio." (Sagredo responds, "Your words, dear Salviati, are so filled with meaning that anything we might say after them would seem shallow.")

The second paragraph of Bell's paper, "On the Problem of Hidden Variables in Quantum Mechanics," appears similarly to be addressed in particular to one person:

> Whether this question [the problem of hidden variables in quantum mechanics] is indeed interesting has been the subject of debate. The present paper does not contribute to that debate. It is addressed to those who do find the question interesting, and more particularly to those among them who believe that [here Bell footnotes Jauch] "the question concerning the existence of such hidden variables received an early and rather decisive answer in the form of von Neumann's proof on the mathematical impossibility of such variables in quantum theory." An attempt will be made to clarify what von Neumann and his successors actually demonstrated.

Bell ends his introduction with a characteristic amused observation at his own expense: "Like all authors of noncommissioned reviews, he thinks he can restate the position with such clarity and simplicity that all previous discussions will be eclipsed."

In fact, that is exactly what happened. But first the Bells went on a trip to California.

When John and Mary Bell arrived at Stanford on November 23, 1963, people were wandering past the terra-cotta roofs and yellow walls, the rows of palms and colonnades, in shock. John F. Kennedy had been shot the day before.

Mary immediately began work with the accelerator group there at SLAC, but John seemed to be alone often, fiddling away with a pencil and a piece of paper. For days he drew little diagrams, all the while looking as if he were trying to solve a particularly vexing word puzzle or mindteaser. All the esoteric particles he had been writing papers about—the pi-mesons that hold the nucleus together and the neutrinos that emerge from its decay—had flown out of his mind: he was consumed by Jauch's impossibility proof and his own suspicion that, as he was to write years later, "what is proved by impossibility proofs is lack of imagination."

Eventually Mary asked him, "Whatever are you doing, John?"

"Well, you know, it's very strange," said John, looking up from his desk for the first time in hours. "I've just been playing around with a simple system of two spin-½ particles" (the name for particles, like electrons, that must "spin" twice around in order to arrive back at the state they started in), "not trying to be very serious, you know, but just to get some simple relations between input and output that might give a local account of the quantum correlations." Mary looked at him, surprised. A local account of quantum correlations? Of all things! "But," John continued, "nothing I have tried has worked. And Bohm's theory is nonlocal, and de Broglie's before him. I'm beginning to feel that it very likely can't be done."

"But why?" asked Mary, curious, bending over his scribbles. She pushed one paper aside, looked at the one beneath it. "Why are you suddenly trying to do this?"

"Josef Jauch is actually trying to strengthen von Neumann's theorem that bans hidden variables. For me, that was just like a red light to a bull." He grinned and Mary laughed.

"So you've returned to von Neumann," she said, smiling.

"Here. Would you take a look at this for me?" he asked, reaching under the top level of papers to a series of calculations carefully written out and interspersed with lines of notes in John's beautiful handwriting. "This is my answer to Jauch."

He handed it to Mary, who pulled up a chair beside his desk and began to read over his calculations, nodding slowly as she read down the page, her bobbed wavy hair falling around her face. She laid the first piece of paper behind the second, and continued reading, in silence. John watched her with a mixture of fondness and expectation on his face.

"Well, really," she said as she looked up, "von Neumann's assumption does look rather foolish when you phrase it like that."

Bell had come to the same conclusion that Grete Hermann had; the same point that Einstein had made to Bergmann and Bargmann. Einstein, of course, never published anything about this, and the offhand presentation of Grete's insight had kept it from broader notice.

"So you don't see anything wrong with my steps?"

"No, it all looks good and clear," she said with a smile. She laid down the paper and said, "So, then—these drawings . . . ?" She gestured at his desk.

"The strange thing that kept on coming up as Jauch and I discussed these matters was how nonlocal Bohm's theory was. In his theory, the

Einstein-Podolsky-Rosen paradox is resolved in the way Einstein would have liked least"—using "spooky" action-at-a-distance. "It's very strange."

"And what you wonder," said Mary, "is whether a hidden-variables theory must be nonlocal in order to agree with the predictions of quantum mechanics."

"Exactly," said John. "So I've been testing the EPR thought experiment—"

"You think Einstein was wrong," said Mary.

"I suspect that action-at-a-distance has not been disposed of yet," said John.

Bell finished his first paper on Neumann's faulty hidden-variables proof, Mary checked it over, and he sent it off to *Reviews of Modern Physics*. It seems likely that David Bohm was the referee.* He suggested that Bell expand on the role of measurement, a subject that occupies fifty pages in Bohm's textbook. Bell tacked on a paragraph: "With or without hidden variables the analysis of the measurement process presents peculiar difficulties, and we enter upon it no more than is strictly necessary for our very limited purpose." He returned the manuscript to the journal.

Here the ghost of von Neumann rises for a third time to prevent—or, at least, once again delay—the general knowledge that he had made a silly mistake. Bell's edited paper, on its return, was misfiled. After a while, the editor wrote Bell to ask for the paper, but sent the letter to SLAC. No one at Stanford felt sufficiently public-spirited to forward this letter on to Bell, whose sabbatical in America had taken him across the country from Stanford to Brandeis, so the paper languished somewhere among piles of paper at *Reviews of Modern Physics* for two years, until 1966, when Bell wrote the journal to ask what had happened.

In 1964, of course, Bell had no idea of all this, and meanwhile, he was feeling more and more confident that it was this requirement of locality that created the essential difficulty to the hidden-variables scheme. One weekend, his ideas coalesced, and it was Bell's turn to make an impossibility proof: no *local* hidden-variables theories.

The equation he came up with was to become famous, known as the Bell inequality. A pair of distant particles may exhibit a certain amount of correlation. The requirements of locality and separability together restrict that amount to beneath a certain level. If the correlations surpass

---

*When a paper is submitted to a science journal such as this one, the journal's editor sends it out to specialists who work in the relevant field. These "referees" read the article and make (anonymous) editorial recommendations. (Thus the full-time editor of a journal does not have to be an expert in the thorny and often changing arcana of every field the journal covers.)

that limit, either locality or separability is violated. Entangled particles violate this inequality with disconcerting frequency: they are flagrantly more correlated than they have any right, by common sense, to be. The fabric of reality requires some form of either nonlocality or nonseparability.

Incredibly enough, after all the *gedanken* experiments and accusations of "metaphysics" of the previous four decades, "the example considered above," wrote Bell of this inequality, "has the advantage that it requires little imagination to envisage the measurements involved actually being made."

Thus, in 1964, the second one of Bell's pair of papers sallied forth in the pages of the very first (and second-to-last) edition of *Physics,* to take its chances in the world. But already a link had been formed: just as Bell, almost uniquely, saw the importance of Bohm's paper the very year it was published, so Bell's paper actually found its critical reader immediately, and this, improbably, in the philosophy department of M.I.T.

A Little Imagination

1969

Abner Shimony

Some consideration of theories involving hidden variables is at present needed to help us avoid dogmatic preconceptions. Such preconceptions not only restrict our thinking in an unjustifiable way but also similarly restrict the kinds of experiments that we are likely to perform.

David Bohm, 1962

. . .

There are three kinds of worlds: the great world, the middle-sized world, and many little worlds. The great world is the world of nature, consisting of the stars, the sun, the planets, the moon, the earth, and all that the earth contains. The middle-sized world is the world of human society, with its nations, governments, armies, religions, factories, farms, schools, families, and everything else formed by human beings. The many little worlds are individual human beings. Each man, woman, and child is a little world, but of course each is shaped and influenced by the middle-sized world of human society, often in strange and surprising ways.

Likewise, each is shaped and influenced by the great world of nature, sometimes in ways that no one could predict.

So begins the story *Tibaldo and the Hole in the Calendar,* a tale of a little Italian boy in 1582 who found that when Pope Gregory reformed the Julian calendar by deleting ten days in the October of that year, his birthday had been deleted too. Because the Julian calendar was accidentally eleven minutes and fourteen seconds longer than the solar year, it had set Easter on an inexorable march toward the summer. When Tibaldo finds that his birthday has fallen through this "hole in the calendar," he sets out on a campaign to get it back. It is a beautiful book, a children's "chapter book" of science and society contained effortlessly within the story of a boy. And if one reads the cover flap, one finds that the book was written by Abner Shimony, a Bostonian theoretical physicist, and illustrated in sepia tones by his son, a Parisian printmaker.

Science is the search for truth, but its searchers are individual people, operating—and sometimes caught—within the webs of the middle-sized world. The opening words of Shimony's children's story would well introduce any story about science, including this one.

Abner Shimony did not know how he had ended up receiving a preprint of this paper of Bell's, and as he sat in his office reading it, the purple ink smearing on his hands, it did not immediately strike him as good fortune. Some friends of his at Brandeis—where Bell had written an early draft of the paper and delivered a talk on it—must have put him on the preprint list. He had not heard of this J. S. Bell from CERN, and already he was finding arithmetical errors that made him suspect it was "just another kook letter."

But this doesn't sound like a kook, he thought. Something kept him reading.

Two years earlier, in the spring of 1962, Shimony had earned his second Ph.D., in physics, from Princeton (while simultaneously teaching the subject of the first one, philosophy, at M.I.T.). When he started studying physics again, his professor gave him the EPR paper and told him to "read it until you understand what's wrong with it." Shimony duly read it and reread it, with the pragmatism of an aspiring physicist and the attention to detail of a philosopher. No fatal flaw came to light. "He wanted to inoculate me," Shimony would later relate with an amused expression, "but it sometimes happens that inoculations cause the disease."

As he read Bell's preprint, the disease developed a complication. The little arithmetical errors faded into triviality and Bell's incredible result, with the beauty and force of a tiger emerging from the long grass where it had lain hidden, leaped up from the page. The premises of EPR that were so plausible—that the world is not spooky but locally causal and that it is not entangled but can be separated into objectively real, unconnected pieces—finally had met the dauntingly successful quantum theory in a place where only one could be right. Shimony thought, *If you have an incompatibility, you'd better do an experiment.*

His heart began to beat faster. He leaned forward in his chair and scanned the page again. As the tiger paced around his mind, its tail swishing, Shimony pondered this experiment that the author Bell had suggested would take only a little imagination to envision. How would it be done?

He remembered a paper written by Bohm that had come out seven years earlier. With one of his best students, Yakir Aharonov, Bohm was musing on one of the aspects of entanglement that had occupied Schrödinger's mind back in 1935. Granted, entanglement exists when the particles are close together, for example in the two electrons of the helium atom, or the two atoms of Rosen's hydrogen molecule. But in their thought-experiment designed to discredit the quantum theory, Einstein, Podolsky, and Rosen assumed, as the theory seemed to indicate, that the particles would remain entangled while infinitely separate. No one had ever proved this was possible. In fact, nothing remained of the EPR argument if, as soon as the particles lost touch, they also lost all memory of each other, decomposing into uninteresting and simple "product states." An entangled state, by contrast, is not the product of two individual states—two particles in an entangled state have *no* individual states.

The paper by Bohm and Aharonov had stuck in Shimony's mind because of the ingenious way the two investigators had gone about this long-fallow problem. They had remembered a brief letter to the editor by Chien-Shiung Wu (a.k.a. Madame Wu) of Columbia University that had appeared in *Physical Review* seven years earlier. With her grad student Irving Shaknov, who died soon after in the Korean War, Wu was working with positronium, the strange pair made up of an electron and its antiparticle. Wu and Shaknov had performed delicate experiments with the high-energy photons (gamma rays, which have even more energy than X-rays) produced when positronium annihilates itself, as it must.

Bohm and Aharonov looked at her data and reanalyzed it with entanglement in mind. The theory suggested that these two gamma rays would be entangled. "Bohm and Aharonov," said Shimony, "avoided the need to perform a new experiment by brilliantly exploiting the results of an experiment which had been performed for an entirely different purpose—an exemplary case of quantum archaeology!"

Bohm and Aharonov had shown that the gamma rays indeed seemed to remain entangled. But was the situation really as extreme as Bell's theorem suggested, with a direct and inexorable conflict between localized, separate particles (each with its own real state) and the quantum theory? Shimony began wondering if any experimentalist he knew would be willing to redo the Wu-Shaknov experiment to greater precision and generality. The purpose would be to test the Bell inequality, which requires really dramatic nonlocal correlations. When Aharonov visited M.I.T., Shimony diffidently brought up the subject. Aharonov—tall, thin, handsome, definitive—said that he and Bohm had already shown all that needed to be shown. For a while, Shimony dropped the subject.

In Manhattan a few years later, in the late 1960s, an astrophysics graduate student at Columbia named John Clauser happened upon Bell's paper in the library of the Goddard Institute for Space Studies. "The same thing happened to me that happened to many people during that time period," Clauser said years later. "It was incredible to me. I didn't understand it, or couldn't believe it." The tiger had leaped into another questioning mind. That evening, Clauser went home to Flushing Bay, where he was living on his racing yacht with the airplanes of La Guardia roaring overhead, and couldn't get Bell's idea out of his head. "I thought, *If I don't believe it, I should be able to give counterexamples.* So I tried to," continued Clauser, "and failed. I realized: This is the most amazing result I've ever seen in my life."

Perhaps too amazing. "I was not yet willing to accept the paper's far-reaching implications," recalled Clauser, "until I saw some experimental evidence that decided between its two significantly different predictions"—in essence, entanglement or "local realism": locality and separability, where the particles have states that are not dependent on each other. "Since Bell's paper is conspicuously vague on the experimental status of its prediction (but crystal clear on everything else), I suspected that Bell was bluffing." While searching for an experiment anywhere in the books, he began thinking about how he would do one himself, rederiving

Bell's result in a more general way that could actually be realized with a real, imperfect, non-*gedanken* experimental apparatus.

"My thesis advisor"—growing impatient with Clauser's extracurricular interest—"told me, 'You're wasting your time.' He wanted me to be a real astronomer." This, however, was not to be.

Meanwhile, Shimony was leaving his tenured philosophy post at M.I.T. for Boston University, to teach physics as well as philosophy. The branch of physics he had focused on for his Ph.D. had been statistical mechanics, and a friend of his at B.U. told him just after he arrived, "We've got a wonderful stat mech grad student for you to work with."

The grad student, Mike Horne, arrived at his door soon after. "I'm not doing stat mech anymore," said Shimony. His thoughts had turned to that purple mimeograph of Bell's paper. But something must have struck him about the tall, soft-spoken Mississippi boy. "I had been in Abner's office exactly five minutes when he showed me Bell's paper," remembered Horne.

"See if you can design an experiment," said Shimony. "EPR's premises are so plausible that maybe something will happen."

Horne found that he could construct a local hidden-variables theory to explain the Wu-Shaknov data—showing, for their purposes, that there was no point in considering experiments like it. The correlation between the gamma rays' photons was too weak. "That," Horne remembered, "was the first thing I gave to Abner that caught his attention." Bell's theorem required an experimental setup that could grant them a closer look at entanglement.

Clauser, meanwhile, unaware of Shimony and Horne, was in hot pursuit of the same goal. A conversation with Madame Wu proved fruitless. Then he heard that Dave Pritchard at M.I.T. was doing an experiment involving the scattering of metal atoms that might fit the bill. When he explained what he was looking for, Pritchard turned to a recently arrived postdoc from Berkeley, Carl Kocher. "Carl, wouldn't your experiment test this?"

"Of course!" said Kocher. "That's why we did it."

When he got back to Columbia, Clauser raced to the library to look in the recent *Physical Review Letters** for the details of Kocher's experiment, which had been performed under the guidance of a highly respected Berkeley professor, Eugene Commins.

---

*Because the explosion in the number of physicists after World War II resulted in a parallel explosion of physics papers, in 1958 *Physical Review* began publishing *Physical Review Letters,* an alternate venue for the "letters"—i.e., short, significant papers.

Clauser found that, like the EPR *gedanken* experiment it was designed to illuminate, Kocher and Commins had measured the entangled particles only when the analyzers* were parallel and when they were perpendicular to each other. It happens that it is only the intermediate cases, when the analyzers are not aligned at right angles, where Bell's inequality shows that all solutions that assume locality and separability fail.

But Kocher and Commins's *source*—two entangled photons of visible light emitted one after the other from a calcium atom in what is called an atomic cascade—seemed perfect for Clauser's needs. If an atom receives a jolt of high-energy light (an ultraviolet photon in the case of the Berkeley experiment), it will absorb that energy and become excited. The excited state is not stable—the atom starts to lose energy again, which it can only do discretely, in the form of a photon. An atomic cascade takes place when the atom loses that energy in two stages, releasing one low-energy photon and then a second, which is entangled with the first.

Kocher and Commins had verified the entanglement of these atomic-cascade photons by measuring their polarization. Polarization, first noticed in the seventeenth century but not named until 1811, is a measure of the inclination or "tilt" of electromagnetic waves. Each of these waves, unlike water waves and sound waves, oscillates in two directions: an electrical component waving up and down, and a magnetic component waving from side to side. The wave travels forward in a direction perpendicular to both of these components. (To visualize these relationships, imagine a fish, whose dorsal fin represents the electric field and whose side fins represent the magnetic field. Like the fish, the wave can tilt without changing the relationship between electric field and magnetic field or the direction of travel, and the angle at which it leans is its polarization.)

Polarization, a wave concept, bears some similarities to spin, a particle concept. Like spin, which always can be described in terms of up and down, polarization can be broken down into two kinds: horizontal and vertical.[†] These horizontal and vertical components are like lengths along

---

*"Analyzer" is a general name for the two pieces of experimental apparatus at either end of an EPR experiment. In the versions of both Bohm and Bell, atoms with a spin of ½ were analyzed by Stern-Gerlach magnets, which measure whether that spin is up or down. Kocher and Commins's version did not use spin-½ particles, so they needed analyzers of a different nature, but the principle is the same.

[†]In fact, spin and polarization can, cautiously, be described as two ways of talking about the same fundamental physical situation; e.g., "spin up" in the photon's direction of motion is the particle description of right-hand-circular polarization in a light wave. (Circular polarization is the result of a syncopation of vertical and horizontal components.)

the *x* and *y* axes of a graph, with polarization like a slope across such a graph.

This tilt, of course, cannot be seen directly. But certain materials called polarizers filter out either the horizontal component of the light or the vertical. Kocher and Commins had used Polaroid sheets, which are a translucent material holding a series of stretched and stained organic molecules in parallel lines. If these molecules are horizontal, they will absorb horizontally polarized light (most reflected light or glare from snow or water is horizontally polarized, so the molecules in Polaroid sunglasses are arranged in horizontal rows to catch it). If they are tilted on their side, they will absorb vertically polarized light. In either case they let the other polarization pass through.

Kocher and Commins found that each pair of atomic-cascade photons was polarized identically—both vertical or both horizontal. *Could this experimental setup test Bell's inequality?* Clauser wondered.

His long-suffering astrophysics thesis advisor, capitulating to Clauser's enthusiasm, said, "Well, why don't you write him a letter and ask him?" Clauser sent off letters not only to Bell but also to Bohm and de Broglie, asking them all what they thought. His proposal sounded viable to all three and none had heard of anyone else who had already solved the problem.

For Bell in particular, this "crazy American student" was the first person to respond seriously to his paper, and he wrote to Clauser: "In view of the general success of quantum mechanics, it is very hard for me to doubt the outcome of such experiments. However, I would prefer these experiments, in which the crucial concepts are very directly tested, to have been done and the results on record.

"Moreover, there is always the chance of an unexpected result, which would shake the world!"

Clauser, who fully expected to find local hidden variables, could hardly have received a more exciting letter. "The McCarthy era was now in the distant past," Clauser recalled. "Instead, the Vietnam War dominated the political thoughts of my generation. Being a young student living in this era of revolutionary thinking, I naturally wanted to 'shake the world.'" Overthrowing the quantum establishment would certainly do it.

In Massachusetts, meanwhile, Richard Holt, a Harvard grad student with bushy hair and an unruly goatee, was painting shelves while listening to the Red Sox lose the 1967 World Series, completely unaware of what he was about to be sucked into. To make space for his thesis experiment, he was cleaning out a tiny room full of disused equipment in the back of Professor Frank Pipkin's bustling basement lab. The little room had previously been used for an experiment in bat sonar, with its walls covered in cork and blindfolded bats flying around obstacles placed in their paths. The apparatus Holt was now moving into the bat cave (as they called it) was built to measure the lifetimes of excited states of mercury. This was done by timing the difference in arrival between the two visible light photons of an atomic cascade—one emitted as the mercury atom entered the intermediate state, and one when it lost energy again. "Bread-and-butter physics," said Holt.

One day, a small, kindly Boston University professor in tortoiseshell glasses, followed by a towering grad student amiably humming into his beard, arrived at the bat cave. Shimony and Horne's yearlong search, in which, Shimony recalled, "we made pests of ourselves looking for a source of correlated low-energy photons," was over. They were determined to woo Holt away from bread and butter.

Holt had already read Bell's paper, on the recommendation of a professor who had thought it was "cute." "I went to the library and *totally* failed to appreciate the significance," he said. But now, "it came back to haunt me," with Shimony urging its importance and Horne telling him it would be fun.

"I was kind of intrigued," remembered Holt. "Then I had to persuade my thesis supervisor that this was worth doing, which I did by saying, 'Well, you know, look, Frank, we'll whip that off in six months and then I'll get back to real physics.' "

The experiment took more than four years.

All seemed set until Shimony's American Physical Society bulletin arrived bearing the strange news that someone at Columbia was planning to conduct the very same experiment. Shimony called Horne, hard at work on their proposal, with the bad news that "we've been scooped." Then he called his former advisor, Eugene Wigner, Dirac's brother-in-law, "and Wigner said, 'Well, *call* this fella! *He* hasn't written it up yet, and *you* haven't written it up yet, so who knows, maybe he'll join in.' " Still upset, Shimony placed the call. When he got off the phone, Clauser was on the team, even "thrilled that somebody was interested," as far as Horne could tell.

The peace-loving Shimony never let Horne know, until almost forty years later, that "there was a hint of shotgun to the marriage." What finally made Clauser want to join Shimony and Horne was the bat-cave experimental apparatus; he had no apparatus of his own and desperately wanted to do the experiment. "This was not Clauser's thesis work for his doctorate," said Horne. "He had to hide it; if he talked to anybody about it at Columbia, they'd say, 'But aren't you supposed to be doing astronomy?' "

After Clauser finished his thesis work, there was a two-week hiatus before he had to defend it for his astronomy Ph.D. He threw himself into Horne and Shimony's paper, and Holt showed him around the apparatus. "It was wonderful to all go at it together," Shimony recalled. "In batting things around, we discovered and refined lots of little points."

One of these little points involved an experimentalist-ex-machina rescue by Holt. Shimony, Horne, and Clauser realized that they had been making their calculations as if the photons always left the atom in exactly opposite directions. But, as Shimony explained, the photons go every which way. Holt knew how to handle what Horne called the mathematical "heavy machinery" of the realistic but complicated calculation.

Meanwhile, Clauser's dream of getting his hands on an apparatus of his own had taken an unexpected turn nearer to reality. He received a job offer from Berkeley, where Carl Kocher's EPR demonstration apparatus still sat somewhere in Eugene Commins's lab. Clauser, of course, would not be hired to study the Bell inequality. Charlie Townes, who had received the Nobel Prize five years earlier for his inspired role in the development of the laser, had offered him a position in radio astronomy, studying the early life of the universe.

So, after a rough draft, and his Ph.D. defense, Clauser unmoored his boat, the *Sinergy,* from her berth in Flushing Bay and set off with high hopes for Berkeley, planning to sail to Galveston, Texas, truck the boat across the Southwest, and put back in the water south of Los Angeles to sail up the fabled California coast.

Meanwhile, Shimony was undaunted ("at least we're making waves") by this turn of events and kept Clauser in the loop by mailing him drafts as he and Horne, with the timely aid of Holt, progressed back in Boston. Various marinas down the East Coast would see a tall, blond sailor hunched over a pay phone arguing about arcanities such as "non-zero solid angles," until the advent of Hurricane Camille put the *Sinergy* onto a truck in Fort Lauderdale. Clauser drove to Los Angeles and then put back into the water at Venice Beach. "The paper was done," he remembered, "by the time I arrived in Berkeley."

Under the title "Proposed Experiment to Test Local Hidden-Variable Theories," it came out in the mid-October *Physical Review Letters.* Bearing their generalization of Bell's theorem "so as to apply to realizable

of 1969 — see p. 410, Clauser

experiments," the four authors declared, "A decisive test can be obtained by modifying the Kocher-Commins experiment."

Commins, however, did not consider his experiment with Kocher a starting point for modification. In fact, he had planned it merely as a "nice demonstration to do for an undergraduate quantum mechanics course." Kocher had even started building the experimental apparatus on a wheeled cart to facilitate this goal. "It turned out to be a little bit more difficult than that," remembered Commins, "but not much. And there was no surprise in the result." Commins was tall enough to look Clauser in the eye when they met, and confident enough to remain unfazed by Clauser's enthused explanation of why the situation demanded further investigation. "There's nothing in that," Commins said. "Don't waste my time."

Help came from an unexpected source: the man who had hired Clauser, not to search for hidden variables, but for the cosmic microwave background of the Big Bang. "Charlie Townes," Clauser remembered with gratitude, "talked Gene Commins—who was still grumbling about all this foolishness—into tolerating the project."

"I was very, very surprised that the whole world made such a big fuss about all of this," Commins said. "Very surprised. Because, you know, it just represents an entangled state. But entangled states have existed in quantum mechanics from the very beginning. If you try to describe the two electrons in the helium atom, they're in an entangled state. It's no big deal."

Of course, Clauser wanted to test if entanglement could persist over a distance some forty-five billion times bigger than the diameter of a helium atom. As Bell had predicted, long-distance entanglement was turning from an impossibility to a dismissible triviality. One of the first things Commins had learned about physics, however, was the existence of someone who thought entanglement of distant systems was not just a big deal, but a deal breaker. Commins was sixteen when his family befriended Einstein's. For the next seven years, until Einstein died, the Commins family would make a monthly pilgrimage to Mercer Street, where Mrs. Commins's piano-playing would accompany Einstein's stately violin. "He was very nice to us," remembered Commins. "He was sort of like a grandfather to my sister and me. But he was very emotional about quantum mechanics, felt it couldn't be right. So he kept trying to poke holes in it, and EPR was an effort of his, back in the thirties.

"Of course Bohr showed he was wrong and that should have been the end of it."

This sort of statement carried a lot of weight with most physicists, even though they would be hard-pressed to explain just what Bohr showed. But Charlie Townes had had an unusual experience that inclined him to take received wisdom, even from so august a source, with a grain of salt. At the same time that Bohm was fighting von Neumann and Bohr with his hidden-variables theory, Townes was in the early stages of developing what would become the laser. Independently, and with crushing definitiveness, both von Neumann and Bohr told him that his project could not work, because of the Heisenberg uncertainty principle, which prohibited the necessary perfect alignment of photons. In his autobiography, *How the Laser Happened,* Townes wrote, "I'm not sure I ever did convince Bohr."

So, with the blessing and advocacy of Townes, and the help of a grad student looking for a thesis project, Clauser began tearing apart Kocher's old apparatus. "Commins told me there was this guy from Columbia who had contacted him and said he was interested in doing the Kocher-Commins experiment," the grad student, Stuart Freedman, remembered. "He wanted to do it to some precision and Commins said to me, '*Boy,* that's crazy.' "

With this glowing recommendation, the open-minded Freedman joined Clauser's project. "I had never really heard of this subject before then. But I thought it was quite interesting," said Freedman, "and I decided, Yeah, it would be possible." Looking back on all the labor, he smiled and recalled, "John was *always* very optimistic, thought *all* we had to do was take this old equipment Kocher had built and fix it up a little bit.

"It turned out that wasn't the case."

Every step—from the entanglement abstractly inherent in Schrödinger's wave equation to EPR, when Einstein imagined just what this implied, to Bell's closer scrutiny, finding a testable conflict, to Horne, Shimony, Clauser, and Holt's blueprint for bringing that conflict into the lab—was a step closer to incarnation. But this story had so far rested in the hands of the theorists. Nineteen sixty-nine was the year in which the experimentalists took command of the age of entanglement.

"How do you tell an experimental physicist from a theorist?" asks Clauser more than thirty years later, in his northern California desert home encrusted with sailing trophies and plaques. Running his finger along the thick spines of schoolbooks, he begins to answer his question: "A *theorist* will have: lots of textbooks (the experimentalist will have some engineering ones, too)." He taps these with his finger. "Lots and lots of *Phys. Rev. Letters.*" In fact, a bookshelf taking up a whole wall is crowded with the pale green journals. "Biographies of the greats, and books written by them." Clauser gestures through the door of his wood-paneled office. "But the *experimentalist* will have"—he turns: here, in the hallway beside the kitchen door, is another floor-to-ceiling bookshelf, packed with rows and rows of narrow, shiny softcover book spines in garish fluorescent colors—"*catalogues.*" He grins. "Anything I need to make, if I don't have the pieces already, I look for it here. I can make anything."

Mail order is usually unnecessary. Outside his rambling garage-cum-machine-shop is a lean-to filled with cardboard boxes overflowing with the remnants of electronic devices, sundry gauges of wire, the flotsam and jetsam of circuit boards, tubes, connectors, magnets, gears, bevels, and cast-off machine carcasses. "To be an experimental physicist," he explains, "you need to be able to make anything. You need a mill and a lathe. But, most of all, to be an experimental physicist, you need a junkyard."

The reason for this is that "the most precious and valuable commodity at any physics department is *floor space,*" he explains. "Everyone has an experiment they want to be doing, and each one needs a room for that. So once an experiment is over, there's a lot of hasty tossing out. Since people don't label the parts of the things they're working on, the people who are cleaning up often don't know how valuable things that they're throwing out might be. But if you *know*—!

"Of course it's not good enough to just have a junkyard: you need to know where things are. Every junkyard owner knows where everything is. He knows that there are two 1932 Ford engines over there, and a '57 Pontiac. . . ."

Let Clauser and Freedman, squabbling slightly, explain their rattling, rumbling, sixteen-foot apparatus, chest-high on its sawhorses. Outside, daily anti–Vietnam War demonstrations bring the National Guard and tear gas to Berkeley while hippies hold flowers and activists throw rocks and the Dirty Word Guy stands upon his right to loud and obscene free speech on the steps of the student union. Freedman and Clauser's room in Birge, the new physics building, is deep down in the windowless second basement, underneath it all.

The experiment is like the House That Jack Built:

> *These are the phototubes that catch the photons.*
> *These are the photons*
>   *that fly through the polarizers*
>   *to be counted by the phototubes*
>   *that catch the photons.*

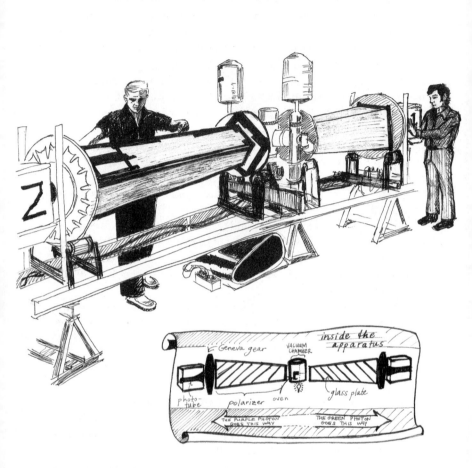

John F. Clauser, Stuart J. Freedman, and their machine

This is the atom that emits the photons
that fly through the polarizers
to be counted by the phototubes
that catch the photons.
This is the lamp that excites the atom
that emits the photons
that fly through the polarizers
to be counted by the phototubes
that catch the photons.
This is the vacuum chamber that enables the lamp
to excite the atom
that emits the photons
that fly through the polarizers
to be counted by the phototubes
that catch the photons.

Clauser starts at the heart of things: "Atom beam. It's dirt simple. It sounds a lot fancier than it is. If you boil water on a stove, you get big, billowy clouds of steam. You could also evaporate metals too, if you get them hot enough.

"Well, if you do the same thing in a *vacuum,* you see something absolutely amazing. Very quickly when you get into even a modest vacuum, the mean free path of the metal atoms"—that is, the distance they're likely to travel without a collision—"becomes bigger than the chamber itself. If you're boiling water, a water molecule comes off and it runs into air, and forms the cloud. But in a vacuum, there isn't any air in the way and it just keeps right on going until it runs into the wall.

"So the easiest experiment to do, is just to take a piece of tantalum* foil, fold it in half, kind of crinkle it open, and you just put a little pellet of something in it. You put this in a bell jar" (a vacuum). "You run a current through it, and the tantalum foil gets hot. If you put a little pellet of copper in there, or aluminum, or calcium—something with a lower melting point—after a while, it's all evaporated, and you look up and you find that you've got copper, or aluminum, or calcium, or whatever it was, all over the walls of your chamber. If you put a sheet with a hole in it [in front of the oven's mouth], and another sheet with a hole in it [in front of that one], you create an atomic beam!" Instead of evaporating all over the walls, the atomic gas is narrowed into a beam. "That's all there is to it."

---

*Tantalum is a refractory metal, which means you can run it up to a very high temperature, because it's very hard to evaporate.

"Our oven is heavier duty than that, of course," says Freedman, showing a block of tantalum slightly bigger than a bar of soap, studded with heater coils. It has a cylindrical hole in the top, which can be plugged once the element to be evaporated into a beam—in this case, calcium—is in place. The evaporated calcium will emerge as a beam from a much smaller puncture hole in front. Clauser and Freedman cut "a nice cylindrical little slug" of about half an ounce of calcium and drop it into the oven, without touching it. Once locked inside its cylindrical brass vacuum chamber, the oven takes anywhere from three to five hours to heat the calcium to its evaporation point.

When the calcium atoms begin to pour out of the puncture hole in the glowing oven—traveling at an average speed of almost 2,000 miles per hour in a comfortable beam of ten billion atoms per cubic centimeter (a *trillionth* the density of solid calcium)—they head straight for the center of the apparatus. Here, in a second cylindrical brass vacuum chamber, a beam of light through a lens in the floor ambushes the atom beam in its flight. Each atom thus excited emits its entangled photons, pale green and then purple.

Finding a light source to trigger this excitation was not easy. "I got into a lot of the technology of making lamps," Freedman recalls, "that I've mercifully had the opportunity to forget, because I use lasers now." But lasers, with their tiny, intense, perfectly tunable beams, would not be ready for several years.

Freedman shone light through a filter so that it carried photons of exactly the right amount of energy to excite the atoms to the planned two-photon cascade. Near ultraviolet in, pale green and purple out. After all the work it took to find the perfect lamps, lenses, polarizers, and phototubes, those wavelengths—2275 Å, 5513 Å, 4227 Å*—would remain stuck in Freedman's mind for the next thirty years.

"Everything had to be very big," says Freedman. "The fundamental problem is that you need a lot of light from the lamp, which meant an enormous polarizer, and, well, enormous polarizers with high efficiency, they don't exist." Clauser had already determined that Polaroid sheets were not efficient enough to test Bell's theorem. Holt, back at Harvard, was using calcite—an elegant natural crystal that, when looked through, shows doubled images of each object. These doubled images occur because calcite separates the light that strikes it into two parallel, oppositely polarized streams. But calcite does not come in the gargantuan size that Freedman and Clauser needed.

---

*An angstrom unit (Å) is a tenth of a nanometer (a ten-billionth of a meter).

Piles-of-plates polarizers are essentially long stacks of ordinary glass. "Being fragile and bulky," Freedman explains in his thesis, they "are normally employed only when more convenient polarizing materials are not applicable." When a pane of glass is tilted just shy of 60 degrees, at what is called Brewster's angle (named after the inventor of the kaleidoscope), it will reflect only light polarized parallel to its surface. The light passing through tends to be polarized at right angles to the reflected light. The piles-of-plates purify this light that passes through, selecting for the perpendicular polarization. A long enough series of such plates—reflection by reflection, pane by pane—will emit at the far end a filtered beam entirely composed of perpendicularly polarized light.

The paper-thin glass designed for microscope slides seemed right, since it would absorb very few of the precious photons in transit. It was so thin, in fact, that the glass arrived in apparently hopelessly wavy warped sheets. Judicious work with an abrasive saw, however, cut out from among the waves twenty flat stretches, graduated in size up to almost a foot square, to accommodate the light beam's spread as it progressed.

In order to measure polarization at different angles (just as in the EPR paper, one might measure either position or momentum, and as in Bohm's and Bell's version, either spin in the x direction or spin in the y direction), the polarizers had to move. Freedman and Clauser turned to Dave Rehder, a tall, laconic, bearded machinist about their age, who had honed his inventiveness at sea on a Navy radar ship, fixing equipment hundreds of miles offshore for a month at a time. Rehder produced everything from little projects like their tantalum oven to the big machine that held the glass plates at Brewster's angle while tilting the long blue plywood boxes (which Clauser called the "coffins") that held them.

The machine did not rotate the coffins continuously, rotisserie style; using something called a Geneva gear, it was set to pause sixteen times per rotation (i.e., every 22.5 degrees), to allow each hundred-second run of the experiment, "like a big clock," says Freedman. "The idea of the Geneva gear came from a book called *Ingenious Mechanisms,* full of industrial-age solutions to engineering problems." A centuries-old watchmaker's invention, Geneva gears are still used in movie projection, converting easily produced rotary motion into a precisely calibrated stop-and-go. A slotted wheel turns continuously, engaging and disengaging a peg on a smaller wheel in a visually hypnotic tango.

Inside each coffin, and attached to its Geneva gear, a frame held the polarizing plates suspended from a taut loop of wire. Rehder, who on his weekends plied a second trade as a commercial fisherman, produced salmon line (a braided stainless-steel cable) and fishing-pole guides to run the lines through.

"We had to do single-photon counting at a time when doing good single-photon counting had just been developed," remembers Freedman. "RCA had just developed these tubes called quanticons, and I got in league with them, in fact learned a great deal about phototubes. They offered me a *job* because I knew so much about phototubes."

Two of these phototubes—one on each end of the piles-of-plate polarizers—caught the photons. Each photon, striking the alkali metal inside the lining of the phototube, was supposed to dislodge an electron. Cesium (its heavenly-blue spectral line discovered a century earlier by Bunsen), only halfway down the periodic table, is the largest atom of all; other alkali metals, like lithium and sodium, are also much bigger than the atoms that surround them on the periodic table. An alkali-metal atom is so big because of a single high-flying electron in its outermost orbit, straying far from the nucleus, whose pull the electron only tenuously feels.

A photon needs only a small amount of energy (called the work function) to free this wide-ranging electron from its atom and send it galloping down the wires to a photomultiplier tube. The photomultiplier, in turn, sends five electrons to the counting computers. Thus "a nice sharp pulse, five, ten, nanoseconds long," as Clauser describes it, "if you do the electron optics nicely," alerts the computers of the arrival of a photon.

But the phototubes had to be cold, to keep the dark current down.

The dark current?

Clauser smiles. "O.K. Ideally it wouldn't exist. Students' and theorists' view of the photomultiplier tube: 'You put photons in. You get counts out. There's some electronics, you get little pulses, you count the pulses, and you know how many photons came in, right?'" He grins at the innocence.

"Reality: You put in a lot of photons, and not many counts come out. You don't put *any* photons at all in, and you get counts coming out.

"The latter are the dark rate." Experimental physics, like life, is messy. "The bigger the photocathode area"—the alkali coat on the inside of the phototube—"the greater the dark current. So ideally you want a small photocathode. The warmer the tube, the greater the dark current, because we have a distribution of electron energies inside. Some of the electrons happen to have high energies; on the tail of the distribution some have energies that are higher than the work function"—the amount of energy it takes to dislodge an outer electron. "They just will escape, all by themselves. How can I tell those apart from those that were ejected by photons? An electron's an electron, and it doesn't take very many electrons escaping to give you one hell of a substantial dark current.

"There are other things. You can get thermionic emissions—you'll spray out electrons off any sharp spots in there. Or among your alkali metals, in your potassium, you might have some potassium-41 in there, and that's radioactive; it's gonna decay and produce electrons.

"All this produces dark current," says Clauser. "*Any*thing produces dark current. There are lots of processes and lots of black arts going on inside the damn tubes."

So they set each little phototube in a horizontal pipe soldered through a Styrofoam-insulated brass tank full of ice. The phototube that caught what Clauser referred to as the "red" photon (*red* in this case is slang for "longer wavelength") needed to be chilled to −78 degrees. This feat required a slush of dry ice and alcohol. "In a Boston winter, when you step off the curb—it looked just like *that*." Clauser laughs: "But *now . . . boy,* you *didn't* want to put your finger in there. And that's what it took to get the dark current down to levels where we could do our experiment."

Every step went slowly; every component was studied, tested, rethought, rebuilt. "People always think you don't have the time to test everything," says Clauser. "The truth is you don't *not* have the time. It's actually the time-saving way of doing it. It's hard, when you're eager to know what Nature's doing: you almost have to train yourself to be *anti-*curious while you're building your hardware. People always want to slap it all together, turn it on, and *see what happens.* But for the first run, you can almost guarantee it's not going to work right.

"There's nothing simple about doing experimental physics, especially if you're doing it with Stone Age hardware, and you're always pushing the edge of the capabilities of whatever equipment you have."

After two years spent building and testing the apparatus, the experiment took 280 hours over a two-month period. "When it was ready, it all ran without any help except for filling the liquid-nitrogen traps," which maintained the vacuum by removing any condensation, "and making sure the capacitor hadn't blown up," as it tended to do,* remembered Freedman. At the end of each hundred-second counting cycle, the machine automatically paused and a sequencer (an old telephone relay that Clauser had rescued) would order one or the other polarizer to turn

---

*Capacitors are a basic component of electrical circuitry that store electric charge.

22.5 degrees in an orchestrated cacophony of domino-like noise and action, vivid in Clauser's mind thirty years later: "These big mama two-horsepower motors would crank over the coffins and the teletype would clatter away," the paper tape pummeling, accordion-style, into a peach basket, spraying its chads across the floor, to the *Ka-chunk* of the serial printer monitoring the quartz crystal that monitored the calcium beam.

Then the apparatus, recorded and realigned, would start its new run, and the lamp would once again excite the atom beam, launching pairs of entangled photons in opposite directions down the angled glass stacks to the phototubes, sending electrons to the counting equipment with their white numbers turning on black wheels, masses and clots of vine-thick wires, plugged and unplugged, snaking over front and back.

Meanwhile, Clauser occasionally got calls from Townes ("We're not getting any astronomy done, now, John"). "But he worked on switching the budgets so I got paid," Clauser remembered with gratitude. Between the two Bell experimenters themselves, however, Freedman remembered "battles. We had real disagreements on how this should be done, and what needed to be done."

One of the first things Clauser told him, recalled Freedman, was about his entry into a paper-airplane contest held by *Scientific American*. Following the strict rules for the amount of paper and glue allowed, "what Clauser devised was the glider that would really fly the farthest, except it looked exactly like a baseball. He glued his paper into a block and then machined it into a sphere on his lathe—he even painted the stitching onto the baseball. He was going to have some baseball pitcher *heave* this." Freedman laughed and shook his head. "They tried to disqualify him from the contest after a long discussion about whether baseballs 'fly' or not.

"And that's typical of John."

"He's a smart guy," recalled Commins later, "but he's pigheaded. An iconoclast. He was very stubborn and wanted everything done his way and sort of made life miserable for poor Freedman. And to add to it, Clauser's a big hulking guy and Freedman's sort of small, and was maybe a bit intimidated."

Outside of the basement of Birge things were easier. "We would go sailing all the time," says Freedman. "And when we were sailing, everything was fine. We had a lot of good times sailing."

30

Nothing Simple About Experimental Physics

1971–1975

Richard Holt

AND THE FIRST RUN WAS FINISHED. Just down the hall from Freedman and Clauser, filling up a room of its own, sat the computer, an IBM 1620 mainframe, in its webs of electronics, printers, and wires that had been rescued from some company's trash by Howard Shugart of the atomic beams department.

Clauser was loading a program on the computer so it could read their results, which had emerged from their apparatus as a perforated paper tape. The program itself was a six-foot-long deck of perforated cards, one card for each line of code.

Down the long, sloping gullet of the card reader, Clauser's deck disappeared, card by card, *grrlllp*. "Hey, Stu," he said as Freedman came in the door carrying the peach basket full of paper tape (which had gotten tangled in its rush from the mouth of the Teletype machine). "I just heard a great story about Commins."

"Oh yeah?" said Freedman, sitting down on a chair between the noisy printer and the big electronics rack that was the computer itself. He started untangling the paper tape.

"It's about Commins up on the roof of this building in the middle of the night with his helium leak detector," said Clauser.

Freedman looked up from the tangled paper tape.

"*Well*," said Clauser, "he applied for permission to do an experiment that involved funneling radioactive gas from up the hill, from the target of the cyclotron, and he was sitting around waiting and not getting any reply at all, and so finally he just decided to go ahead and *do* it. So he runs this pipe fifty feet in the air, gets it all set up—and it *leaks.*" Clauser was bouncing back and forth between the console and the card reader while telling his story. "And he thinks, Shit, I don't want to draw attention to this, so he has to fix the leak at *night.*" The last card disappeared into the reader; Clauser loped back to the console. "And he fixes it, and finishes his experiment—still no word from the honchos—and gets a nice little result, publishes, and then, *finally,* he hears from them." Big grin: " 'Your permission is denied.' "

Freedman laughed too. "If *Gene* is willing to ignore the committee, that shows just how absurd it is. . . . Alright, paper tape's ready."

"Load away," said Clauser, and very slowly the paper-tape reader worked its way through the spool, spitting out punch cards at the other end.

Returning to their room, they began to read their results from the neat matrix of holes punched through numbers on the punch cards. Clauser plotted out a graph, following the changing number of photon coincidences. The coincidences—the simultaneous detection of the green photon (the first photon to be emitted in the atomic cascade) in one phototube, and its purple mate in the other—rose and fell in number depending on the ever-changing alignment between the two polarizers. When the polarizers are parallel, there are many coincidences; there are fewer when they are at slightly different tilts; there are almost none when they are at right angles to each other.

Quantum mechanics predicts for this relationship a smooth, snake-like undulation—a sine wave, with a rounded peak at the origin, when the polarizers are parallel, sliding quickly past 45 degrees to its trough at 90 degrees, then rising back up to another rounded peak at 180 degrees, when they are parallel again. The photons are correlated, regardless of the distance between them; and with parallel polarizers, learning the position of one is enough to know the polarization of the other.

But the curve that emerged under Clauser's pencil was scalloped, like the windows in a mosque—*exactly what you'd expect if there were local hidden variables, if there were no entanglement, if quantum mechanics was wrong.* The photons seemed to be following separate patterns. Not the perfect straight-lined zigzag that would be a totally conclusive result, but still extremely suggestive.

"Wow," said Clauser, staring at this cusped curve. As he recalled later, "At that point, I had just been calculating predictions from a whole bunch of ad hoc hidden-variable theories that people had come up with, and it looked like some of the stuff I'd been calculating: *Maybe we've found it!*"

"O.K., let's check it through," said Freedman, his brow furrowed, comparing the numbers from the punch card with the points on the graph.

" 'There's always the chance you might *shake the world*,' " Clauser quoted (as he often did) from Bell's letter back in 1969.

Freedman grinned, and raised his eyebrows: "We'll see." He shook his head, looking again first at the punch card, then at the graph.

Just then Commins walked into the room. "How's it going, Stu, . . . John? Finished your first run-through yet?"

Freedman looked at Clauser; Clauser looked at Freedman. Freedman picked up Clauser's graph: "Look at this."

Commins frowned thoughtfully at it. "You've checked this over?"

"Yeah, twice," said Clauser.

"Of course, we need more data," said Freedman.

"Definitely," said Commins, laying the graph down on the table.

Clauser picked it up again. "We're pretty marginal, but it looks like we're within the error bars."

Commins was not excited. "Well. Why don't you start that back up and get some data to compare to this?" He jerked a thumb behind him at the now peaceful machine. "I'll see you two soon, and we'll see about this."

"I mean, yeah," said Clauser as Commins was leaving. "We've obviously got to do some more testing. But this graph!"

Freedman looked over at Clauser with a grin: "Don't worry, John, if the effect's there, we'll find it with a longer experiment."

Gingerly they slid a new lump of calcium into the oven; poured more dry ice foaming into the slurry around the red photomultiplier; refilled the liquid-nitrogen traps; loaded more paper tape into the Teletype; turned off the lights, and started the whole thing on another round.

They had both become so used to the machine's periodic sudden cacophonies that they didn't react a minute and a half later when the motors came on with a roar, the Teletype clattered, and the big coffins slowly turned their one-sixteenth of a rotation. Then, amid the familiar chaos came a new sound, of glass against glass, and then glass against metal.

"Did you hear that?"

"Ooooh, that did not sound good."

Clauser stopped the machine as Freedman turned on the lights. They opened the big blue plywood lid of the coffin. Like a huge transparent spine, the thin sheets of glass lay neatly in their row—all except one. Askew, instantly distinguishable, its metal frame was directly in the path of the photons.

"Aw *shit*," Clauser said as he pulled the frayed salmon line from out of the innards of the coffin. "Damn *wire's* broken." This had left its polarizer sheet flopping around inside, sometimes obstructing the photons, sometimes out of their way, skewing hours and hours of data.

Freedman eased the loose plate out of the coffin. "*That's* why we got that funny-looking curve."

"Back to square one," said Clauser. He grinned. "But if we get the quantum-mechanical result, I'll quit physics."

"I have to get the quantum-mechanical result," Holt joked to Shimony back in Boston, "because if my experiment comes out for local hidden variables I'll get the Nobel Prize but Harvard won't give me a Ph.D."

Harvard might have been hidebound, but Holt's thesis supervisor, Frank Pipkin, was not, with his droll attitude toward things other people took too seriously and his inspiring enthusiasm (which overcame his terrible lectures, faithfully attended by the hard core of the physics department). "Pipkin was an agnostic about these fundamental things," remembered Holt. "He preferred to do experiments wherever they came up and not waste too much time agonizing over things." He supervised fifteen different grad students, with experiments ranging over the increasingly divergent realms of elementary particle, nuclear, and atomic physics. Still, "you would walk in the door after a week or so and he would pick up right where you left off," remembered Holt. "Just a very, very smart guy."

Inspired by the civil-rights movement, Holt was teaching math and physics in a segregated school in North Carolina when he got a call from Pipkin, telling him about the mercury-cascade apparatus, which had been designed and left behind by one of Pipkin's recent students. Pipkin knew he would be interested in such a small, neat piece of equipment: "I liked tabletop physics," said Holt. "In those days you got a lot of papers with a hundred names on them; I didn't want to be part of that." And in contrast to its West Coast cousin, Holt's apparatus took up less than a cubic foot.

Instead of using an ultraviolet lamp to excite an atomic beam in a big vacuum chamber, Holt's experiment *was* a lamp. Sealed into a glass tube full of mercury vapor was an electron gun, which would excite the vapor like a neon sign. The photons that the lamp emitted—a slightly paler green and deeper purple than those in the Berkeley experiment—were entangled (unlike the Berkeley experiment, but like Bertlmann's socks, opposite polarizations were correlated). Clauser, who watched Holt at work before he left for Berkeley, recalled the creation of this glass tube as "a nontrivial glassblowing feat."

The mercury was more complicated than calcium. Five times as heavy, it also has an array of common isotopes (different versions of the same atom), ranging in weight between 204 and 196 hydrogen atoms. Holt's source was the relatively rare mercury-198. So, "then you have to distill into the glass tube this ultra-pure separated isotope of mercury. You can't even *see* it; the only way you know it's there is by running a radio-frequency discharge on it," so its spectral line can be seen through a spectroscope.

As at the Cavendish Lab in England, the string-and-sealing-wax approach was encouraged at Harvard's Lyman Lab. "Some things I had to have built by the machine shop," Holt recalled, "but we were encouraged to build our own whenever possible." Pipkin was a penny-pincher, but the apotheosis of this approach was the slight-framed Kenneth Bainbridge, at this point nearing seventy, famous as the director of the test of the first atomic bomb. He was missing part of a finger, which he had cut out when a bit of radioactive material had fallen on it. Bainbridge used to buy cheap soup pots in Harvard Square for his experiments.

As in Freedman and Clauser's experiment, phototubes stood at each end of Holt's apparatus, and "one very sad day one slipped out and crashed." Pipkin brought Holt upstairs to where his high-energy-physics group worked. This was one arena in which penny-pinching was unnecessary. "They had photomultipliers by the box load! 'Pick the one with the best specs,' " said Pipkin.

Holt sometimes wished for an accomplice, but the basement lab surrounding him was full of people working on interesting experiments and able to offer practical advice when he popped out of the bat cave with a problem. The grad student who had preceded Holt with the earlier incarnation of this apparatus "actually brought a sleeping bag to the lab, but I definitely did not want to go as far as that." Instead, he got into a routine of sleeping late and then working to all hours.

"People called me the hidden variable."

Freedman slowly opened the door to the dark Berkeley basement lab room. They had fixed the salmon line, rerigged the system to run more smoothly, and the apparatus was now on another run. "Don't worry about the light," said Clauser from behind the machine. "The red-photon counter* has lost it again and I'm shutting the whole thing down." Freedman turned on the lights, discovering Clauser bent over the offending overheated phototube, a perturbed but interested look on his face as he examined it, like a towheaded California farm boy who has just found a frog and is turning it over in his hands, ignoring its squirming. "It's all overexcited," he said. "Time for a vacation."

Freedman nodded. "Man, that red phototube has been such a pain."

"I know," said Clauser. "I'll blame *it* if Holt and Pipkin beat us."

"Yeah, well, that won't be the end of the world," said Freedman. "Science is cooperative, right? We're all just trying to find out what's going on."

"Yeah, yeah," said Clauser. "But I'd like to find out what's going on *before* Harvard does!" He grinned, but he was thinking, *It would be much better if we didn't have competitors, if we didn't feel rushed.* He looked down again. *Stupid phototube.* "Abner called me today from *Italy* asking if we'd gotten results yet."

Shimony was at a conference in Varenna.

"Oh, right, wasn't he going to talk about our experiment?"

"Yeah, and Bell gave a new more general inequality, by which Abner was able to show that our experiment will test not just local hidden-variables theories, but *any* theory that depends on locality and realism." Realism, in this context, is roughly the same as objectivity (existence independent of the mind of the experimenter) and separability (existence independent of distant things). "Big stakes."

Freedman was nodding with interest. "It would be good to get that in my thesis."

"Yeah. Abner said Bell gave a great speech. He started, 'Theoretical physicists live in a classical world, looking out into a quantum mechanical world.' " They both looked at their huge machine. "Experimentalists, too," said Freedman. Clauser laughed. "Yeah. And since nobody knows where

---

*I.e., the one that counted the longer-wavelength photon, which in this case was actually green.

the boundary is, said Bell, that makes the quantum theory seem a little provisional. 'It seems legitimate to speculate on how the theory might evolve. But of course no one is obliged to join in such speculation.' "

Freedman laughed. "I'd like to hear Bell speak."

"Yeah, he must be ten feet tall."

"Anyway, John, I've come up with something." Freedman looked earnest under his Neil Youngish mop of hair.

"What is it?" said Clauser, sounding less interested than he was.

"Well, the biggest violation of the inequality would occur at 22.5 degrees and 67.5 degrees." When the polarizers were separated by these angles, the quantum-mechanical sine wave would be farthest from the hidden-variable zigzag. "You can make a much simpler inequality as it applies to the experiment just using them." He handed Clauser his jottings on a piece of paper.

Clauser peered at them, and then began to nod.

"We might want to do another run *just* at these two angles, since they're the critical ones," Freedman suggested.

"Sounds good."

"I was also thinking," said Freedman slowly, ". . . well, I know that Dick Holt's not getting such a good count rate—we talked on the phone the other day and you know he doesn't have the equipment we do, his thesis advisor's a real cheapskate—and I was thinking maybe I should tell him about my new inequality, might help him out a little."

Clauser looked at Freedman as if he had just spoken in Russian.

Then he shrugged. "Well, I don't know how our team's going to win if you give secrets to the other side."

The web of experimentalists who wanted to work with entanglement was beginning to spread, meanwhile, beyond the immediate colleagues of Clauser, Horne, Shimony, and Holt. Their paper, under discussion at the informal but grandly named Brazos Valley Philosophical Society in southeastern Texas late in 1969, had captured the interest of Ed Fry, a good-natured young experimentalist and new professor at Texas A&M.

"Immediately that evening," Fry said thirty-five years later, he realized what he wanted to do. Lasers were just becoming available, and one of these could be tuned to wavelengths around 5461 Å—a bright emerald green. As it happened, that was a color that could trigger an atomic cascade in mercury. A violet photon would emerge, followed by, and entangled with, an ultraviolet one. The purity of the laser beam would make the experiment drastically neater, the entanglement much simpler to produce.

Fry and Jim McGuire, a theorist friend who had brought him to the philosophical club that evening, were thrilled with the idea; neither knew how unfashionable this subject was. "We fed off each other's excitement and enthusiasm," remembered Fry.

Their excitement and enthusiasm were quickly dampened by the discovery that no one wanted to fund Fry's beautiful experiment. The reviews of his grant proposal "were a clear window on the culture," as he remembered. "There were specific references to time and money already being wasted at Berkeley and Harvard."

Further experimental exploration of entanglement, it seemed, was not to be. "I was devastated."

"Dick Holt and Pipkin didn't tell anyone their results," Clauser remembered. "We didn't know *what* was going on with them. So we just stumbled along with our experiment, and finished and published it in 1972," in *Physical Review Letters*. "And it was a very interesting result," said Freedman. "It excluded hidden variables by a large margin. It was fairly conclusive." The highly calibrated classical mechanics of the Clauser-Freedman machine had displayed a fully entangled, spooky result.

After Freedman finished his thesis, in May 1972 ("It was the biggest thesis ever to come out of this lab," he recalls. "*Everything* about this project was big"), he accepted a position at Princeton, and he went to Harvard in order to finally meet Holt in person. He first found his way to Pipkin's office a few floors above the subterranean bat cave. There was no one in the office, so Freedman sat down to wait. Then he spotted Holt's thesis lying on the desk, neatly stapled into its thin posterboard cover. He picked it up with curiosity and started reading.

Holt's abstract began with the statement that he had performed "the linear polarization correlation measurement proposed by Clauser et al." Next (as in Freedman's own thesis) came Freedman's version of Bell's inequality, and then Holt gave his results in comparison to it. Freedman looked at this number twice, and his eyebrows rose. Holt had set up the inequality in a slightly different way than Freedman had, but his result was almost as far *inside* the bounds of the inequality as Freedman's had been in violation of it. "This value is in strong disagreement with the quantum prediction," concluded Holt. In direct opposition to the Berkeley results, entanglement failed.

Freedman read on, fascinated, his eyebrows now lowered: What was going on here? he wondered. It wasn't a problem with count rate. He had a statistically significant result, measuring at the two angles that the Freedman inequality called for.

The two angles. Freedman remembered his phone call, relaying his simplified version of the inequality to Holt as a favor. In fact, after that, Holt constructed what he called "a sort of Rube Goldberg apparatus" that measured the photon coincidences only at 22.5 degrees and 67.5 degrees. When an engineer at the lab tried to persuade him to use a Geneva mechanism (as the Berkeley team had done), Holt listened politely, but "being young and headstrong, I thought I could do it"—he laughed as he told the story years later—"simpler."

Sitting in Pipkin's office, Freedman thought: *Either he's found Clauser's "world-shaking result" . . . or I've caused him a tremendous amount of harm.* Any systematic error is harder to see when you measure in only two places.

Pipkin came in, and Freedman stood up to shake hands, feeling a little strange. "I see you've found Dick's thesis," said Pipkin.

Freedman nodded.

"We keep debating whether or not we should publish. No one has come up with any satisfactory suggestions for what might have gone wrong. But whenever he wants to publish, I'm advising caution, and whenever I think it's our duty to publish, he's wary. Here's his most recent version of the preprint." He handed the paper over to Freedman. The byline said, "F. M. Pipkin and W. C. Fields, Harvard University." Freedman laughed.

Just then Holt came in the door. "W.C.!" said Pipkin, with an ironic grin. "I'd like you to meet Stu Freedman."

"Hey, Stu, nice to finally meet you," said Holt, proffering his hand.

"Yeah, Dick, I'm working at Princeton now and I thought I'd come by."

"I see you've found us out," said Holt, gesturing toward the crimson-covered thesis in Freedman's hands. "We don't know what to do with it. I ran into Bob Pound—one of the professors here—in the hall the other day, and he said, 'So you're gonna publish that thing?' and I said, 'Well, we're sort of thinking of it,' and he sort of looked off into the distance and said, 'Ahh, he was a promising young physicist. . . .' "

Pipkin laughed out loud, as did Holt. Freedman still felt a little uncomfortable.

"So no one can find anything wrong?" he asked.

"Yes, Fields, I take it you're not so keen to publish today," said Pipkin.

"Oh well, you know, if at first you don't succeed, try, try again. Then quit. No use being a damn fool about things," said Holt, grinning and shrugging.

"If you were W. C. *Pauli* right now instead of W. C. Fields it might be more helpful," suggested Pipkin.

"Do you think Pauli would have ageless words of wisdom to equal *that*?" asked Holt.

Holt had spent the previous two years looking more or less fruitlessly for errors in the experiment. "I became pretty good at thinking of bizarre things that could be going on," he remembered. A clear culprit never emerged. Muddying the waters further, the next year, in 1974, a team testing Bell's inequality in Sicily published results concurring with Holt's result, sharply disagreeing with the quantum-mechanical predictions.

The Sicilian experiment was actually a version of the Wu-Shaknov experiment about which Mike Horne had decided back in 1968, "There was no point in ever considering that experiment again. It was not adequate. You could reproduce it with a hidden-variables scheme." (That is, the entanglement displayed in Wu's experiment was weak enough that it could also be explained in a nonquantum way, by unspooky local hidden variables.) Clauser had come to similar conclusions, and though Wu's student Leonard Kasday had argued that two added assumptions allowed the experiment to demonstrate strong entanglement, Shimony found these glaringly inadequate.

In Texas, Fry had found the confusion "totally reenergizing." His experiment could no longer be deemed such a waste of money and time. In 1974, with "enough money to buy the laser," Fry started work, with the encouragement of Clauser, but originally without the help of his student Randall Thompson, whom Fry was trying to protect from involvement in a project that it seemed could only be bad for his career.

Meanwhile, "we were standing out all alone with no confirmation," remembered Clauser of himself and Freedman. "It was a funny position for me to be in, too, because I had expected just the opposite to be true. I was pretty confused by it all." In words that are echoed by many others who have tried to deeply understand the situation that is, in the words of John Bell, "revealed or *concealed* by quantum mechanics," he continued a quarter of a century later, "Still am. I still don't understand it."

Horne and Shimony were also still coming to grips with what lessons to draw from these experiments. Horne became a professor at a little college outside of Boston called Stonehill, and "we were together all the time," Horne remembered. "We argued and argued and argued. And Clauser and I argued and argued, mainly by telephone, trying to come to an agreement" about the logical underpinnings of their version of Bell's inequality and what, in fact, they had actually proved. "All three of us discussed it, all through '72, '73, '74—a forever paper."

"In formulating special relativity," explained Clauser, "Einstein noted that one cannot ask the universal questions 'when?' and 'where?' in a precise fashion, without first defining the associated primitive entities, 'time' and 'distance.' " Famously, Einstein found that the best definitions were operational ones: time is what you measure with a clock, and distance is what you measure with a ruler.

The Einstein-Podolsky-Rosen paper, in trying to restore objectivity to quantum mechanics, had tried to answer the question "what?" in the same way. EPR had written, "If, without in any way disturbing a system, we can predict with certainty the value of a physical quantity, then there exists an element of reality corresponding to this quantity."

The analysis of Bell, however, had proved this EPR definition of an "element of reality" inadequate. Entanglement exists, and requires guidelines for a new answer to the question "what?" Clauser recalled that he and Horne were thinking along these lines: "Does the chair exist when you're not there to look at it? Maybe we could define *what* as 'something you could put in a very large box' where *chairness* can be measured . . . whatever properties that might involve, like: it has varnish. Define the properties and you've got the object, right?

" 'No!' says quantum mechanics. You can't measure the properties and expect them to have anything to say about the actual object.

"This," said Clauser, "is what I found infuriating." (Horne's take was, "Well, that's the way it is.") "It's the whole Bishop Berkeley question of 'If a tree falls in the forest and no one hears it, does it make a sound?' Quantum mechanics answers the bishop, 'No.'

"If you close the box, is the chair still there? How about a shoe? You'd be pretty surprised if your shoes changed color in the box coming home from the shoe store. How about ions? We make ions by chipping electrons off atoms—they are *things*, we assume. You can see even just one of them. Put it in a potential well"—an electron trap—"and shine light on it, it will fluoresce. And if you put another one in with the first one, the fluorescence will double. And so on. So if you put an ion in your box, is it in the box when you close the lid?

"What if you have two photons in an entangled state"—Clauser raised his eyebrows with a grin—"and you put one in the box you're holding and one in the box I'm holding? Are they there? You can measure the impact of a single photon on an object. It seems like it's a thing. Is *it* in the box when you close the lid?

"It's easy to be confident about the chair, the shoes, even pretty confident about the ion, but the photon is harder. Are they all things? If you're a scientist, you've got to be precise in your definition.

"The varnish is there on the chair whether or not you 'measure' it, or whether you are a good or bad varnish measurer. I may or may not know what I have measured . . . (in fact, it's arrogant to say that you actually know, like 'I just measured the *spin* of this electron': what you mean is you got a certain result in a Stern-Gerlach device. You have no idea what you really measured).

"Now, suddenly, with photons, their polarization depends on *how* you measure it and what you're looking for. But if I'm wrong in any one of these places, can I be right about the chair or the shoes?   ,

"What physicists tend to say to reassure you about the chair and the shoes is, 'Don't worry about the fuzziness—it's so small it doesn't affect the macroscopic world.' But that's just scare tactics with numbers! Atoms are inescapably there. [In 1990] IBM glued down atoms in a silicon substrate, spelling out 'IBM,' and you can read it. Just because the guys in the thirties couldn't see 'em didn't mean they weren't there."

Back in 1974, as Clauser and Horne continued their discussion over three thousand miles of phone line, Clauser performed a new experiment, which Shimony declared to be "of capital importance." Contemplating the objectivity of photons ("we need to know that answer to make an airtight Bell's theorem case"), Clauser made sure that they could in fact behave as particles, either reflecting from a half-silvered mirror or passing through it, never splitting in half like a wave.

But this ostensible object, when in an entangled state, has no definable properties on its own. "It's not a question of, Is the varnish made from pine trees? but *Is* it varnish? What is varnish? Does varnish exist? Do *trees* exist?" Clauser threw his hands in the air: *Beats me!*

"So it turns out I'm in deep *caca*," Clauser said, summarizing the indefinite results of his and Horne's paper, "when I try to define the interrogative 'what.'." They finished the paper in 1974, acknowledging "many valuable discussions with A. Shimony," and the Clauser-Horne version of the Bell inequality they set forth there is still considered the gold standard. In fact, no experiment has yet tested it, and strangely enough, experiments of this sort in the last twenty years, though greatly improving on the Freedman-Clauser experiment in speed and beauty, are arguably further from testing this inequality than the huge, ponderous machine now languishing in the "boneyard" attic of Old LeConte Hall at Berkeley.

And what of Holt's experiment? Clauser repeated it, "on scrounged equipment," discovering that the transition in mercury was much harder to work with than their calcium. "It was *grim*. I think I ended up with four hundred hours of counting time, or something horrible." This was twice as long as the Freedman and Clauser experiment, or Holt's. "So this went on for weeks and weeks." But he eventually got the quantum-mechanical result, demonstrating entanglement, in 1976. A few months later, Ed Fry published his own long-dreamed-of experiment, and Clauser "heaved a sigh of relief": Fry, too, demonstrated the presence of entanglement and the absence of any local, realistic explanation. With the emerald-green beam of a laser, Fry and his student Thompson had collected their data in less than an hour and a half, the opposite of grim.

The experimentalists were beginning to get a grasp on entanglement.

31

In Which the Settings Are Changed

1975–1982

Alain Aspect

Hanging next to the door of Bell's office at CERN was a poster of a long-necked Modigliani lady in a hat; her eyes and the eyes of Bell himself watched the twenty-seven-year-old Alain Aspect, a genial, mustachioed graduate student talking eagerly about boxes of water.

It was early in 1975, and Aspect had just returned to Europe from a three-year stint of French "national service," teaching in Cameroon. Soon after his return, he had suffered what he described as a *coup de foudre.* "In October 1974," he remembered, "I read John Bell's famous paper 'On the Einstein-Podolsky-Rosen Paradox,' and it was love at first sight. This was the most exciting subject I could dream of." Immediately he decided to make Bell's theorem the subject of his Ph.D. thesis at his alma mater, the University of Paris-South, in Orsay.

Meanwhile, Clauser was trying to get a job. "I must have applied to at least a dozen different places, and at all of them I was totally rejected." Universities were uneasy about hiring a professor who would encourage the next generation to question the foundations of quantum theory. Finally Clauser found an opening at the Lawrence Livermore National Laboratory, in the hills east of Oakland, researching plasmas (David Bohm's first love).

"I don't know anything at all about plasma physics," Clauser announced at his job interview. "But I do know a lot about doing experimental physics. I'm a very talented experimental physicist."

"You can learn plasma physics," was the reply. He was hired in 1976 and stayed there for a decade. At Livermore, his self-proclaimed skills as an experimentalist were well used. But almost wasted was an equal, though unproclaimed, skill. Clauser had the gift—rare even in the teaching profession—of explaining complicated subjects to students clearly, vividly, and patiently. The university career where he could have combined these skills did not materialize in the three decades since he first applied for such a job.

"Back in the sixties and seventies, reputable physicists did not ask questions about quantum mechanics," explained Fry in 2000. "I think that Clauser took the brunt of this attitude—in part, I believe, because he was actually *doing* the experiment, not just talking about the theory."

Fry himself had better luck with academia. In the midst of performing his experiment, he was granted tenure. Thirty years later, by then the head of the physics department at Texas A&M, he learned that this open-minded institutional decision was thanks to an intervention from Frank Pipkin, Holt's adviser at Harvard. Realizing that the tenure committee was about to reject the Bell experimenter, one of Fry's friends asked Pipkin to come to College Station, Texas.

"If you had sent me just Ed's file to look at, I would have rejected him very quickly," Pipkin told the committee. "However, after spending a day in his lab I can tell you that this guy is a winner and I would bet on his success." Pipkin's renown in atomic physics won over the skeptical committee.

Bell himself was acutely aware of the stigma attached to the experiments his work had inspired, but thus far Aspect was not. Before heading to West Africa in 1972, Aspect remembered, "I had a quite good education in classical physics, and I knew my education in quantum physics was extremely *bad*." The classes he had taken on the subject comprised equation-solving with little discussion of physical meaning, let alone inculcation of any stigmas.

So for his three equatorial years in Cameroon, Aspect taught himself quantum mechanics, using a recent textbook by the great French physicist Claude Cohen-Tannoudji. This book had two strengths: "First, it is real physics," said Aspect. "Second, it is neutral with respect to foundations. No brainwashing, no 'Bohr solved all of that.' " As a result, "I was able to solve the equations but nobody had washed my brain.

"I was totally convinced by Einstein and Bell," he said. But what experiment to do? In rereading Bell's 1964 paper, Aspect realized that its last lines told him "there was still an important test to be done."

He raced to Geneva to tell Bell his idea.

Bell had ended his paper on a cautionary note. If there was enough time for a light-speed signal to correlate the particles, then entanglement would lose much of its mystery. Conceivably, quantum mechanics might work, wrote Bell, only when "the settings of the instruments are made sufficiently in advance to allow them to reach some mutual rapport" by exchange of signals at the speed of light. "In that connection, experiments of the type proposed [in 1957] by Bohm and Aharonov, in which the settings are changed during the flight of the particles, are crucial."

The practical problem with this experiment was that the huge, fragile, piles-of-plates polarizers at either end of the Freedman-Clauser experiment could not move into their settings quickly. Aspect had come up with a beautiful (and, importantly, frugal) alternative idea. Its main ingredient was water.

"Each polarizer," Aspect explained to Bell, "would be replaced by a setup involving a switching device followed by two polarizers in two different orientations." At any given time, the switch would allow a path to only one of the two polarizers. "The switch would rapidly redirect the incident light from one polarizer to the other one," leaving no time for light-speed signals to facilitate any kind of "mutual rapport" between distant ends of the apparatus. He turned to the blackboard, where he wrote the appropriate inequality for the situation "if the two switches work at random and are uncorrelated."

Aspect's "switches" were two glass boxes full of water, over forty-two feet apart from each other, on either side of the beam of cascading calcium that produced the photons. Each box of water carried a sound wave, far higher than the human ear can hear. (Transducers on either side of the boxes converted electrical signals into this ultrasonic wave.)

Sound waves, unlike light waves, need a medium—hence the silence of outer space. They move by repeatedly compressing and then relieving pressure in their medium. Aspect's ultrasonic wave cycled between a swiftly oscillating high, striping the water in an alternating pattern of dense and thin; and a flat low, leaving the water unaffected. The striped pattern acted as a diffraction grating, bending the light to send it to the side polarizer; without it, the light traveled straight through to the main polarizer. The wave cycled between striped and flat quickly—"the switching between the two channels would occur about every ten nanoseconds," Aspect explained, four times faster than the photons would travel the twenty-one feet separating the oven from the water.

It would not, he admitted, be the ideal scheme, "since the change is not truly random, but rather quasiperiodic. Nevertheless, the two switches on the two sides would be driven by different generators at different frequencies," meaning that the two boxes would oscillate at different rates, and in practice the rates would drift. "It is then very natural," said Aspect, "to assume that they function in an uncorrelated way."*

When Aspect finished his eager presentation, he stood silently awaiting a reply. Bell asked his first question with a trace of irony: "Have you a permanent position?" Aspect was only a graduate student, but—because of the uniqueness of the French system, and in drastic contrast to his counterparts in America—his position at the École normale supérieure was actually permanent. Even with this advantage, it was not easy.

"There will be serious fights," Bell warned him. But the stigma was not the only thing he worried about. "One should not spend all his time on concepts. You are an experimentalist, which keeps your feet on the ground, so you are not in so much danger. For me, I am a theorist and this subject must remain my hobby.

"If you spend all your time thinking about it, you are in danger of becoming crazy."

Freedman drifted away from Bell physics, but he found his thesis experiment haunting him even thirty years later. "The Bell experiment was a null experiment—that's an experiment that measures no deviation from what you expected—and in my time here, I've done twenty-four null experiments, finding that those things that you didn't expect to be there are, in fact, *not* there. So this was sort of how I started my career."

The "constant theme" of Freedman's career was, as he said in 2000, "It's really a big help if you know what the right answer should be: if you don't get it, you might suspect that there's something wrong with your equipment—and that's probably the case.

"I have quite a reputation for doing this, for stepping into a field

---

*This experiment (published in *Physical Review Letters* for Christmas 1982) was so difficult to carry out that Aspect and his student, Jean Dalibard, listed the machinist, Gérard Roger, as an author.

where there's something very exciting going on"—Freedman grinned—"and leaving it with nothing interesting." The last word on the subject was far from spoken, however.

Holt, too, left Bell and EPR behind for the "bread-and-butter physics" his apparatus had once been designed to do. He went on to a career of measuring atomic lifetimes via cascade photons, measuring spectra with lasers, and measuring energy levels in atoms too complicated for quantum mechanics (essentially any atom beyond hydrogen). Looking back, "I would say that I played a relatively minor role in the CHSH [Clauser-Horne-Shimony-Holt] business," Holt said. Then he grinned. "But I did set the world on fire with my wrong result."

As it had with Freedman, the experience started him contemplating how science proceeds. "There's an interesting scientific principle that a wrong answer can be much more stimulating to the field than just sort of finding the answer that's in the back of the book. A wrong result gets people excited. Worried.

"Obviously, you don't *really* want that to be happening—it's O.K. for a *theorist* to come up with a speculative new theory that gets shot down, but *experimentalists* are supposed to be very careful and their error limits are supposed to be realistic. Unfortunately, with this experiment, whenever you're looking for a stronger correlation, any kind of systematic error you can imagine typically weakens it and moves it toward the hidden-variable range. It was a hard experiment. In those days, at any rate, with the kind of equipment I had, and . . . well, what can I say?" He laughed with a shrug. "I screwed up."

Clarity, however, about which experiment is right is not the same as clarity about the quantum mechanics these experiments were designed to elucidate. "The thing is," Holt said, "I'm a scientist and I sort of want to believe what Nature says is the answer and not what I just think I know ahead of time, and I've always thought that quantum mechanics was just so wonderful because it's a surprise—it's sort of this hidden knowledge that we high priests of science in a way"—he laughed—"can find. . . .

"Which isn't to say I want to keep it secret," he explained. "But if everything were just obvious, if you could just look around and see how the universe is, well, that wouldn't be very interesting. Quantum mechanics is very subtle; that's its fascination.

"For me, well—*all* of my interest in physics has been because I personally wanted to know the answer to these questions . . . and," he said simply and a little wistfully, "I still don't know the answers. . . . That's the frustrating thing. . . . I think it's going to be a long time before people know what quantum mechanics means. All these modern experiments *clearly* show that you have to at least provisionally accept the quantum-mechanical way of thinking about states of unobserved things. And yet it's a very unsatisfying kind of thing.

"Quantum mechanics—how you talk about it, that is—still remains unfinished business. But I believe something will come up, from an unexpected direction. . . . We may not have to solve the problem as originally posed at all—the problem will just disappear one day because we'll find out we were asking the wrong question."

In 1975, Mike Horne thought he was leaving Bell physics behind, too. He and Shimony had become fascinated by a beautiful new experimental apparatus, a neutron interferometer, just invented by Helmut Rauch in Vienna. In contrast to Clauser's experiment exhibiting the particle nature of light, Rauch's machine dramatically displayed the wave nature of matter.

In the first years of the nineteenth century, as Horne vividly describes it, the great physicist Thomas Young "showed experimentally that the addition [or superposition] of two equally bright light beams can make darkness, and, under slightly different conditions, can make light four times brighter than either original beam." Horne smiles: "That is, 1 plus 1 equals 0, and, under other conditions, equals 4." This is called interference, and it signifies the presence of a wave.

But Rauch was displaying these hallmark wave phenomena with particles of matter. Beams of neutrons, particles produced from the seething hot core of a nuclear reactor, were interfering with each other like waves.

The neutron interferometer offered these neutrons two alternative V-shaped paths, like a child choosing to throw a ball to his friend by bouncing it off the floor or off a low ceiling. Whether, after entering the interferometer, a neutron hit the "floor" or the "ceiling," its final destination was the same, so two neutrons, starting at the same point but taking alternative paths, would meet again afterward. These alternative paths seen together traced out a diamond shape. When they met, the neutrons would interfere with each other.

Dramatically, even a *single* neutron in the interferometer would interfere with itself. This, like so many quantum-mechanical conundrums, is impossible to picture. It is as if one neutron had traveled down both paths simultaneously.

Horne and Shimony had been immersed for a decade in the mysteries of two-particle entanglement; now they were distracted by this magical single-particle effect. "Abner and I both thought, This is going to be a very valuable device; people are going to play with it for *years*," remembered Horne. Just for starters, Horne and Shimony realized, it could be used to show that a spin-½ particle, like a neutron, returns to its starting position only when it is spun in place two complete rotations. The exciting possibilities of this machine were not lost on others interested in elementary quantum mechanics. Before Shimony and Horne could publish their paper, Herb Bernstein, just down the Massachusetts Turnpike from them in Amherst, had also proposed the experiment, and Rauch lost no time in performing it in Vienna, with a young Austrian physicist Horne's age named Anton Zeilinger.

Zeilinger turned up again soon afterward at a Bell physics conference in Erice, Sicily. Horne remembered that of the fifteen or twenty in attendance, including Bell and Shimony, "only one person came who wasn't talking about two particles—he was talking about *one* particle—and that was Anton."

Horne was now also interested in one particle. "We hit it off right away," he remembered, the soft-spoken Southerner and the charismatic *Österreicher*, two tall, bearded men bent together, deep in conversation. "We spent many days talking. He was initiating me into the details of neutron interferometry," Horne remembered, "and I told him about things he was not particularly in on at the time. . . ."

"This was my first real encounter with the international scientific community," remembered Zeilinger. "There, I heard for the first time about Bell's theorem, about the EPR paradox, about entanglement, and the like. Needless to say, I did not get any real understanding of what this was all about, but I got the hunch that it was something very important." Fascinated, he tried to learn from Horne everything he could about entanglement.

When Horne returned home, he went straight to Cliff Shull's lab at M.I.T., where he had heard rumors of neutron interferometry. Not much more than half Horne's height, the beloved and distinguished Shull had essentially founded the field of neutron diffraction—that is, utilizing the wave nature of the apparently particulate neutrons. (Shull's Nobel Prize for this work was still, in 1975, another twenty years in the future; his collaborator, Ernie Wollan, would not live long enough to receive it with him.)

"I hear you're building a fundamental quantum toy called a neutron interferometer and I'd like to play!" Horne said to Shull. "Can I play?"

"Sure," said Shull. "Take that desk right there."

"So I just sat down," Horne remembered, "and from then on I went over there every Tuesday—the day I didn't teach classes at Stonehill—and often on the weekend and every holiday and all the Christmas breaks, and all the summers for about the next twelve years." And they were "a fun twelve years" for someone who wasn't a theoretician or an experimentalist—"I'm just sort of a middleman—I'm just *around*. So when they wanted to move lead bricks, I joined in with them moving lead bricks." But when they wanted insightful help in planning the details of experiments, the easygoing Horne was good at that, too.

Soon after Horne set up shop in Shull's lab, Zeilinger showed up too, with his family, intent on doing neutron interferometry and buying, as a laughing Horne remembered, "the biggest American car he could get— some sort of Oldsmobile land cruiser, a humongous station wagon the size of a yacht."

Next to arrive (actually he was returning to his alma mater) was Danny Greenberger, a witty, stocky New Yorker from the Bronx who had been on the trail of the neutron interferometer since a few years before it had been invented. He had wanted to learn the effect of gravity on the interference of a neutron—an experiment that was performed* almost immediately after the neutron interferometer was turned on.

He had met Horne and Zeilinger at one of the first neutron interferometry conferences, and "we hit it off wonderfully," Greenberger remembered, "so I started going up regularly to M.I.T., and Cliff supported all three of us." Shull was "a lovely person," and Greenberger called that decade visiting at his lab "the high point of my professional career, in terms of having fun and doing interesting things with great colleagues." It would also lead, by a decade-long path, to the discovery that entanglement of *three* particles was stranger than anything that had gone before.

---

*By Roberto Collela, Al Overhauser, and Sam Werner (and thus known as the "COW" experiment) in Ann Arbor, Michigan. Werner subsequently made the University of Missouri a center for beautiful neutron experiments.

Entanglement
Comes of Age

1981–2005

Clockwise from top:
Anton Zeilinger, Daniel Greenberger, and Michael Horne's hat

32

Schrödinger's Centennial

1987

John Bell's self-portrait

BELL HAD BECOME ONE of the premier theorists of one of the premier centers of physics in the world. He was "called the Oracle of CERN," Bertlmann remembered: "there was a certain aura around him and his office." This reputation had little to do with Bell's theorem. "If you went to CERN," remembered Nicolas Gisin, who in 1981 had just received his Ph.D. in quantum mechanics nearby at the University of Geneva, "... and asked at random about John's contributions to physics, his work on the foundation of quantum physics would barely be mentioned."

The people who came to the Oracle—teetering on the wobbly visitor's chair in his office, or meeting him at exactly four o'clock in what Bell referred to as the CERN "canteen" for a cup of *verveine* tea, encountered Bell's "goodwill in all respects," as one of his collaborators described it. But Bell was not only a man of goodwill. "There was with John also this feeling of energy and strong engagement under the usually peaceful surface. To do physics was a challenge of basic importance, and there were matters that could at rare occasions increase his temper to a point that was felt like a small eruption."

Bertlmann has a sketch Bell drew of his own profile—winter hat, beard, glasses, and all—tossed off in a few confident, spare, and exactly right lines. "He wrote like that, too," said Bertlmann. "He could frame a sentence perfectly, straight from his stomach. People were scared of him—he could be very fierce. To me, he was like a father, we would joke around.

"But he never told me about his work with the quantum mechanics; though I knew he dabbled in it a bit, I was a typical physicist—not interested (at CERN they thought he was a little mad with the quantum mechanics). That shows a very strong character—he knew instinctively what the right time was and to wait for that right time.

"Immediately after the socks paper came out," in 1981, "I was an expert in quantum mechanics," remembered Bertlmann with irony. "So I read up quickly." He found that John Bell was not the only person a little mad with the quantum mechanics. "What amazed me was the way Einstein and Schrödinger felt about it. Not, 'we have these different models to try.' They felt it in their *blood.* They were searching for an inner truth. I was *shaken* by their letters.

"And you know, in those days, it was thought of as a sin to question the orthodox quantum mechanics." At CERN, this attitude remained strong throughout the eighties.

One evening in 1987, the Bells and the Bertlmanns were sitting on the flagstone patio outside the CERN cafeteria amid groves of young saplings. "The sun was falling down where the Jura met the Alps," Bertlmann remembered, "and this lovely red light was everywhere—the long wavelengths of the sun matching John's red hair. It was a beautiful evening." John and Mary were about to go home for dinner; Reinhold and Renate were about to go back to his office, where Renate's artwork was spread out in one companionable corner. It was a peaceful paused moment, everyone languidly stretched out on their chairs feeling the last warmth of the sun on their skin.

"John, whether or not they end up giving it to you," said Bertlmann, "I think you deserve the Nobel Prize for your quantum mechanics."

John did not move, his eyes closed. "No, I don't," he said matter-of-factly. He mentioned a highly technical experiment that his friend Jack Steinberger, a slight-framed, good-natured experimentalist, had recently performed at CERN. "It's like Steinberger's experiment—it is a very beautiful experiment, very beautifully done, but it's a null experiment, and you don't get the Nobel Prize for a null experiment." He was silent, thinking about Steinberger, and then said, "His work is so good, he should get the Nobel Prize, and he will. But for me—there are the Nobel rules as well—it's hard to make the case that my inequality benefits mankind." He squinted at the sun on the mountains.

Bertlmann said, "I disagree with you."

Bell looked over at his friend with an amused expression on his face.

"It's not a null result. You *have* proved something new—nonlocality. And for that I think you deserve the Nobel Prize."

Bell's forehead furrowed, his chin on his chest. He seemed pleased at Bertlmann's insistence, but then—raising one sunset-reddened arm, his legs splayed out in front of him—he shrugged his shoulders. "Who cares about this nonlocality?"

The sun sank lower and lower and the mountains got darker and darker, and the moment was over.

Steinberger received the Nobel Prize Bell had predicted for him less than a year later—not for his beautiful null experiment, but for work from almost thirty years before. Even sooner, however, Bell would meet someone who was to play one of the starring roles in a drastic change of physicists' attitudes toward nonlocality.

In August 1987, in one of the University of Vienna's colonnaded and embellished rooms, at a conference celebrating the hundred-year anniversary of Schrödinger's birth, Bell found himself on a panel with another charming and confident Austrian—Zeilinger, whom he had last met in Sicily a dozen years before as an enthusiastic neutron physicist coming to grips with entanglement for the very first time.

Insisting that physics should explain *how* events happen, Bell said that quantum mechanics was missing something crucial. He found himself faced, as usual, by polite complementarian disagreement. Zeilinger, though unusually appreciative of Einstein, Schrödinger, and the man sitting beside him for their clear understanding of "the radical changes in our worldview that quantum mechanics necessitates," liked the austerity of the Copenhagen interpretation. He was drawn to the idea that the wavefunction might be simply a description of knowledge (or even, in soon-to-be-popular terminology, *information*), at which point the paradoxes mostly disappear, and it depends on the theorist whether any reality or causes should be sought in the nonlocality or nonseparability beneath this curtain of processing information. Zeilinger, for his part, was beginning to suspect that the lesson of quantum mechanics was that "there is no difference between epistemology and ontology: being and knowing are intertwined."

"Bell was quite conservative, and wanted to be as careful as possible,"

Zeilinger later said. It was a point of view that he had great respect for. "But I," he went on, "am a romantic, and I wanted to be as radical as possible." Rather than suppressing the extreme nature of the Copenhagen interpretation, as the generation of physicists after the death of Bohr had tended to do, Zeilinger recognized and embraced it.

He doubted individual events would ever be explained, a perspective that was anathema to Bell. But in Zeilinger, Bell was introduced to the new wave of quantum theorists who, whatever their views on the reality or ineffability of the world that the wavefunction described with such mathematical precision and physical vagueness, would center their whole careers on affirmative and joyful investigations of Bell's long-neglected nonlocality. All Zeilinger needed was the perfect source of entangled particles, and 1987 was the year he would find it.

33

Counting to Three

1985–1988

Michael Horne

Mike Horne plays a stand-up bass that belonged to his brother, who never came home from Korea. In the summers between semesters at Ole Miss, before he took up the bass, he had drummed for a lounge trio. Now he applies that sense of timing and syncopation to the bass, which is almost a percussion instrument as the bassist's fingers walk up and down the neck in complicated rhythmic patterns.

Anton Zeilinger is also a bassist, in what Bohr would call a complementary way. As a boy, he learned to play classical music on his bass with a bow, its voice speaking deeper than a human's. While Horne had internalized the striding and stepping rhythms the instrument could produce as his fingers walked its neck, Zeilinger knew its voice. "He knows what it's going to sound like before he plays it," says Horne, "but his timing is too square for jazz. While, me, I can 'walk' but I can't 'talk.' " (Danny Greenberger, in physics if not in music the third member of the trio, loves to play standards of the World War II era on his piano.)

One winter's day, a few years before Zeilinger's appearance with Bell on the panel at the Schrödinger Centennial, the two complementary bassists were sitting in Shull's lab when they saw a poster advertising a conference the next summer near the Soviet border:

FIFTIETH ANNIVERSARY OF EPR,
JOENSUU, FINLAND,
JULY 1985.

"Gee, have you ever been to Finland?" asked Zeilinger.

"No," Horne said, "I've never been to Finland."

"Well, let's go to Finland!" said Zeilinger.

But after a decade of beautiful one-particle experiments, "We don't have anything to say about two particles," said Horne. "You know this is about two particles, right?"

They were silent.

Then Zeilinger said, "There must be a connection . . ."

Horne began to nod. "Between two-particle polarization correlations and neutron interferometry."

That afternoon, they found that two connected diamond-shaped interferometers could test Bell's inequality. If two entangled particles departed in opposite directions, entering two opposite interferometers, the results would display entanglement in a way similar to those of the Clauser-Freedman experiment.

If the experimenter looks for interference, he will find that each particle traversed both sides of its interferometer, interfering with itself. But if the experimenter instead decides to measure which path ("floor" or "ceiling") one of the particles took through its interferometer, he will know, without measuring it, that the other particle took the opposite path through the other interferometer.

This insight took Horne and Zeilinger to Finland and Greenberger to another conference at the World Trade Center in New York City, but at neither place did anyone come up with a source that could produce entangled particles that moved in exactly opposite directions, as was necessary for the double diamond to work. "We couldn't do neutrons," Horne explained. "We had no way of making a pair of neutrons: neutrons just come out of the reactor; and we couldn't use the cascades like Clauser and Holt did because the atom that emits the photons is left behind," carrying off some of the momentum, so the photons don't leave back-to-back. "So we returned to the old positronium annihilation as a possibility, where nothing is left behind as a remnant." This was particularly interesting because positronium sometimes decays into *three* gamma rays.

"Has anybody ever tried to look at what happens with Bell's theorem for three particles?" Greenberger asked a little while later.

"I don't know," said Horne. "No, I'm sure they haven't."

"I'm gonna work on that," said Greenberger.

"Good," said Zeilinger, whose curiosity had been piqued by the anomalous positronium decay.

Cliff Shull had hoped that Zeilinger would succeed him when he retired in 1987, "but M.I.T. had other plans," remembered Greenberger. "They wanted to rip the whole lab apart." So Zeilinger returned to Vienna. But he arranged for a Fulbright scholarship, for which Greenberger "was the only person in the world who fit."

One day, Horne, now alone, was wandering through Shull's lab when he saw a recent *Physical Review Letters* lying on someone's desk. Thumbing through it, he came upon their double-diamond experiment, turned from theory to fact by Rupamanjari Ghosh, working on his Ph.D. with one of the pioneers of quantum optics, Leonard Mandel, in Rochester, New York.

Ghosh and Mandel had discovered the entanglement latent when light strikes a crystal, in a beautiful process laboring under the ungainly name of "spontaneous parametric down-conversion." This down-conversion turned out to be the ideal source to perform the two-particle interferometry experiment that Horne, Zeilinger, and Greenberger had been seeking. When a beam of light shines through a certain kind of crystal, a very tiny number of its photons split in two (it is not understood why), each carrying half the energy of the parent. If the light is ultraviolet, the daughter photons are visible as a lower-energy, many-colored nimbus emerging on the other side of the crystal. Mandel and Ghosh discovered that each pair of daughter photons is entangled, both in their colors and their directions of flight.*

Horne called Zeilinger: "Go get this paper; they're doing our experiment. And it looks like a thing that you might want to get into."

Up to that point, as Horne remarked, Zeilinger "had never plugged in a laser." But he saw the excitement inherent in such an easy source of entanglement, and he picked up the phone and called Rochester. "There's undoubtedly a lot of art, a lot of trade secrets, in setting up those down-conversions. You'd never know, looking at a *Physical Review* letter, exactly

---

*The same discovery was made simultaneously in Maryland by Yanhua Shih, in his Ph.D. dissertation under Carroll Alley.

how they do it," explained Horne. "What laser do you recommend I buy?" Zeilinger asked Mandel. "Where do you get these crystals? How do you cut them? How do you illuminate them?" Zeilinger had found his source, and in a decade he would be taking the world of physics by storm.

Meanwhile, for the spring and summer of 1987, Greenberger was sharing Zeilinger's office in the Atominstitut in Vienna, looking at interactions between three particles, "and they were *very* complicated.* Slowly," however, "I started getting a feeling for it." He discussed strategy with Zeilinger, got long-distance suggestions from Horne, "and every morning I'd come in and say, 'I have a better Bell's theorem than yesterday.' "

Horne was less than riveted by Greenberger's new obsession. "Did Abner ever tell you that much of the seventies and eighties was spent refereeing papers? The journals sent a *mountain* of them. I'd seen a gazillion inequalities." So when Greenberger came home "and we'd visit, say, in his kitchen or my kitchen, and he'd say, 'I have a new inequality,' I was not attentive enough, because I was through with that stuff."

One morning in Vienna, Greenberger came into the office with a furrowed brow. "Anton, I am so confused—there's nothing left. We've got such a good Bell's theorem that there's *no* freedom at all for the particles." There was no inequality needed anymore; just a single yes-or-no question.

Bell's inequality, stated roughly, is that a local hidden-variables explanation of entanglement requires certain kinds of outcomes to occur more often than they actually do. Greenberger, with the help of Zeilinger and Horne, had found that such a local, objective, separable explanation of *three*-particle entanglement requires something that never happens. In an ordinary two-particle Bell's-theorem experiment, the case against local hidden variables is built up slowly, over thousands of measurements. But in a three-particle case, a single measurement would judge against objective local realism in favor of quantum mechanics. The inequality had vanished.

"Well, that can't be right," said Zeilinger.

"We started thinking about it," remembered Greenberger, "and that's when we realized what we had done. But it had to hit us over the head."

In 1989, at a conference in Sicily in honor of Heisenberg's principle, aptly titled "62 Years of Uncertainty," David Mermin listened to Greenberger

---

*Part of the reason for this is "the three-body problem": a physicist can predict the path of any two "bodies" in orbit—e.g., the earth and the sun—but if he also considers a third body, like the moon, the paths are no longer perfectly predictable.

"making incomprehensible remarks about why it might be interesting to think about a particle decaying into a pair of particles each of which then decayed into a pair of its own." Mermin, who a few years earlier had written such a lucid description of Bell's theorem that it was praised by Feynman, had been a friend of Greenberger's for years. (They had met as "Cornell English department husbands.")

At the "62 Years of Uncertainty" conference, Greenberger started by announcing that he was going to show something that Mermin "was quite sure was impossible. Danny said he was going to show a version of Bell's theorem in which all the relevant probabilities were 1 and 0." (A probability of 1 means something will certainly happen; a probability of 0 means it will never happen.) "So I stopped listening. I decided that there had to be a mistake somewhere and I was tired anyway. So I paid no attention."

Also present was a mathematical philosopher from Britain named Michael Redhead. With two colleagues, he then wrote a sixty-page paper purporting to improve upon the mathematical rigor of the now three-year-old Greenberger-Horne-Zeilinger result, which its authors had never gotten around to publishing. A preprint of this paper had two salutary effects. First, it spurred Greenberger into publication—three pages of dot-matrix font titled "Going Beyond Bell's Theorem." Second, it annoyed Mermin into alertness. "Mathematical rigor, it seemed to me, was *irrelevant* to the argument," he said. "I mean, either Danny had a physical point or he didn't have a physical point. *All* of the Bell's-theorem arguments were really very simple from the point of view of mathematics, and very subtle from the point of view of physics." And somewhere in the sixty pages, Mermin finally realized that Greenberger, Horne, and Zeilinger really had come up with a striking simplification of Bell's theorem.

As it happens, Mermin—whose day job at Cornell is solid state physics—writes an occasional column for the magazine *Physics Today*, ubiquitous in physics departments and labs. Under the title "What's Wrong with These Elements of Reality?," his piece for June 1990 described Greenberger, Horne, and Zeilinger's "new and beautiful twist" on the EPR paradox, its spookiness more vivid than ever before.

In analogy to EPR, measuring two particles gives enough information to predict the results for the third. "In the absence of spooky actions at a distance or the metaphysical cunning of a Niels Bohr," notes Mermin, "the two faraway . . . measurements cannot 'disturb' the particle" we are about to measure. Hence EPR would accord an "element of reality" to the result for this measurement.

Now, with three particles, we are dealing with six elements of reality:

each particle's spin measured horizontally and vertically. "All six of the elements of reality have to be there," observes Mermin, "because we can predict in advance what any one of the six values will be, by measurements made so far away that they cannot disturb the particle that subsequently does indeed display the predicted value.

"This conclusion," he notes parenthetically, "is highly heretical," because, as far as quantum mechanics is concerned, two components of the spin of the same particle cannot exist at the same time—to assert that they do is a form of local hidden variables.

"Heresy or not," continues Mermin, "since the result of either measurement can be predicted with probability 1"—i.e., 100 percent of the time—"from the results of other measurements made arbitrarily far away, an open-minded person might be sorely tempted to renounce quantum theology in favor of an interpretation less hostile to the elements of reality."

As in Bell's version of EPR, however, the elements of reality do not survive "the straightforward quantum mechanical predictions for some additional experiments, entirely unencumbered by accompanying metaphysical baggage. In the GHZ [Greenberger, Horne, and Zeilinger] case the demolition is spectacularly more efficient.

"Farewell elements of reality!" trumpets Mermin. "And farewell in a hurry. The compelling hypothesis that they exist can be refuted in a single measurement."

The first thing Mermin did when he understood the Greenberger-Horne-Zeilinger argument was to write a letter to John Bell about it.

Bell wrote back: "This fills me with admiration."

34

"Against 'Measurement' "

1989–1990

AT THE "62 YEARS OF UNCERTAINTY" CONFERENCE in 1989, John Bell, born one year after uncertainty, gave the speech of his career. Mermin rated it "close to being the most spellbinding lecture I have ever heard. (The only competitors are Richard Feynman's 1965 Messenger Lectures at Cornell,"* where, in six lectures, he walked his nonscientific audience through the natural laws.) Titled "Against 'Measurement,' " it was, as it turned out, Bell's parting shot against the science he had lived with all his life. It remained, in his opinion, inexcusably vague in its formulation, its physicists inexcusably complacent about this vagueness. Published a year later in *Physics World*, "the article," Mermin comments, "conveys his brilliance and his wit, but not, of course, the music of his voice."

"Surely, after sixty-two years, we should have an exact formulation of some serious part of quantum mechanics," Bell began.

> By "exact" I do not of course mean "exactly true." I mean only that the theory should be fully formulated in mathematical terms, with nothing left to the discretion of the theoretical physicist—until workable approximations are needed in applications. By "serious" I mean that some substantial fragment of physics should be covered. Non-relativistic "particle" quantum mechanics . . . is serious enough. For it covers "a large part of physics and the whole of chemistry" [as Dirac liked to say]. I mean too, by "serious," that "apparatus" should not be separated off from the rest of the world into black boxes, as if it were not made of atoms and not ruled by quantum mechanics.
>
> The question, "should we not have an exact formulation?," is often answered by one or both of two others. I will try to reply to them: *Why bother? Why not look it up in a good book?*

*Later published as *The Character of Physical Law*.

Perhaps the most distinguished of "why bother?"ers has been Dirac.

Bell explained Dirac's system of differentiating between "second-class difficulties"—problems that, in his opinion, should be solved as soon as possible*—and "first-class difficulties" not ripe for solution at the present time.

Dirac gives at least this much comfort to those who are troubled by these questions: he sees that they exist and are difficult. Many other distinguished physicists do not. . . . When they do admit some ambiguity in the usual formulations, they are likely to insist that ordinary quantum mechanics is just fine "for all practical purposes." I agree with them about that:

ORDINARY QUANTUM MECHANICS (as far as I know) IS JUST FINE FOR ALL PRACTICAL PURPOSES.

Even when I begin by insisting on this myself, and in capital letters, it is likely to be insisted on repeatedly in the course of the discussion. So it is convenient to have an abbreviation for the last phrase:

FOR ALL PRACTICAL PURPOSES = FAPP.

Bell could imagine the impatience of these physicists who understood the theory "in their bones," but

is it not good to know what follows from what, even if it is not really necessary FAPP? Suppose for example that quantum mechanics were found to resist precise formulation. Suppose that when formulation beyond FAPP is attempted, we find an unmovable finger obstinately pointing outside the subject, to the mind of the observer, to the Hindu scriptures, to God, or even only Gravitation? Would not that be very, very interesting?

Why not look it up in a good book?

---

*A technical footnote: Among Dirac's "second-class difficulties" were the infinities of quantum electrodynamics.

But which good book? In fact it is seldom that a "no problem" person is, on reflection, willing to endorse a treatment already in the literature. Usually the good unproblematic formulation is still in the head of the person in question, who has been too busy with practical things to put it on paper. I think that this reserve, as regards the formulations already in the good books, is well founded. For the good books known to me are not much concerned with physical precision. This is clear already from their vocabulary.

Bell's list of words that "have no place in a formulation with any pretension to physical precision" included all the words that physicists use all the time that "imply an artificial division of the world," like "system," "apparatus," and "environment." "Observation" and "information" are similarly imprecise:

Einstein said that it is theory which decides what is "observable." I think he was right—"observation" is a complicated and theory-laden business. Then that notion should not appear in the formulation of fundamental theory. "Information"? Whose information? Information about what? On this list of bad words from good books, the worst of all is "measurement." It must have a section to itself.

Bell quoted several sentences from Dirac's *Quantum Mechanics* in which the concept of "measurement" appears in "the fundamental interpretive rules" of the theory. For example, Dirac writes, "A measurement always causes the system to jump into [a specific state] of the variable that is being measured."

The "system," in this context, is whatever object we are measuring—for example, a particle. The "variable" is the attribute we are measuring—for example, the position of the particle. Dirac's statement, one of the fundamental rules of quantum mechanics as described by one of its most matter-of-fact, literal practitioners, says that to measure the position of a quantum particle is not to find out where it is, but to cause it to be somewhere. This "jump" into specificity, the famously vague "collapse of the wavefunction," is related to Bohr's "quantum jump," when an electron disappears from one atomic energy level to emerge in another. Jumps are not described by the quantum theory. They merely happen, pushing us briefly off the map of what we understand.

"It would seem," Bell concluded,

that the theory is exclusively concerned about "results of measurement" and has nothing to say about anything else. What exactly qualifies some physical systems to play the role of "measurer"? Was the wavefunction of the world waiting to jump for thousands of millions of years until a single-celled living creature appeared? Or did it have to wait a little longer, for some better qualified system . . . with a Ph.D.? If the theory is to apply to anything but highly idealized laboratory operations, are we not obliged to admit that more or less "measurement-like" processes are going on more or less all the time, more or less everywhere? Do we not have jumping then all the time?

The first charge against "measurement" in the fundamental axioms of quantum mechanics is that it anchors there the shifty split [Bell was echoing Heisenberg's term for the same thing—the cut, or *Schnitt*] of the world into "system" and "apparatus." A second charge is that the word comes loaded with meaning from everyday life, meaning which is entirely inappropriate in the quantum context. When it is said that something is "measured" it is difficult not to think of the result as referring to some pre-existing property of the object in question. . . .

Even in a low-brow practical account, I think it would be good to replace the word "measurement" . . . by the word "experiment" . . . [which] is altogether less misleading. However, the idea that quantum mechanics, our most fundamental physical theory, is exclusively even about the results of experiments would remain disappointing.

In the beginning natural philosophers tried to understand the world around them. Trying to do that, they hit upon the great idea of contriving artificially simple situations in which the number of factors involved is reduced to a minimum. Divide and conquer. Experimental science was born. But experiment is a tool. The aim remains: to understand the world. To restrict quantum mechanics to be exclusively about piddling laboratory operations is to betray the great enterprise.

But how is measurement to be squared with the equations of quantum mechanics, which are famously silent on the subject? Bell dived into three "good books" for the answer, starting with the famous and hallowed *Quantum Mechanics* of Lev Landau, written with his longtime collaborator, Evgenii Lifshitz.

I can offer three reasons for this choice:

(i) It is indeed a good book.
(ii) It has a very good pedigree. Landau sat at the feet of Bohr. Bohr himself never wrote a systematic account of the theory. Perhaps that of Landau and Lifshitz is the nearest to Bohr that we have.
(iii) It is the only book on the subject in which I have read every word.

This last came about because my friend John Sykes enlisted me as a technical assistant when he did the English translation. My recommendation of this book has nothing to do with the fact that one per cent of what you pay for it comes to me.

Landau and Lifshitz agree with Dirac that "a measurement always causes the system to jump" into a specific state. Moreover, the possibility of describing the attributes of an atom "requires the presence . . . of physical objects which obey classical mechanics." This is not merely because the atom is too small to see otherwise, but because the atom actually has no specific attributes unless it interacts with something else that has. The classical apparatus briefly, and without explanation, loans to the quantum atom its decisiveness, describability, and specificity, and we call this a "measurement."

But the apparatus itself, a machine or computer, is made of indecisive, quantum atoms. How does their fogginess add up to the normal "classical" behavior of the machine? The question is unaddressed in Landau's textbook, beyond the statement that the relationship of quantum theory to the outdated, yet necessary, classical physics is "very unusual among physical theories."

Landau's "vaguely defined wavefunction collapse, when used with good taste and discretion, is adequate FAPP," Bell continued. "It remains that the theory is ambiguous in principle, about exactly when and exactly how the collapse occurs, about what is macroscopic and what is microscopic, what quantum and what classical. We are allowed to ask: is such ambiguity dictated by experimental facts? Or could theoretical physicists do better if they tried harder?"

He turned to another *Quantum Mechanics,* this one written by an old CERN friend, Kurt Gottfried.

Again I can give three reasons for this choice:

> (i) It is indeed a good book. The CERN library had four copies. Two have been stolen—already a good sign. The two that remain are falling apart from much use.
> (ii) It has a very good pedigree. Kurt Gottfried was inspired by the treatments of Dirac and Pauli. His personal teachers were . . . Julian Schwinger, Viktor Weisskopf . . . [et al.]
> (iii) I have read some of it more than once.

This last came about as follows. I have often had the pleasure of discussing these things with Viki Weisskopf [who arrived at CERN as its director at the same time as Bell did]. Always he would end up with "you should read Kurt Gottfried." Always I would say "I have read Kurt Gottfried." But Viki would always say again next time "you should read Kurt Gottfried." So finally I read again some parts of KG, and again, and again, and again. . . .

A particle is mathematically described by a wavefunction. This wavefunction is almost always a summation of many apparently incompatible states, all simultaneously true, and interfering with each other in a way that boggles common sense. Gottfried explains in his book how this is to be understood. "Intuitively un-interpretable" mathematical terms involving interference between macroscopic states must be removed, he writes, or the theory will be "an empty mathematical formalism." This new summation, created when the weirdest terms are removed, is then interpreted as a simple list of options. That is, one would look at mystifying statements like "The cat is both dead and alive" or "The particle is both here and there," and read them as understandable ones: "The cat is either dead or alive" and "The particle is either here or there."

"But this," declared a skeptical Bell, "suggests that the original theory, 'an empty mathematical formalism,' is not just being approximated—but discarded and replaced."

After disposing of one more "good book" in similar form, Bell finished by describing two attempts to bring us "towards a precise quantum mechanics." Dispensing with the shifty split, these two theories can handle "system" and "apparatus" alike. The first, the de Broglie–Bohm nonlocal-hidden-variables theory, posits a world of real particles that have real positions and momenta regardless of whether or not they are observed, and are guided by the mysterious, spooky pilot wave.

In the second, the Ghirardi-Rimini-Weber theory, the collapse of the wavefunction is turned from metaphysical puzzlement to mathematical precision. GianCarlo Ghirardi and his colleagues doctor the Schrödinger equation so that it will "collapse" into specificity of its own accord. For a few particles, the effect is small, but this allows the theory to handle larger things better than ordinary quantum mechanics does: "pointers very rapidly point," said Bell, "and cats are very quickly killed or spared."

Bell staved off his audience's automatic dismissal of these theories by closing with Feynman: "We do not know where we are stupid until we stick our necks out."

At least two physicists left "62 Years of Uncertainty" with the hearty sound of "Don't be a sissy!" ringing in their ears. Hearing these words over the crowd noise of a conference reception, Mermin looked over to see "John Bell, encouraging a younger physicist not to let the scope of his speculative investigations be overly constrained by the wisdom of his elders."

The next June, in the green and welcoming hills of Western Massachusetts, two Amherst college professors, George Greenstein and Arthur Zajonc (pronounced Z*y-ons*), brought a dozen quantum physicists interested in the foundations of the subject together to live in a fraternity house for a week (with a professional chef). Greenstein, the son of a physicist, was a gentle, white-bearded astronomer with an unusual gift for translating science for the layman; Zajonc was a contemplative physicist with an interest in the big picture. The attendees included Bell, Greenberger, Horne, Zeilinger, Shimony, and Mermin. Since the best part of any conference is always the conversation over coffee or beer, the chance meeting in the hall, the argument over dinner, Greenstein and Zajonc decided to organize their conference to be nothing but these events. "No prepared talks, no schedule, no proceedings," remembered Mermin, "just wonderful conversations."

During that idyllic week, many details of the "proper" Greenberger-Horne-Zeilinger paper were hammered out—this time with all three present, plus the help of Shimony as co-author and the critical input of Mermin, who had just sent his "What's Wrong with These Elements of Reality?" to the editor. Halfway through the conference, Zajonc arrived at lunch with the new *Physics Today* in his hand; Mermin's article had made its dramatic entrance, surprising physicists across the country with the drama of three-particle entanglement. (When Horne arrived home, there was a postcard from Clauser, who did not know anything about the GHZ paper until he read *Physics Today*, even whether it was in preprint or reprint stage. "You old fox!," he wrote. "Send me a (p)reprint of GHZ. Mermin seems to think this is super-hot stuff.")

In Amherst, Kurt Gottfried got his chance to respond to Bell's "wonderfully barbed attack" on his book in "Against 'Measurement.' " "I was delighted that the most profound student of quantum mechanics since the Founding Fathers, and an old friend from CERN, had paid close attention to what I had written." But he found "John's critique of orthodoxy to be rather less overwhelming than his superb rhetoric."

Thus started the central debate of the conference, in which, as Mermin remembered, "Bell claimed that in some deep way quantum mechanics lacked the naturalness that all classical theories possessed." There was no problem with the interpretation of *classical* physics. For example, as Gottfried granted Bell, "Einstein's equations tell you" their own interpretation. "You do not need him whispering in your ear." In quantum mechanics, on the other hand, Gottfried admitted, even the greatest classical physicist "would need help: 'Oh, I forgot to tell you that according to Rabbi Born, a great thinker in the yeshiva that flourished in Göttingen in the early part of the 20th century, [the amplitude-squared of the Schrödinger equation] is' " to be interpreted as a *probability,* despite all indications to the contrary. Bell felt it was obvious that something profound was missing from quantum mechanics; "Kurt," Mermin said, "never felt this in his bones." Shimony remembered that "Bell pressed Gottfried severely about confusing FAPP with a full theoretical solution," while Viki Weisskopf, the two men's old friend from CERN, acted as moderator. Gottfried's answers met neither Bell's standards nor his own, starting him on a decade of honing his thoughts on the subject that would culminate in an entirely rewritten textbook, which actually begins with the Einstein-Podolsky-Rosen paradox.

"Of all the people who played in this area, he seemed to *give* the least," Horne commented about Bell in Amherst. "He was totally dissatisfied with quantum mechanics, and very strongly committed to an objectively real view of the world." Greenstein remembered, "It was a really knockdown, drag-out fight. And it *mattered,* to everybody, what they were arguing about." He was in particular struck by Bell's "burning intensity and seriousness in the midst of his very gentle demeanor. An almost otherworldly, very different kind of person." Mermin and Gottfried compared Bell's wrath against superficiality to that of an Old Testament prophet.

Late one afternoon, as the conference spread out for a picnic, Shimony "propounded on the rules and strategy of croquet," remembered Greenstein, "with the same thoroughness he has brought to the study of quantum mechanics," and Mermin and Weisskopf played piano four hands. But Greenstein noticed the combatants Bell and Gottfried standing apart. With concern he drew near, only to see them companionably comparing their cameras.

Mermin and Gottfried, on the long drive back to Cornell, "agreed that John was truly unique in the world of physics, as a personality and as an intellect—at once scientist, philosopher, and humanist. He was a person to whom deep ideas mattered deeply. . . . He belonged, also, to that small company of physicists whom either of us would walk miles to hear lecture on any topic whatsoever."

Three months later, on October 1, 1990, Bell died of a stroke in his kitchen in Geneva, so suddenly and unexpectedly that Bertlmann couldn't even come to the funeral. He was only sixty-two, never having given anyone except Mary a hint of the terrible migraine headaches that he had suffered from all his life.

Over the years between 1964 and 1990, Bell's theorem had gone from an unknown result to a respected one—albeit one with roots in an unloved quantum subculture. Unlike Clauser and Freedman's 1972 experiment or Fry's 1976 one, Aspect's in the early eighties had achieved some fame. About entanglement over long distances, it was felt that all had been said and done that needed to be said and done. But a decade later, at the turn of the millennium, quantum optics labs in Innsbruck, Geneva, and Los Alamos (to name a few) would be full of new excitement about the possibilities of using entanglement in increasingly dramatic, long-distance, and potentially useful ways. For those who now work in these newly wide-open fields, it is a frustrating and poignant irony that, if only Bell could have lived out his biblically allotted three-score years and ten, he would have seen such wonderful things begun by his half-secret hobby.

Just before he died, however, Bell took a trip to Oxford. He met there the postdoctoral student whose idea about entanglement—though neither one knew it at the time—would set in motion the sea change.

35

Are You Telling Me This Could Be Practical?

1989–1991

Artur Ekert

IN THE LIBRARY of Oxford's Clarendon Lab sat Artur Ekert, a postdoc who might have been mistaken for a young Yul Brynner (the famous shiny pate would come later), reading EPR for the first time. Born in Poland of a family from all over Europe, Ekert had known where he wanted to go to college ever since a friend had given him a few papers written in 1985 by Oxford's most brilliant and reclusive physicist, David Deutsch.

"At the time, there were probably two or three people in the world who were interested in quantum computing," Ekert remembered. "David was one of them. And he wrote those few papers which *no one* actually really read seriously at the time." A quantum process could simulate any classical computation, Paul Benioff showed in 1980. A year later, Feynman famously threw down the gauntlet to computer scientists in his "Simulating Physics with Computers" speech—Bell's inequality shows that only a quantum computer would be able to fully imitate Nature, because only a quantum computer can capture entanglement.

Though the young Feynman had dramatically speeded up the computing in the Manhattan Project (done on cumbersome multipliers and sorters) by using computers working in parallel, it was Deutsch who took the third step toward the quantum computer, describing the undreamed-of computing speed inherent in what he called quantum parallelism (the superposition principle of wave mechanics, in which two waves added together produce a new wave), where many possibilities can be simultaneously explored.

Deutsch was a wispy and fragile-looking man who was awake only at night, rarely left Oxford, and felt that quantum mechanics and, in particular, the quantum computer prove that the universe is constantly splitting into many different worlds. Quantum parallelism meant, to Deutsch, the power of a machine that could compute in all parallel worlds simultaneously. He became an adviser and friend to the sunny, down-to-earth Ekert.

Ekert's hobby was cryptography. "I just liked public-key cryptosystems—I was fascinated by them." In the 1970s, "a lovely piece of number theory" developed by Pierre de Fermat in the seventeenth century had been co-opted by the British Secret Service for cryptography—"a *big* surprise," said Ekert, laughing, "to many mathematicians who thought the field was so pure it would never be tainted by any practical application." The system, called RSA after the three civilians at M.I.T. who rediscovered it (a few years after the classified version), elegantly solves the problem of the eavesdropper.

An unbreakable code is, after all, only as secure as its key—and the key has to be distributed, by means of a courier or over some line of communication, never fully safe from eavesdropping. The idea behind RSA is that the key does not need to be top secret if the mathematical function (part of an equation) that encodes the message is very hard to reverse. For example, it is easy to multiply two numbers together—these two numbers are called "factors" of the bigger number they produce. It is much harder to start with only the bigger number and work backward to find out what the factors were.

A slightly more complicated hard-to-reverse function was provided by Fermat's discovery of a tidy relationship between prime numbers and modular arithmetic. (Modular arithmetic is essentially a form of division where the important part of the answer is not the number of times one number goes into another, but the *remainder:* 11 modulo 2 = 1, because 2 goes into 11 five times with a remainder of 1; alternatively, 11 modulo 5 also equals 1, because of the same argument.)

In RSA the sender, "Alice," encodes the message to "Bob" in a series of long numbers. Taking one of these at a time, she raises it to the power of an encrypting exponent, then divides it by a very large number created by multiplying two prime numbers. Neither the exponent nor the very large number needs to be secret. The *remainder* of this division is the code. Given the code and the public key, it is almost impossible to recover the plaintext by factoring the product of the two primes, even with a herd of computers, particularly if the very large number is on the order of two hundred digits. Amazingly enough, however, if that code is raised to the power of the decrypting exponent (the private key that only Bob knows), and divided by the same very large number, its remainder is the plaintext. The problem of the eavesdropper can be staved off.

But, though hard, factoring large numbers into primes is not physically impossible; it just takes a hugely long time using existing computers. If any malicious eavesdropper ever worked out a faster way to factor large numbers, the RSA code would fall apart.

So it happened that Ekert, in the quiet of Clarendon's library, sat back in his chair in shock when he read for the first time the famous EPR definition of "an element of reality": *if, without in any way disturbing a system, we can predict with certainty the value of a physical quantity, then there exists an element of physical reality corresponding to this physical quantity.* "It just clicked in my brain: wow—this is about eavesdropping!" An eavesdropper just wants to learn the value of the code "without in any way disturbing the system." If the code shows traces of prying, then Alice and Bob won't use it, and the eavesdropper's work would be in vain.

"Local realism," Ekert realized, "allows you somehow to incorporate the definition of perfect eavesdropping into the formulation." His mind was whirling: "Ha!—but I *know* it was refuted!" An entangled code would be an un-eavesdroppable code. He leaped up and began to search the Clarendon Library for Bell's 1964 paper.

What if Alice and Bob shared a long series of entangled photons?

"If the Clauser-Horne-Shimony-Holt inequalities are violated" (meaning that Alice and Bob's photons remain entangled), Ekert began to realize, "then there's no chance that the eavesdropper touched the particles on the way." Any prying would destroy the entanglement. The inequality is "like a signature that the particles were not touched."

"I was very happy about it," Ekert remembered, but then he found that, aside from the intellectually fearless Deutsch, "hardly *anyone* wanted to talk about this; people were just very dismissive about the whole thing."

Then Bell himself arrived to give a speech at Oxford. After the speech, Ekert, overflowing with excitement, came up to Bell and explained his idea, in his likable accent (a combination of several different European languages, including the Queen's English).

Bell stared at the elated young graduate student. He had always firmly discouraged students from delving into the foundations of quantum mechanics; a subject so unfashionable and esoteric could only make them unemployable. He had never expected to hear anything like what Ekert was saying. "Are you telling me that this could be of practical use?" Bell asked.

"I said, 'Yes, I think it can.'

"And he said, 'Well, it's *unbelievable.*' "

They did not have long to talk. "I was just a student and there were many people who wanted to whisk him away," said Ekert, but Bell left Oxford knowing that a new chapter in the history of entanglement was about to begin.

36

The Turn of the Millennium

1997–2002

Nicolas Gisin

I F YOU HAD LIVED high above Lake Geneva as one millennium turned into another and you scanned the city and its surroundings once in a while with your high-powered Swiss binoculars, you might have noticed some interesting activities of certain physicists—on bicycles, in Fiats, making their way to Swisscom telephone central stations. This phenomenon was not entanglement, exactly, but it was definitely *correlation*, of the kind that might be achieved with a few mobile phones. Up there in your castle, you might have wondered what the source of this suspicious correlation might be, and why physicists would want access to the phone lines of the entire city of Geneva.

Other odd things had been happening in other places. In Innsbruck, there was a mysterious, fully coded transmission of a photograph of the divinely spherical *Venus* of Willendorf, the world's first-known sculpture. And in Los Alamos, physicists using an equation that they referred to as "the inequality" repeatedly caught an amorphous and pernicious eavesdropper known only by her semimythological code name, Eve.

Obviously something strange was afoot.

In the second half of the nineties, Anton Zeilinger had emerged as the world leader in bringing entangled theory into experimental life. Down-conversion—where a crystal splits high-energy laser photons into a halo of lower-energy entangled photons—made it possible. No longer a career killer, these investigations into entanglement now drew brilliant postdocs and graduate students from all over the world to his lab in Innsbruck, in the Austrian Alps. "It was easy in the early nineties to make a list of *great* things that could be done, now that there was such a convenient source of entangled pairs," explained Horne, characteristically downplaying his own role. "Anton's claim to fame is that he went and *did* them." Reliably, each year news of one or even two landmark explorations of entanglement would emerge from Zeilinger's Bauhaus barracks in the midst of the snowy mountains.

Quantum teleportation, which Zeilinger and his team achieved in Innsbruck in 1997, is slightly subtler than its spectacular name. Alice in Wonderland has a photon whose state she wants to send to Bob in Milwaukee. She entangles that photon with her member of an already entangled photon pair. This causes Bob's member of the (previously entangled) pair to fall into the desired state. The coyness of quantum mechanics allows this perfect result to happen only about a quarter of the time, but the rest of the time the photon is a trivial flip or phase shift away from perfection. (Flipping the photon or shifting its phase are operations Bob can easily perform. As an analogy, imagine Alice has sent an unmarked paper cutout to Bob through the mail. When Bob opens the envelope, he does not know if he is looking at it upside down or back-ward.) To complete the otherwise spooky process of quantum teleporta-tion, Alice will have to tell him whether or not he should flip the result to have a photon identical to the one she started with. Without "classical communication," e.g., a phone call, nothing is gained.

In 1998, entanglement-swapping (or teleportation of entanglement, as Zeilinger sometimes calls it) was the next piece of theoretical magic that Zeilinger and his team incarnated. This time Alice in Wonderland has two entangled photons; so does Bob in Milwaukee. If they meet in Innsbruck, each carrying only one of their photons, and entangle them together, these newly entangled photons lose all memory of their erst-while partner-photons, and vice versa. Meanwhile, the two forgotten photons back in Wonderland and Milwaukee are now entangled with each other, even though they have never met.

In 2000, Zeilinger sent an encrypted photograph (over optical fibers from "Alice," in this case a computer in one building, to "Bob," a computer a few buildings away) of the "Venus" of Willendorf. The encrypted photograph was composed of random-looking speckles in garish hues, but after "Bob" decrypted this he was left with an almost perfect image of the little round fertility goddess, in shades of clay brown against a black background.

Meanwhile, among the red rock and pine trees of the Sangre de Cristo Mountains of New Mexico, Paul Kwiat, an endearingly birdlike man in glasses and suspenders with boundless energy and encyclopedic knowledge, led his team in attempting various eavesdropping strategies on their Alice and Bob. The Clauser-Horne-Shimony-Holt inequality did indeed reliably signal the presence of the wily "Eve," just as Ekert had imagined back in 1991.

" 'I am a quantum engineer, but on Sundays I have principles,' " Nicolas Gisin began his speech. It was a decade after Bell's death, and the occasion was a celebration of Bell in the form of a conference in Vienna organized by Bertlmann and Zeilinger, who had just moved there from Innsbruck.

"That is how John Bell opened his 'underground colloquium' in March 1983," Gisin explained, smiling, in his Swiss French accent:

> words which I will never forget! John Bell, the great John Bell, presented himself as an engineer—one of those people who make things work without even understanding how they function—whereas I thought of John Bell as one of the greatest *theoreticians*.
>
> In March 1983 the Association Vaudoise des Chercheurs en Physique organized their annual one-week course in Montana—an excellent combination of ski and physics—on the foundations of quantum mechanics. For one of these reasons peculiar to the community of people interested solely in these foundations, John Bell was invited, but without any time slot for a presentation. With some friends we, i.e., the Ph.D. students, managed to convince him to give us an evening lecture, after dinner, while the professors enjoyed the local wine. . . .
>
> The talk took place in the basement, the ceiling was too low and the students sat on the ground: a perfect underground atmosphere. When John Bell started you could "hear the silence": *I am a quantum engineer, but on Sundays I have principles.*

With gray hair and a gaunt, intelligent face, Gisin looked like a wizard transplanted into the real world and dressed in a mufti. He had started his career in a way that Bell would have cautioned him against: with degrees in physics and in mathematics, he wrote his Ph.D. thesis in 1981 on an alternative version of Schrödinger's equation that did not require an observer or apparatus to "collapse" the wavefunction. (For this, the de Broglie Foundation, whose stated mission is to encourage the questioning of quantum dogma, gave him a prize.) With a young family to support, Gisin went to work for a telecom start-up. Shimony had advised him that an optical career could bridge the industry and the academy. Gisin continued to publish his heretical papers on the side, while enjoying the dawn of the fiber-optic revolution in telecommunications.

By 1988, the erstwhile mathematical quantum philosopher was also an expert on the hardware-software interface of fiber optics, and the University of Geneva asked him to head the optics section of its Applied Physics group, essentially a research arm of the Swiss telephone company. Gisin transformed this lab into a cutting-edge center of quantum cryptography. His goal was to bring quantum cryptography out of the lab, and by 1997, he and his group were demonstrating entanglement between photons separated by six miles as the crow flies. The entangled photons started in the Swisscom central station in downtown Geneva, and traveled down a total of nearly nine and a half miles of the commercial phone lines before meeting either "Alice," in a little station southwest of the city in Bernex, or "Bob," north along the lake in a village called Bellevue. The two sides of Gisin's life had come together: the quantum philosophy and the telecom engineering, the principle and the practical.

In these experiments, classical communication had an even greater role than usual: "When you do this experiment, first you have to make sure—is the guy there?" explained Gisin with a smile. "Maybe he's going there with his bicycle—is he late, did he go out for a coffee? These kinds of trivialities have also to be organized and just get a little more complicated than when everyone is in one lab. But it's fun also."

Codes, too, played a nonquantum role, as the only way to enter the locked telecom central station. "Then you have thirty seconds to give your name and all that to the police" over the intercom, Gisin explained. "If you fail to do so, or go out to smoke a cigarette, and come in without respecting these rules, the police is there in a few minutes. That makes sense because once you are in there, you have access in these fibers to all the data—the boring data, and . . ." he paused, and then said, "the other data."

In less than two decades, he had watched fiber optics go from a lab-based toy to a web of glass and light spanning the globe, and Gisin expected to see entanglement follow a similar path, now that it had sloughed off the cloak of vagueness and negativity it had worn for so long.

"Quantum mechanics [has] now existed for about eighty-five years," said Gisin, "and has mainly been considered as a theory of paradoxes, of mathematics, and strange counterintuitive ideas. So it was actually looked at from a quantitatively negative point of view—with rules like, *you cannot measure this and that simultaneously; you should not draw pictures of elementary quantum processes; you cannot clone a photon* . . . all these are negative rules."

Gisin saw entanglement-based quantum cryptography as the turning point. "In 1991, Artur Ekert's discovery . . . changed the physicist's world: entanglement and quantum nonlocality became respectable." What was negative became positive, ushering in "a kind of psychological revolution among physicists (and actually only, of course, a fraction of the physicists—young ones mainly, also computer scientists) who began to realize that quantum mechanics being so *different* from classical physics also opens up the possibility for doing something that is radically *new*.

"I think this is the really important thing, this psychological revolution. It is changing the view that physicists have about quantum physics in general."

"I am a quantum engineer, but on Sundays I have principles." That statement had given Gisin, as a new Ph.D., a wider sense of the possibilities of his career. His Vienna speech dedicated to Bell, titled "Sundays in a Quantum Engineer's Life," was a portrait of an experimentalist taking a weekly Sabbath to think about the big picture, which he described in the form of conversations between a physicist and an engineer. These archetypes symbolized not only Gisin and his team, but also the unconventional Swiss cognitive scientist, philosopher, and physicist Antoine Suarez, who after meeting John Bell in 1988 had become possessed with the idea of a dramatic experiment to "investigate in depth the tension between quantum mechanics and relativity."

Relativity is adamantly a local theory about separable, real objects, while entanglement seems to deny that these attributes can actually all coexist in nature. Yet no one has been able to catch the two in an out-and-out contradiction. To speak of the locality (or lack thereof) of quantum mechanics is to speak of this still unplumbed and mysterious relationship.

On a really nice Sunday, in Gisin's story, "the physicist thought about testing the speed of what Einstein called the spooky action at a distance!"—the nonlocal "collapse of the wavefunction," where the vast and vaporous quantum wave, upon measurement, becomes a minuscule particle in a specific place. In entanglement, it seems as if the measurement of one particle collapses the wavefunction of the other. If cause must precede effect, and if Alice's measurement of her entangled particle affects Bob's entangled particle at a distance, then the correlation should disappear if they measure their particles at exactly the same time.

But the theory of relativity makes *at exactly the same time* an extremely loaded phrase. One of Einstein's great insights, of course, was that time is merely "what we measure with a clock" and not absolute, godlike, and unalterable, as we tend to think of it. Further, clocks in motion tick more slowly than clocks standing still. The faster a clock moves, the slower it ticks, until, at the speed of light, time stands still. Depending on an observer's state of motion, and thus how quickly his clock was ticking, he might report events as occurring in a different order than would another observer "standing still."*

Physicists refer to observers in different states of motion as being in different "reference frames": the classic example compares "the moving-train reference frame" with "the platform reference frame." And the concept of simultaneity, two things occurring "at exactly the same time," has meaning only for observers in the same reference frame.

Gisin continued:

> The engineer likes the challenge of aligning his system such that both measurements take place "really at the same time." This is far from obvious, knowing that Alice [in Bernex] and Bob [in Bellevue] are connected by almost 20 km of optical fibers over a straight line distance of more than 10 km!
>
> But, the engineer has heard of relativity and asks: "In which reference frame should I align the experiment?"
>
> "Well, hum, I do not really know!" admits the physicist. "Let's try the most obvious choices: the reference frame in which the Swiss Alps are at rest! And also the reference frame of the cosmic background radiation (center of mass of the Universe)!" . . .
>
> "Okay," says the engineer, but "what exactly should be aligned?

---

*But recall that "standing still" actually means moving pretty quickly as the earth rolls around the sun.

The beam splitters?* The detectors? The computers? The observers?"

The physicist is amazed. Now that he dared to consider the assumption that the collapse is real, so many questions arise, and each hypothetical answer can, in principle, be tested!

The following Sunday, our physicist goes for a walk with his friend David Bohm. . . . They spoke about the engineer's question: What should be aligned? Clearly, it should be the device that triggers the collapse. But what could one reasonably assume as the trigger? After a few minutes of silence, the physicist states: "It must be the detector! That's where the irreversible event happens!"

"Possibly," replies Bohm, "but I bet it is the beam splitters." (Indeed, in Bohm's pilot wave model, the irreversible choice is made at the beam splitters; in this model the detectors merely reveal this choice). . . .

This experiment has really been performed in 1999 in Geneva. The referee reports are interesting, ranging from fascination to desperation. . . . No doubt that for many physicists the collapse is taboo (but certainly not for John Bell). . . .

"The experimental result established impressive lower bounds on the speed of the spooky action," thousands of times the speed of light, Gisin noted. He reminded his listeners that, after all, "for a long time the speed of sound was the fastest measurable speed, while light was assumed to be instantaneously everywhere."

There comes yet another sunny Sunday. The physicist rests in his armchair and thinks: *All this is quite exciting! But what if we let relativity enter the game even deeper? What if the detectors are in relative motion such that each detector in its own reference frame analyzes its photon before the other? It would seem then that each photon-detector pair must make [its] choice before the other!*

"This renews the tension between quantum physics and relativity," says the physicist to himself. . . . Abner Shimony, a good friend of John Bell, termed this tension as "peaceful coexistence," because the tension does not lead to any testable conflict.

"However," continues loudly the physicist although he is alone,

---

*The beam splitters send the entangled photons in opposite directions (in this case, one to Bernex and one to Bellevue).

"once one assumes that the collapse is a real phenomenon, and once one considers specific models, then the conflict is real and is testable." . . . If both measurements happen before the other, then the quantum correlation should disappear, however large the speed of the spooky action! Very excited, the physicist starts to evaluate some orders of magnitude.

This was Suarez's dream experiment, for which he had been fundraising and seeking an experimentalist since 1992. His search took him to Innsbruck to talk with Zeilinger and Kwiat (who was there at the time), but they did not have the long-distance capabilities necessary for such an experiment. Finally, in 1997, he was introduced to Gisin, with his "lab" encompassing all the fiber-optic phone lines of Geneva, and the experiment began.

Suarez realized that about 112 miles per hour would be a high enough speed to see the effect. " 'A Ferrari can do it!' " exclaimed Gisin, recounting the train of thought in his speech. " 'So, my dear engineer, let's perform the experiment!'

" 'But,' complains the engineer, 'do I really need to activate this bloody moving detector! This requires liquid nitrogen and is a real mess!' "

Once the engineering was made feasible, "this experiment was also performed in Geneva, in the spring of 1999," reported Gisin. "The two-photon interferences were still visible, independently of the relative velocity between Alice and Bob's reference frames." Alice, in her reference frame, measures her photon first; from Bob's point of view, he has measured his photon first; yet the correlation is still present.

"What would John Bell have thought of this?"

Later, Gisin explained privately: "Indeed, quantum mechanics just predicts that these correlations are there, but it doesn't really describe how they come about. If you have a natural picture that one happens first and influences the other (which is the kind of vague idea that I guess many physicists have, just to be comfortable with this correlation), this kind of naïve explanation simply doesn't hold in the face of our experimental results.

"So you have to abandon this kind of mental picture, but then— which picture should you have? I don't have a good picture, I have to say. I find these results quite puzzling and I'm not very comfortable with the idea that we should just give up any picture at all. Most physicists will always have some kind of mental picture to complement the mathematical tools." Gisin's response to puzzlement was to plan further experiments, to explore every nook and cranny of entanglement that he is able.

"The main lesson I learned from John Bell led me, on the one side, to become a quantum engineer, and, on the other side, not to forget about the principles," he concluded his speech. "There is nothing wrong with metaphysical assumptions, but the good ones are those that can be tested."

37

A Mystery, Perhaps

1981–2006

THE IDEA OF QUANTUM COMPUTING met the idea of entanglement almost immediately, back in 1981, at an M.I.T. conference on "The Physics of Computation." Feynman gave the keynote address to the assembled computer scientists.

"Might I say immediately, so that you know where I really intend to go, that we always have had—" Feynman stopped, and looked around with an expression of joking paranoia: "(secret, secret, close the doors!)—we always have had a great deal of difficulty in understanding the world-view that quantum mechanics represents." His expression was frank. "At least I do, because I'm an old enough man that I haven't got to the point that this stuff is obvious to me. Okay, I still get nervous with it. . . ." Feynman was famous for iconoclasm, but this was still not what the audience was expecting. Quantum theorists had rarely been so candid on this subject since the deaths of Einstein and Bohr.

"You know how it always is, every new idea, it takes a generation or two until it becomes obvious that there's no real problem. It has not yet become obvious to me that there's no real problem. I cannot define the real problem, therefore I suspect that there's no real problem, but I'm not *sure* there's no real problem. So that's why I like to investigate things. Can I learn anything, from asking this question about computers, about this *may-* or *may-not-*be mystery as to what the worldview of quantum mechanics is?"

Well, could a classical computer simulate a quantum system? "If . . . there's no hocus-pocus, the answer is certainly 'No!' This is called the hidden-variable problem." He proceeded to walk the audience through his version of Bell's theorem. "I've entertained myself always by squeezing the difficulty of quantum mechanics into a smaller and smaller place, so as to get more and more worried about this particular item. It seems to be almost ridiculous that you can squeeze it to a numerical question that one thing is bigger than another"—a Bell inequality.

"What I'm trying to do," he told the computer scientists, "is get you people . . . to digest as well as possible the *real answers* of quantum mechanics, and see if you can't invent a different point of view than the physicists have had to invent to describe this.

"In fact"—Feynman looked up at his audience, his brow wrinkling—"in fact, the physicists have no good point of view." He grinned at the surprised faces in the crowd. ". . . So, I would like to see if there's some other way out. . . . I don't know"—he shrugged—"maybe physics is absolutely okay the way it is."

He paused; he wanted to know for sure, not maybe. He had been discussing the subject with the organizer of this conference, the great computer scientist Ed Fredkin. "The program that Fredkin is always pushing, about trying to find a computer simulation of physics, seems to me to be an excellent program to follow out. He and I have had wonderful, intense, and *interminable* arguments"—another grin—"and *my* argument is always that the real use of it would be . . . the challenge of *explaining* quantum mechanical phenomena. . . .

"And I'm not happy," finished Feynman in a rush of words, "with all the analyses that go with just the classical theory, because Nature isn't classical, dammit! And if you want to make a simulation of Nature, you'd better make it quantum-mechanical, and, by *golly*, it's a wonderful problem because it doesn't look so easy."

In 1993, while a postdoc at Los Alamos, Seth Lloyd, the ponytailed, mountaineering scion of a western Massachusetts family of madrigal singers, doctors, chamber musicians, and university presidents, proposed a breakthrough design for a quantum computer, one that could actually be built with current technology. At the turn of the millennium, Lloyd's once and future colleague Isaac Chuang, himself also a rock climber and violinist, was one of the people who actually managed to perform a quantum computation using the molecules in a vial of liquid as one of these Lloyd machines. (The molecules factored the number 15 into 3 and 5.)

Two men from Bell Labs with large mustaches and unusually fine writing styles, Peter Shor and Lov Grover, found algorithms for a quantum computer to run that are far, far faster than any classical computer could ever be—Shor's, in 1994, for factoring large numbers, and Grover's, in 1996, for searching a database. "If computers that you build are quantum," Shor explained (in a *Science News* poetry contest), "Then spies of all factions will want 'em. / Our codes will all fail, and they'll read our email, / Till we've crypto that's quantum, and daunt 'em."

All our banks, all the Internet, all of a hundred daily operations of little or huge importance depend on modern cryptography, mostly RSA. Deutsch and Ekert explain: "Any RSA-encrypted message recorded today will become readable moments after the first quantum factorization engine is switched on. . . . Confidence in the slowness of technological process is all that the security of the RSA system now rests on. . . .

"No one can even conceive of how to factorize, say, a thousand-digit number by classical means; the computation would take many times longer than the estimated age of the universe. In contrast, quantum computers could factorize thousand-digit numbers in a fraction of a second."

If, as the great teacher John Wheeler put it, *it* comes from *bit*—if things (its) are made of information (bits)—then, as the universe stores and processes that data, it is computing. In Lloyd's book, *Programming the Universe,* he reminds his readers, "The universe is a quantum system, and almost all of its pieces are entangled." So, he explains, if it's a computer, it's a *quantum* one. Uncertainty provides the seeds from which new detail and structure emerge, and through entanglement, "quantum mechanics, unlike classical mechanics, can create information out of nothing."

This pervasive entanglement is both the blessing and the bane of quantum-computer scientists. If the quantum computer entangles with its environment, it will produce random results (that is, results that are correlated with things the computer programmer doesn't want and can't control). Quantum error-correcting involves fighting unwanted entanglement with more entanglement: a shell of entangled particles that protects the important computing particles inside.

"When we started working with quantum computing seriously, we learned a lot about the nature of entanglement," explained Ekert in 2005. "In the last six years we learned much more about the whole entanglement business, about the structure of entanglement, than we learned in the previous seventy. So there's been *enormous* progress, mostly mathematical, in trying to quantify entanglement, in trying to find ways of measuring it, finding ways of detecting entanglement."

When people ask, "What's entanglement good for?" Ekert tells a story. James Clerk Maxwell, the greatest theorist of the nineteenth century (who spent his career at Cambridge University, where Ekert now works), was once asked, "What's electricity good for?" Ekert repeats Maxwell's reply: "Well, *I* do not know but I'm pretty sure that Her Majesty's government will tax it soon."

Michael Nielsen, the coauthor with Isaac Chuang of the first graduate-level textbook on the subject (*Quantum Computation and Quantum Information*), explains, "Quantum information science has revealed that entanglement is a quantifiable physical resource, like energy, that enables information-processing tasks: some systems have a little entanglement; others have a lot. The more entanglement is available, the better suited a system is to quantum information processing." He remarks on the parallel between now and the eighteenth century, when steam engines and other new machines spurred a deeper understanding of energy, and helped lead to the laws of thermodynamics that govern it. Now the quantum computer spurs us on to a deeper understanding of another resource, entanglement, and will, we hope, lead us to insights into its currently inexplicable ways.

Those of us unburdened by physics graduate degrees are always wanting to know the answers to questions that have no answers—hence Bell's ironic name for his book, *Speakable and Unspeakable in Quantum Mechanics*. Consider the question of Rutherford's friend, who had just read of his lectures: "After such reading my mind always fastens on the Proton—I wonder what *you* have in your head—in your mind's eye—when you think of it. I don't know whether anyone has invented any guts for it." Now, years later, there is a quick, superficial, and correct answer to that question: the "guts" of a proton or neutron are quarks. But Rutherford's friend's question remains essentially unanswered. Is there any way to understand what a quark is? These things are not particles, they are not waves; in some ways they are utterly deterministic, in others, completely random; they always confound instinctive understanding.

"All these fifty years of conscious brooding have brought me no nearer to the question, 'What are light quanta?' Nowadays every Tom, Dick, and Harry thinks he knows it, but he is mistaken." Einstein wrote this to his best friend, Besso, in 1951, four years before he died. *Bohr's* mind could bend around complementarity, duality, uncertainty, but, still, Rutherford—who was always on Bohr's side in the arguments with Einstein—once thundered at him, "Bohr, my boy, you are too complacent about ignorance."

The great experimental physicist I. I. Rabi described this problem in an interview with Jeremy Bernstein. "There's the miracle of doing something like making an electron-positron pair!* One actually creates a remarkable thing like an electron. It's a marvelous thing. I don't see how it's made. It just appears. It's a kind of materialization—the ghost shows up in reality. There it is. You can calculate how many electrons will be

---

*Matter and antimatter emerging from the energy of a collision.

produced, and with what probability. But how was it born? What was it made of? It's this kind of question that, as an experimenter, I would like to see answered.

"The theory doesn't answer the sort of question that led me into physics in the first place.

"I wanted to know what the thing really was."

And so, of course, did Bell. He wanted to know not how a particle responds to a "measurement," but what it really is. While still at Stanford in 1964, after writing his paper on the nonlocality of hidden variables, Bell wrote another paper in collaboration with a professor there named Mike Nauenberg. Their paper, "The Moral Aspect of Quantum Mechanics," ends with these resounding paragraphs:

It is easy to imagine a [quantum state, described by a huge Schrödinger equation] for the whole universe, quietly pursuing its linear evolution through all of time and containing somehow all possible worlds. But the usual interpretive axioms of quantum mechanics come into play only when the system interacts with something else, is "observed." For the universe, there *is* nothing else, and quantum mechanics in its traditional form has simply nothing to say.

It gives no way of, indeed no meaning in, picking out from the wave of possibility the single unique thread of history.

These considerations, in our opinion, lead inescapably to the conclusion that quantum mechanics is, at the best, incomplete.*

We look forward to a new theory which can refer meaningfully to events in a given system without requiring "observation" by another system. The critical test cases requiring this conclusion are systems containing consciousness and the universe as a whole. Actually, the writers share with most physicists a degree of embarrassment at consciousness being dragged into physics, and share the usual feeling that to consider the universe as a whole is at least immodest, if not blasphemous.

However, these are only logical test cases. It seems likely to us that physics will have again adopted a more objective description of nature long before it begins to understand consciousness, and the

---

*"This minority view is as old as quantum mechanics itself, so the new theory may be a long time coming. . . . We emphasize not only that our view is that of a minority, but also that current interest in such questions is small. The typical physicist feels that they have long been answered, and that he will fully understand just how if ever he can spare 20 minutes to think about it." (This is Bell's footnote.)

universe as a whole may well play no central role in this development. It remains a logical possibility that it is the act of consciousness which is ultimately responsible for the reduction of the wave
packet [in other words, "the collapse of the wavefunction"]. It is
also possible that something like the quantum mechanical state
function may continue to play a role, supplemented by variables
describing the actual—as distinct from the possible—course of
events ("hidden variables") although this approach seems to face
severe difficulties in describing separated systems in a sensible way.

What is much more likely is that the new way of seeing things
will involve an imaginative leap that will astonish us.

In any case, it seems that the quantum mechanical description
will be superceded. In this it is like all theories made by man. But to
an unusual extent its ultimate fate is apparent in its internal structure. It carries in itself the seeds of its own destruction.

— superseded !

Back in Vienna

2005

ANOTHER CONFERENCE is in full swing under the soaring arches of the University of Vienna, in the city where Pauli and Schrödinger were born and raised. Zeilinger walks by in a tweed coat with a backpack on his back. With him are Horne in a porkpie hat and Greenberger, who is saying, "We're like very intelligent flies living in a two-dimensional universe which is actually a line drawn by a drunk."

Sitting on a bench are Shimony and Gisin, Gisin with a question for the man Greenberger calls "Mr. Quantum Mechanics": "It is now banal for two photons which have never met to be entangled. Do you think John Bell would have guessed this?"

"Bell had no idea that was possible," says Shimony.

Bertlmann sweeps by. The careful observer can see a green-and-black-striped sock under the cuff of one trouser leg and a red-and-black-striped sock under the other cuff.

Gisin tells Shimony that it was his review article with Clauser in 1978 that pulled Gisin into this field. "I read it in a hotel in India, drinking tea." He was on a trip to go scuba diving with his brother, an expert diver.

In 1997, Clauser wrote a paper in honor of Shimony's birthday, titled "De Broglie–Wave Interference of Small Rocks and Live Viruses." With one of Ed Fry's students, he had developed an atom interferometer, and the technique they had used he believed could someday demonstrate the quantum nature of something as large, in quantum terms, as a virus. He sounded the call to the neutron interferometers and other fundamental quantum investigators to pay attention, and Zeilinger was listening.

His group has just sent a knobbly fluorinated fullerene molecule (a sixty-carbon-atom soccer ball brandishing fluorine atoms from every seam) through a two-slit experiment, dramatically displaying the wave nature of matter. Zeilinger is fast chasing down Clauser's dream of interfering "small rocks and live viruses," believing that a careful enough experiment can show anything existing in a placeless quantum-mechanical limbo. Careful experiments are expensive, of course, and the bigger an object is, the less willing it is to display quantum behavior. "The boundary between quantum and classical is just a question of money" is Zeilinger's slogan. It seems that the University of Vienna is listening.

Gisin also has an amazing new experiment. Two entangled photons start out together. One enters a box full of noisy, uncorrelated photons while the other bypasses it. The photon that emerges from the box is still entangled with the photon that went around it. "This is a change," Gisin is saying to Shimony. "I was told, 'Entanglement is very ethereal, very fragile. Don't study it.' But entanglement is *robust*. We routinely send an entangled photon down a kilometer of fiber!"

A student says, "Schrödinger and de Broglie would love this stuff!"

Says Gisin, "Yes, but they would still ask, Where's the boundary? When does it become classical? Anton would say that there is no boundary; we are quantum. . . . I do not know what that means." He pauses and then says, "I think that as we continue, we will find it. I don't believe that *I* will find it, but my experiments—"

"They contribute," says the student.

"They contribute," says Gisin.

Though Gisin stopped scuba diving after his brother tragically drowned, Ekert has, as of a few years ago, become a serious scuba diver himself, spending more and more time in Singapore, where he holds a second professorship. The conference photo on the long and wide front steps is taken with Ekert standing behind everyone else, present but totally invisible to the camera's eye, enthusiastically discussing entanglement with some students. In the photo, Kwiat is seen climbing on the building, hanging by one arm from a square pillar in the last row. (He has just moved from the plateaus and canyons of Los Alamos to the plains of Illinois, and the title of his speech is "Scaling Mt. Entanglement, or What to Do if You Live Somewhere Where All the Spatial Derivatives Are Zero.") He looks back at Ekert: "Being a hidden variable, I see."

Deep in conversation under one of the soaring arched windows of the university are two young physicists, members of the new generation studying and arguing about quantum theory's foundations. They are quantum-information theorists, who focus on the deeper lessons that can be drawn from quantum cryptography and quantum computation.

The older of the two, a charming and energetic Texan named Chris Fuchs, works at Bell Labs. He has just published his samizdat text, *Notes on a Paulian Idea*, which has been for a long time circulating on the Internet—a book of his e-mail correspondence full of the stylish prose with which he argues, cajoles, and teaches everyone else in the field. The Paulian idea of its title encapsulated in the epigraph, quoted from the great man himself: "In the new pattern of thought we do not assume any longer the *detached observer* . . . but an observer who by his indeterminable effects creates a new situation."

The book begins with a beautiful foreword by Mermin. As he explained in his piece for the Bell conference a few years earlier, "Until quite recently I was entirely on Bell's side on the matter of knowledge/information"—words, Bell said, that should not appear in any serious formulation of physics. "But then I fell into bad company. I started hanging out with the quantum computation crowd, for many of whom quantum mechanics is self-evidently and unproblematically all about information." Because of the arguments of the old and new guard of the Copenhagen interpretation—his friends Gottfried and Fuchs, respectively—Mermin has become much more sympathetic to its tenets.

"The past decade has seen the growth of intense interest in applications of quantum mechanics to information processing, brought about by its deep intellectual richness, in fortuitous but sociologically significant resonance with our cultural obsession with keeping secrets," Mermin writes in the foreword. "Chris Fuchs is the conscience of the field. He never loses sight of the real aim of these pursuits, and if you yourself thought it had to do with secure data transmission, RSA code cracking, fast searches, fighting decoherence, concocting ever more ingenious tricks, and such, you should look up from your beautiful algorithms or candidate qubits [quantum bits for quantum computers] for a few hours every now and then to browse through these pages."

Quantum mechanics is a law of thought in a world that is sensitive to the touch is Fuchs's motto, which resounds throughout the work. "It is that *zing!* in the system—that sensitivity to the touch—that keeps us from ever saying more than ψ"—the answer to the Schrödinger equation— "of it," as he explains in his speech at this conference. "This is the thing that is real about the system. It is our task to give better expression to that idea, and start to appreciate the doors it opens for us to shape and manipulate the world." He suspects that "we will someday be able to point to some ontological content"—content about *being* as opposed to the *knowing* that the theory clearly covers—"in the quantum theory. But that ontological statement will have more to do with our interface with the world—namely, that in learning about it, we change it—than with the world itself." Then he laughs. "Whatever that might mean."

His tall, long-haired, blond friend standing by the window with him in the black AC/DC T-shirt is named Terry Rudolph. Now a professor at Imperial College London, he grew up in Australia. He is more than capable of holding his own in conversation with a reincarnation of Pauli. But he is not a reincarnation of Heisenberg. "I belong to the crazy group that believes Einstein was right," he says. In his senior year in college, as a twenty-year-old math and physics major at the University of Queensland who was feeling lukewarm about physics, "I went to this lecture because I thought the professor was going to talk about the exam. Instead he gave us Mermin's paper on Bell's inequality. I almost failed the exam because I was so sure it was wrong. I went off for two weeks trying to see where the mistake was.

"It's the most important, profound feature of modern physics that this is the way the world is."

When Fuchs says that he thinks the "ontological content" of the quantum theory will be that it reacts to us, Rudolph says, "It could still be that there is an underlying reality. Because, you know, if the wavefunction is knowledge, then you still have the question, 'knowledge about what?' "

"I like the word *belief* better than *knowledge,*" says Fuchs. "The unique thing about the quantum world is that it never lets people's opinions get so aligned that they can jump to the conclusion of a freestanding reality. This is captured by the fact that the best that can be said of almost all measurements can only be couched in statistical language."

"Alright, statistical language about *what,* if not a freestanding reality?" asks Rudolph.

"Belief about the *answers* to the *questions* we can ask, about what will win our bets for us. Who could want more from a physical theory?" says Fuchs. Then he relents: "Yeah, yeah, I know you want more . . ." With an impish grin he says, "But I suspect you can't get it."

"It's too anthropocentric for me," says Rudolph. "It's all to do with me and how *I'm* describing the world. But I actually believe"—his Australian accent gets stronger—"that if we had never evolved, you know, if the earth had been a bit closer to the sun and these human monkeys had never come out, that the universe would have still been here, and that *something* would still have been going on. And whatever *that* theory is, it certainly doesn't depend on the betting strategies of a bunch of glorified monkeys."

Fuchs and Rudolph's argument is one of the central ones of the quantum theory, and has been so, in various forms, ever since the beginning. But the point on which they agree is even more interesting.

"Almost all the formal structure of the quantum theory . . . is not really about physics at all," Fuchs wrote in 1998. "It is about the formal tools for describing what we know." Many oddities of quantum theory show a strong similarity to information theory, a powerful set of ideas developed alongside the theory of computation to explain the transmission of information. Rudolph and Fuchs agree that what we call the quantum theory is actually mostly information theory, theory about our knowledge of quantum entities rather than about the entities themselves.

If information theory could be shorn from quantum mechanics, what would be left?

"What we need to do is remove from quantum mechanics all those information-theoretic aspects that are due to how we perceive the world and find a quantum mechanics that conforms to the world when I'm not around," is how Rudolph puts it.

"The distillate that remains—the piece of quantum theory with no information-theoretic significance—will be our first unadorned glimpse of 'quantum reality,' " is how Fuchs puts it.

How would someone go about this task?

"One approach is to see, if I impose very simple information-theoretic restraints on a classical theory, how much of quantum mechanics the resulting theory looks like," explains Rudolph, essentially playing with hidden-variables models, and seeing where they fail, as John Bell once did. "You know, there's no underlying principle to quantum mechanics in the same way there is for relativity. In relativity, you've got a simple principle that the laws of physics look the same to all inertial observers, and *bang*, you can derive huge amounts from that. So here's an information-theoretic constraint: the observer can never know precisely the position and the momentum of a particle. Bang. And we look and see how constrained the resulting theory must be and how similar to quantum mechanics it must be."

"The task," agrees Fuchs, "is not to make sense of the quantum axioms by heaping more structure, more definitions, more science-fiction imagery on top of them, but to throw them away wholesale and start afresh. We should be relentless in asking ourselves, From what deep *physical* principles might we *derive* this exquisite mathematical structure? Those principles should be crisp, they should be compelling. They should stir the soul.

"When I was in junior high school, I sat down with Martin Gardner's book *Relativity for the Million* and came away with an understanding of the subject that sustains me even today. The concepts were strange to my everyday world, but they were clear enough that I could get a grasp of them knowing little more mathematics than arithmetic. One should expect nothing less for a proper foundation to the quantum. Until we can explain the essence of the theory to a junior high school or high school student—the essence, not the mathematics—and have them walk away with a deep, lasting memory, I well believe we will have not understood a thing about quantum foundations."

A year after Rudolph read Mermin's paper on Bell's inequality, he finished his honors thesis in physics, making the first major step toward becoming a physicist. He decided to celebrate by wandering around the world for a year. He wanted to see Africa, where his grandmother and mother had once lived; Europe, where his aunt and uncle made their home among the gorgeous peaks of the Austrian Tyrol, and North America, where he himself would ultimately get his Ph.D. and work at Bell Labs, and where he would meet Chris Fuchs. When he was about to leave Australia to set out on this journey, his mother told him an incredible story.

Her own mother had been a very innocent twenty-six-year-old Irish Catholic who had become pregnant after a relationship with a brilliant and much older man. The young woman gave her baby to the father of the child, as he had asked. But after two years of missing her daughter, when she came across the toddler with a nanny in a park in Dublin, she snatched her out of the baby carriage, and took her across the globe to South Africa.

And it was thus that the twenty-one-year-old Rudolph, who had only a year earlier encountered the idea of entanglement for the first time and as a result dedicated himself to studying physics, learned that his grandfather was Schrödinger.

GLOSSARY

acausality—A state in which effects have no causes.

accelerator—A machine that accelerates charged particles to near the speed of light, letting them collide with a fixed target or with each other, to study what is created out of the energy of their collisions.

action-at-a-distance—The problem was defined by Isaac Newton in the late seventeenth century: gravity seemed to be a force of one body acting on another far away, with no intermediary. This seems, as he wrote, "inconceivable."

But if so, then how else could forces that worked at a distance, like gravity, electricity, and magnetism, be transmitted? In the mid-nineteenth century, Michael Faraday, one of the greatest experimentalists of all time, saw that a clue to this mystery lay in the behavior of iron filings lying on a piece of paper over a magnet. The filings line up in rows radiating from pole to pole, tracing out "lines of force." Faraday compared these lines of magnetic transmission "to the vibrations upon the surface of disturbed water, or those of air in the phenomenon of sound"—cases where the action is not instantaneously at a distance, but by chains of local contact. He extended his "lines of force" idea to gravitation, electricity, and light (which he declared a vibration in the electrical lines of force).

His visualizable phrase, "lines of force," is now usually replaced by the more opaque term "the field."

algorithm—A given set of rules a computer follows to solve a given problem.

alpha particle—A helium nucleus (two protons and two neutrons), which can act as a single particle with +2 charge.

angular momentum—The momentum of an object constrained to move in a circle. For *intrinsic angular momentum* see *spin*.

atom—Though not, as its name means in Greek, indivisible, it is the smallest piece of an element that still retains its nature.

Bell's theorem / Bell's inequality—The discovery that quantum mechanics is compatible with either, but not both, of the following ideas:

      1. locality
      2. some form of independent, separable reality.

The world, at the quantum level, is strangely and inexplicably correlated. Some, defying special relativity, say that the moral of the story is nonlocality; others say quantum things are not real when not observed. And there is a murmur of a suggestion that the central reality, at the quantum level, *is* entanglement: that relationships between quantum "things" are more fundamental and objective than the things themselves.

The "inequality" of Bell's theorem is a mathematical statement restricting the amount that two unconnected particles can be correlated, an apparently trivial truth that is nonetheless not respected by quantum mechanics.

Bose-Einstein condensate—When bosons (see *bosons* and *fermions*) are cold enough—when they do not have enough energy to jiggle around—they synchronize, slipping into the same quantum state, and creating a relatively huge, undivided quantum entity.

bosons and *fermions*—Bosons have even-numbered spin, meaning they carry intrinsic angular momentum in multiples of $\hbar$ (Planck's constant divided by 2π). Fermions have half-integer spin, carrying intrinsic angular momentum in odd multiples of $\hbar/2$.

A fermion will obey Pauli's exclusion principle, tending to pair with another fermion of opposite spin. An electron, for example, is a fermion, with spin ½. A pair of fermions make a boson (spin 0 or 1). Three fermions make a fermion (spin ½ or ½). And so on.

Photons are bosons that are not built from anything else. These and their kin transmit the four forces (electromagnetic, strong, weak, and, presumably, gravity). These bosons are not conserved—they can be created and destroyed endlessly.

Cartesian coordinates—The common x, y, and z on a graph. In the seventeenth century, René Descartes introduced this system of identifying a location by its distance from a point of origin.

CERN—European center for nuclear and particle physics research, near Geneva. (The *C* in CERN actually stands for *Conseil*—the name has changed but the acronym remains.)

CH inequality—In 1974, Clauser and Horne (in conversation with Shimony) produced an improved inequality that was more stringent but harder to test in the lab.

CHSH inequality—The version of Bell's inequality that can be tested in the lab. From Clauser, Horne, Shimony, and Holt (1969).

classical—Relating to the physics before quantum theory, or physics that is consonant with pre-quantum ideas.

complementarity—Bohr's view, first stated in 1927, that the quantum paradoxes are inherent, and that quantum-mechanical results must always be described in classical—but contradictory—terms. In particular, a description in space-time precludes a causal description, and vice versa. If causality is maintained by the principle of the conservation of energy and the principle of the conservation of momentum, then the *uncertainty principle* emerges: a description of a quantum-mechanical object in terms of position (space-time) cannot be made at the same time as a description in terms of momentum or energy (causality).

conservation of energy—Energy is neither created nor destroyed; it merely changes form.

correspondence principle (until 1920 called the *analogy principle*)—Bohr's ground rule that quantum mechanics should correspond to classical mechanics in the domain where classical mechanics is successful, where actions are large compared to the quantum.

determinism—As the Marquis de Laplace wrote (around the end of the eighteenth century) of what became known as his Demon: "We may regard the present state of the universe as the effect of its past and the cause of its future. An intellect which at any given moment knew all of the forces that animate nature and the mutual positions of the beings that compose it, if this intellect were vast enough to submit the data to analysis, could condense into a single formula the movements of the greatest bodies of the universe and that of the lightest atom; for such an intellect nothing could be uncertain and the future just like the past would be present before its eyes."

diffraction—A fundamental property of waves, from the Latin meaning "to break in two." A row of waves striking a barrier with a gap in it roughly the size of their wavelength will break their forward progress, radiating from the gap in half-circles.

down-conversion (also called *spontaneous parametric down-conversion*)—Two low-energy entangled photons are produced from one high-energy photon when it passes through a certain kind of crystal.

electron—A particle that carries a negative electric charge exactly equal to the much bigger proton's positive electric charge; electrons circle the nucleus, creating the atom's size, shape, and electrical properties by their number and distribution in space. Electrons in metals form an "electron gas" flowing around all the nuclei; this flow can be directed down a copper wire to give us the electricity we use.

elementary particle—In the thirties, this was a short list: electron, photon, proton, neutron. It grew longer as pions and muons fell from the sky and neutrinos came shooting out of the nucleus. Now protons, neutrons, and mesons are thought to be made out of *quarks,* even smaller and more elementary particles. Gluons arrived to glue the quarks together—they are elementary too.

Meanwhile, because of $E = mc^2$ (Einstein's equation of 1905, which means that matter can be transformed into energy and vice versa), if you smash a particle hard enough, you create other particles out of the energy of the collision. The Dirac equation of 1928 predicted the existence of antimatter—positive electrons and negative protons. If a positron and an electron meet, they annihilate each other in a burst of light—matter becomes energy. The reverse is even more amazing: a photon spontaneously *becomes* a positron and an electron—light becomes matter! Thus the search for a particle that is truly "elementary" and indivisible seems less and less hopeful.

entanglement—A condition of two or more bits of matter or light behaving, though separated, as if intimately connected. The word is due to Schrödinger's English- and German-language replies to EPR in 1935—"entanglement" in August; "*Verschränkung*" (which in English might be better conveyed as a "cross-linkage") in December.

EPR—The abbreviation for Einstein, Podolsky, and Rosen's 1935 paper, "Can Quantum-Mechanical Description of Physical Reality Be Considered Complete?" They answer "no": reality should not involve either nonlocality (a.k.a. "spooky action-at-a-distance") or nonseparability of objects that are far apart (a.k.a. "entanglement"). For an outline of EPR's reasoning, see the "Longer Summaries" after this Glossary.

excited state—The state of an atom or nucleus that has absorbed electromagnetic radiation and gained energy. See *ground state.*

fermions—See *bosons* and *fermions.*

field—The word *field* in physics, which seems so intimidating, is actually not markedly different from the word as it is used in common English. A field, whether or not it is visible, is a continuous expanse defined by what grows or happens there: a field of poppies, a field of study, a playing field. Often these fields are perceptible only by their effects. For example, we can't see the field of pressure differences in the air, but we can feel the wind and see the grass bending. Radio waves, light, X-rays, and their kin are waves in the electromagnetic field.

frequency (symbolized by the Greek letter v, called *nu*)—Cycles per second; for a wave, the number of peaks that pass a given point in a second. For a small range of the electromagnetic spectrum, our eyes are sensitive to frequencies, which we perceive as the colors of the rainbow: blue is high frequency, red is low frequency. Sound waves, on the other hand, are waves of pressure difference in the air—high-frequency sounds are shrill, and low-frequency sounds are deep.

gedanken *experiment*—A hybrid German-English word for "thought-experiment": an application of logic and the laws of physics to elucidate an imaginary physical scenario.

GHZ—Greenberger, Horne, and Zeilinger's paper on Bell's theorem without inequalities, first published in 1988. They show that when *three* particles are entangled, then local realism is refuted decisively by a single measurement. (By contrast, when only two particles are entangled, the refutation provided by Bell's inequality requires a long time and many thousands of measurements to be decisive.) (See the "Longer Summaries" below.)

ground state—The stable, lowest-energy state of a particle.

h—See *Planck's constant.*

$\hbar$ ("h-bar")—Planck's constant divided by 2π.

hidden variables (also known as *hidden parameters*)—"To know the quantum mechanical state of a system implies, in general, only statistical restrictions on the results of measurements. It seems interesting to ask if this statistical element be thought of as arising, as in classical statistical mechanics, because the states in question are averages over better defined states for which individually the results would be quite determined." So begins John Bell's famous paper "On the problem of hidden variables in quantum mechanics." The hypothetical "better defined states" would then be described by unknown quantities in addition to the wavefunction. These are the "hidden variables"—"'hidden'," explains Bell, "because if states with prescribed values of these variables could actually be prepared, quantum mechanics would be observably inadequate." While the final sentence of the EPR paper ("We believe, however, that such a theory is possible") is often interpreted as a call for hidden variables, Bell showed that only if they were nonlocal could they reproduce the results of the quantum theory.

interference—A fundamental property of waves, discovered by Thomas Young in the early years of the nineteenth century. When a wave interferes destructively with itself, its peak and trough cancel each other out: the result is flat water, an acoustic dead zone, or unexpected darkness or absence. Waves interfere constructively when two peaks coincide, creating a wave bigger than the sum of their heights.

ion—An atom shorn of one or more of its electrons.

light-quantum (plural: *light-quanta*)—Einstein's name for what became known as a *photon*—the elementary particle of light (or of any other section of the electromagnetic spectrum, from kilometer-long radio waves to gamma rays the size of an atom).

local—Not distant. Examples of local connections are a sound wave hitting your eardrum, a light wave hitting your retina, or someone pushing you. Instantly arriving home after clicking your ruby slippers, mind reading, influencing faraway events by the power of positive thinking, etc., would be nonlocal events.

locality—The condition that things can influence other things only by a chain of local connections, and thus influences cannot travel faster than the speed of light.

local realism (also called *Einstein's separation principle*)—Until quantum mechanics, this idea was the goal of scientific explanation: that the world can be divided into individuals and that these cannot affect one another without contact, whether direct or indirect. ("In front of me stand two boxes," Einstein wrote Schrödinger in 1935. "The separation principle: The second box is independent of anything that happens in the first box.")

matrix—An array of numbers arranged in rows and columns and treated as a single entity.

mechanics—The science of how things move, acted upon by forces.

meson—See *pi meson; mu meson.*

momentum (abbreviated "*p*")—The speed of an object in a particular direction times its mass. (Position is abbreviated "*q*.")

mu meson or *muon*—A heavy electron. Very different from a pi meson or pion. The confusion started because the muon was discovered experimentally only three years after Hideki Yukawa predicted the "heavy quantum" (pion), and initially it seemed to satisfy Yukawa's prediction. Actually the pion (a spin-0, photonlike particle) decays into the muon (a spin-½, electronlike particle).*

neutrino—A tiny particle related to the electron, which appears to be (nearly) massless.

neutron—A very small but heavy neutrally charged piece of matter, which resides in the atomic nucleus.

neutron interferometer—A wonderful device showing the wave nature of matter, in which a single neutron can seem to have gone down two different paths in order to interfere with itself, as a beam of light does.

nonlocality—The idea that small bits of matter and light are influenced by each other without any means, faster than the speed of light, by what Einstein mockingly termed "spooky action-at-a-distance."

*nonseparability (*also called *entanglement)*—The idea, predicted by quantum mechanics, that small bits of matter and light, even if physically separated from each other, somehow can remain united.

*nucleus (*plural: *nuclei)*—The heavy heart of an atom, which is otherwise almost empty. It is made up of protons and neutrons.

observable—A basic attribute of a quantum-mechanical system. (The terminology emphasizes the importance of *measurement,* and the accompanying agnosticism about what went before the measurement. Like the story about the physicist, mathematician, and philosopher on a train through Scotland, who see a field full of sheep: The physicist says, "Look—the sheep in Scotland have black faces." The mathematician adds, "In this field, at least." The philosopher: "On the side facing the train." Quantum mechanics takes the point of view of the philosopher.)

p—Abbreviation for *momentum* (position is "*q*").

particle—In classical mechanics, a particle was an imaginary point that had mass (i.e., weight) but no size. It could move, but because it was a mere point, it could not rotate. The word has come to mean "things smaller than an atom," whether or not these things are mathematical points. See *elementary particle.*

particle physics—The science of the relationships between elementary particles. See *elementary particle.*

phase—Two waves are "in phase" when their crests line up perfectly—this causes constructive interference. If waves are perfectly "out of phase," they will destructively interfere, canceling each other out.

*"Who ordered that?" said I. I. Rabi.

photon—See *light-quantum.*

pi meson or *pion*—In 1935, Hideki Yukawa saw that the force holding the nucleus together must be much stronger than gravity or electromagnetism, the two known forces. As light-quanta carry the electromagnetic force, he proposed "heavy quanta" for this strong force. These heavy quanta, which became known as mesons, appeared in 1947, and Yukawa, entirely educated in Japan, rapidly became the first Japanese to receive the Nobel Prize for physics. He had actually taken his bride's last name three years before the prediction of mesons that would make that name so famous.

Planck's constant (known as "*h*")—The ratio of the energy of light (in terms of quanta) to the frequency of light (in terms of waves). It mathematically bridges the gap between waves and particles. This ratio is constant, and shows up in all equations of quantum physics. It is a tiny number, 6.626×10^{-34} kg-m^2 / sec (i.e., kilogram-meters-squared per second, or joule-seconds).

plasma—The fourth state of matter, which makes up much of the universe (e.g., stars). Plasma occurs when a gas is heated beyond the ability of its atoms to hold on to their electrons, which begin to flow in a synchronized, collective motion. (Metals also contain this "sea of electrons," but their atomic nuclei are arranged in a crystalline pattern, rather than floating freely like the electrons.) Even a fire on earth oscillates between gas in its colder sections and plasma where it's hottest.

polarization—The "tilt" of the electromagnetic field, an attribute of all electromagnetic waves. Unlike water waves and sound waves, electromagnetic waves undulate in *two* directions: the *electric* field waves in one direction (say, up-down), the *magnetic* field is perpendicular to that (left-right), and the wave travels in yet a third direction that is perpendicular to both (forward).

To visualize this, you could think of the electromagnetic wave as a fish: the electric field is its dorsal fin; the magnetic field is its side fins. If the fish tilts, that doesn't change the relationship between its fins or the direction it travels, and the angle at which it leans is its polarization.

position (symbolized by the letter *q*)—The location in space. In classical physics, the position of a particle is specified by three numbers; in quantum physics, by a wave that in many cases extends arbitrarily far in all directions. This fact results in the uncertainty principle. See *quantum state.*

positivism—The belief that observable quantities are all that is meaningful to talk about.

positron—An antielectron; it has a positive charge.

positronium—An evanescent particle that is composed of an electron and an antielectron (positron): they circle each other before self-annihilation in a burst of light (gamma rays).

proton—About the same size and weight as the neutron, also located in the atomic nucleus, but positively charged.

ψ—The Greek letter *psi* of the Schrödinger equation stands for the wavefunction describing a quantum entity.

q—Abbreviation for *position* (momentum is "*p*").

quantization—The fact that energy and matter cannot be divided endlessly into smaller pieces: there is the quantum, the ultimate, indivisible piece (prefigured by the ancient Greek concept of the "atom").

quantum cryptography—A way to secretly distribute a key for use in making an unbreakable code. There are two main versions of this quantum key distribution.

The earlier one, developed by Stephen Weisner and later Charlie Bennett and Giles Brussard, is based on the Heisenberg Uncertainty Principle. Ekert's entanglement-based version of 1991, created independently of this American and Canadian precedent, entails a key created from entangled particles shared between "Alice" and "Bob" (the conventional names for the sender and receiver of codes) (see the "Longer Summaries" following this Glossary).

quantum electrodynamics (also known as *QED*: "Physicists," David Wick writes, "love to see the mathematicians cringe.")—The refinement of quantum mechanics devoted to the interaction of charged matter (e.g., electrons) and light. While plagued by mysterious infinities that show up in the calculations, physicists know how to subtract these away to make incredibly accurate predictions.

quantum mechanics—The science, as of 1925, of the doings of the most elemental pieces of matter and light.

quantum state—What can be known about a quantum object.*

An ordinary classical state can be, in theory, completely known. All a dog's attributes, for example, can be listed: his location in three dimensions of space and one of time, his momentum, his energy, et cetera. But a quantum state, described by the Schrödinger wave (or psi) function, is maximally known when about half of these attributes (or "observables") are known (if position, then not momentum; if time, then not energy).

quark—The constituent particle of protons, neutrons, and the mesons that bind them.

qubit (pronounced "*Q*"-*bit*)—A quantum bit. A bit ("binary digit") is a single piece of information: either 0 or 1. A qubit can be 0, 1, or a *superposition* of 0 and 1. (The word was coined in 1995 by Ben Schumacher of Kenyon College.)

radiation—Waves or particles emitted from an atom. This word has a different emphasis depending on the source of the radiation.

1. An ordinary atom merely loses energy when an electron drops from a higher energy level to a lower one and emits *electromagnetic* radiation, in this case visible light.

2. It can also lose energy if the atom emits the radiation from its nucleus. In this case, it will be much higher energy radiation composed of *gamma* rays, which are photons even shorter and harder than X-rays.

3. A radioactive atom actually loses pieces of itself—in the process, decaying into another, lighter, more stable atom—when it emits *alpha* radiation composed of protons and neutrons, or *beta* radiation composed of electrons.

radioactivity—The decay of large, unstable atomic nuclei. Though spontaneous, its probability for a given atom is well known. See *radiation,* sense 3.

realism—"If one asks what, irrespective of quantum mechanics, is characteristic of the world of ideas of physics," as Einstein wrote, "one is first of all struck by the following: the concepts of physics relate to a real outside world . . . independent of the perceiving subject."

"It's only when in the course of some high-brow argument, pursuing some ideological position or other," John Bell said once in an interview, that physicists "might come to question the existence of reality. But I think that in their hearts they know that reality is there. The problem is to form some sharp conception of

---

*The word *state* is so vague that many of the founding fathers, particularly Schrödinger, complained about its use. For clarity, Kramers used the term *physical situation* instead of *state.*

reality at the atomic scale, which is consistent with the methods that we know work predicting phenomena."

RSA—A form of cryptography developed in 1977 by the M.I.T. professors Rivest, Shamir, and Adelman. "Bob," the recipient of a coded message, publishes a "public key," which rests on a mathematical function that is easy to run in one direction and hard to run in reverse without an extra key that only Bob knows.

A metaphor often used to give a sense of how RSA works: Bob sends "Alice" a box with an open lock to which only he has the key. She puts her message in the box and closes the lock. Now even she can't get the message back out, and neither can any code-breaker, only Bob.

Schrödinger's cat—See *superposition.*

separability, or *independence*—In the analysis of science historian Don Howard, Einstein's argument against quantum mechanics (often described as "it does not satisfy both locality and realism") can be sharpened to "quantum mechanics does not satisfy both locality and separability." Separability would then be, in Einstein's words, the "assumption of the mutually independent existence (the 'being-thus') of spatially distant things." Separable things have definite states or characteristics of their own. Einstein goes on to explain that without such an assumption, "physical thought in the sense familiar to us would not be possible. Nor does one see how physical laws could be formulated and tested without such a clean separation. . . ."

He connects this separability with locality thus:

"For the relative independence of spatially distant things (A and B), this idea is characteristic: an external influence on A has no *immediate* effect on B; this is known as the 'principle of local action.' " Both interrelated principles, Einstein notes, are best applied in a field theory.

The opposite of *separability,* then, is entanglement, where distant things do not possess distinct states of their own.

The opposite of *locality* is action-at-a-distance, where things possess distinct states of their own, which then can be changed at a distance.

spectral line—A narrow band of color from the rainbow spectrum. Any element may be identified—when heated by a Bunsen burner and viewed through a prism or spectroscope—by its characteristic series of these colored lines.

spin—Quantum particles, even when totally at rest, still have attributes of a turning object. This is also called "intrinsic angular momentum" (ordinary angular momentum is the momentum one feels when turning or moving along a curve), and its value can only be in integer or half-integer multiples of $\hbar$ (Planck's constant divided by 2π). Since magnetism is produced whenever there is an electric charge moving, the "spin" of an electron makes it also a tiny magnet, and much magnetism in matter finds its root in spin.

An example of a spin-½ particle is an electron. The value of their intrinsic angular momentum is either $+\frac{1}{2}\hbar$ or $-\frac{1}{2}\hbar$. Photons are spin-1; pions are spin-0.

state—The condition and attributes of a given object. In classical physics, this can be specified by giving the values of a list of "observables"—the object's position and momentum, for a start.

statistical mechanics—The study of how free-moving atoms create the observable properties of (in particular) gases. (For example, the temperature of a gas is merely the average of the kinetic energies of all its atoms or molecules.) Ludwig Boltzmann, fighting atomic doubters on all sides in Vienna, and J. Willard Gibbs, serenely isolated at Yale, developed this subject in the late nineteenth century and the first few years of the twentieth. Equally isolated at his patent office (and in ignorance of the work of his predecessors), Einstein redeveloped the whole discipline from scratch in 1900–1904. This work laid the foundations not only of his famous 1905 paper on atomic size, but also of his principal contributions to the quantum theory, all of which, notes Pais, "are statistical in origin" (see *Subtle Is the Lord*, p. 56).

Stern-Gerlach experiment—Silver atoms that pass through an inhomogeneous magnetic field (created by oddly shaped magnets) have a quantized response to the field: they are deflected either up or down, with no range of responses. Designed and performed in 1921 by Otto Stern and Walther Gerlach.

superposition—Two simple waves, added together, produce another wave. This leads to *interference*.

Superposition is one of the main features of quantum mechanics. "Schrödinger's cat" is the classic reductio ad absurdum. It describes a situation where "an indeterminacy originally restricted to the atomic domain becomes transformed into macroscopic indeterminacy, which can then be *resolved*"—Schrödinger emphasized—"by direct observation.

"That," he explained, is what "prevents us from so naively accepting as valid a 'blurred model' for representing reality"—a "blurred model," in which the whole world is fundamentally wave-like.

A radioactive atom, for example, is usually in a superposition of both having emitted a neutron and not having emitted a neutron. The thought-experiment runs as follows: Schrödinger's cat is trapped in a box with a vial of poison that is triggered by a quantum event, the emission of a neutron. After a certain amount of time, the atom will definitely be in a superposition of both having emitted the neutron and not having emitted the neutron. Does that mean that the poison is consequently in a superposition of being both inside and outside its bottle? Does that mean that the cat is, as a result, in a superposition of having been killed and also remaining alive?

But anyone who opens the cat box will see either a dead cat or a live one. Does the act of measurement then "collapse the wavefunction," forcing the cat to either die or live?

uncertainty relations—Heisenberg's 1927 formula $\Delta p \Delta q \geq \frac{1}{2}\hbar$. I.e., the uncertainty in momentum (p) times the uncertainty in position (q) is greater than or equal to a small but nonzero number (*Planck's constant, h,* divided by 4π). . . . This equation, if you think about it, means that neither uncertainty can be zero—i.e., neither quantity can be fully known.

ν—The Greek letter *nu,* the abbreviation for "frequency."

wavefunction—A solution to the Schrödinger equation, describing the state of a given quantum entity. Symbolized by ψ, the Greek letter psi, it is also known as the psi-function (particularly by those who are agnostic about the reality of waves).

x, y, and z coordinates (also called *Cartesian coordinates*)—A set of numbers that together give the address or position of something on a three-dimensional graph. Drawn on a two-dimensional surface, the *x* axis runs left to right, the *y* axis runs up and down, and the *z* axis runs perpendicular to the page.

LONGER SUMMARIES

EPR (Einstein-Podolsky-Rosen paper, 1935)

Since Heisenberg and Bohr were always saying that quantum mechanics is in itself complete, the Einstein-Podolsky-Rosen paper started by offering an innocuous definition of a complete theory: "*Every element in physical reality must have a counterpart in the physical theory.*"

They continued with a way to judge that "physical reality": "*If, without in any way disturbing a system, we can predict with certainty the value of a physical quantity, then there exists an element of reality corresponding to this quantity.*"

Then followed what Einstein, Schrödinger, and Podolsky referred to as a paradox (though Rosen did not like the word).

1) Given the uncertainty principle, which says that, for example, knowledge of momentum precludes knowledge of position (or, as Bohr said in his speech at Solvay in 1927, "the demand of causality"—summarized by the statements of the conservation of energy and momentum—precludes "the space-time description"), either:

    a) "Quantum mechanics is incomplete" —Einstein.

or:

    b) "These properties are not simultaneously real" —Bohr.

2) Let's assume that Einstein was wrong, and analyze the situation as if quantum mechanics were complete.

3) Notice that by measuring *one* of Rosen's correlated hydrogen atoms, we can learn things about the other, without touching it.

    a) Even when the atoms are separated, we could measure the momentum of one (p) and instantly, because of symmetry, know what the momentum of the other one would be ($-p$).

    b) But we could have instead measured the first atom's position. We can divine the position of the second atom from the two-particle wavefunction and the known position of the first.

    c) The second atom, alone and untouched, is apparently prepared to tell us *either* its position *or* its momentum, depending on what we measure on the first. But without action at a distance, what we call an element of reality cannot be (instantly) affected by a measurement on another, faraway element of reality.

4) We come to the conclusion that, if there is no action at a distance and quantum mechanics is complete, position and momentum *do* have simultaneous (if inaccessible) reality, contrary to what the theory seemed to suggest.

5) This means that assuming the *negation* of what (in this summary) we have called the Einstein point of view "leads," as Einstein, Podolsky, and Rosen wrote, "to the negation of the only other alternative"—the Bohr point of view (see 1a and 1b).

6) Hence we are left with the Einstein point of view: quantum mechanics is incomplete.

GHZ (Greenberger-Horne-Zeilinger paper, 1988; Mermin *Physics Today* article, 1990; Greenberger-Horne-Shimony-Zeilinger paper, 1990)

Start with three particles—a, b, and c—which emerge back-to-back-to-back from a common source. Add three detectors—A, B, and C—each at a great distance from each other and from the source of the particles. Each detector has two measurement settings (say, horizontal or vertical) with only two possibilities for the result of each measurement (+1 and −1). If the particles have spin ½, then the detector is probably something like a Stern-Gerlach magnet, measuring horizontally along the *x* axis or vertically along the *y* axis, and a "+1" result means *spin up*.

Watching each detector alone we find that these results (+1 or −1) occur equally often. After some observation (or a glance at the quantum-mechanical state of the three particles), however, interesting correlations emerge. Take the case when detectors A and B are set to measure along the *y* axis, and C along the *x* axis. Our results in this case *always* contain an even number of "spin downs"—for example, (−1, −1, and +1) or (+1, +1, and +1)—and thus always multiply to +1. This knowledge, coupled with observations of the results at two of the detectors, lets us predict with certainty results at the third.

"In the absence of spooky actions at a distance or the metaphysical cunning of a Niels Bohr," notes Mermin, "the two faraway *y*-component measurements cannot 'disturb' the particle" we are about to measure along the *x* axis (see Mermin, *Physics Today*, June 1990, 9). Hence EPR would accord an "element of reality" to the result for this horizontal measurement of particle "c." A similar setup would allow us to predict the results for the other two—or for any of the three along the *y* axis, for that matter. The table on the following page summarizes these results, along with those of another run, to which we will return.

We are dealing with six elements of reality: each particle's spin measured horizontally and vertically. "All six of the elements of reality have to be there," observes Mermin, "because we can predict in advance what any one of the six values will be, by measurements made so far away that they cannot disturb the particle that subsequently does indeed display the predicted value.

"This conclusion," he notes, "is highly heretical," because, as far as quantum mechanics is concerned, two components of the spin of the same particle cannot exist at the same time—to assert that they do is to assert the presence of local hidden variables.

"Heresy or not," continues Mermin, "since the result of either measurement can be predicted with probability 1"—i.e., 100 percent of the time—"from the results of other measurements made arbitrarily far away, an open-minded person might be sorely tempted to renounce quantum theology in favor of an interpretation less hostile to the elements of reality."

As in Bell's version of EPR, however, the elements of reality do not survive "the straightforward quantum mechanical predictions for some additional experiments, entirely unencumbered by accompanying metaphysical baggage," continues Mermin.

"In the GHZ case the demolition is spectacularly more efficient.

"Suppose, heretically, that the elements of reality really do exist. . . ." We can't measure all six at the same time, but we can look at the quantum-mechanical predictions for the product of multiplying all three together, which is constant:

SETTINGS AT DETECTORS:			QM PREDICTIONS:	
A	B	C	AN ODD NUMBER HAS THE VALUE:	THE PRODUCT OF THE RESULTS:
horizontal	**vertical**	**vertical**	+1	+1
vertical	*horizontal*	**vertical**	+1	+1
vertical	**vertical**	*horizontal*	+1	+1
horizontal	*horizontal*	*horizontal*	−1	−1

We could symbolize each element of reality by the letter of the detector that revealed it, using bold for a detector aligned vertically and italics for horizontal (so that "particle 'a' measured by detector A aligned vertically" would be **A**, while "particle 'a' measured by detector A aligned horizontally" would be *A*). Then the first three results of the table give us these three equations:

$$A\mathbf{B}\mathbf{C} = +1$$
$$\mathbf{A}B\mathbf{C} = +1$$
$$\mathbf{A}\mathbf{B}C = +1.$$

If we multiply these equations together we will get another prediction for the same system:

$$\mathbf{A}^2\mathbf{B}^2\mathbf{C}^2 ABC = +1$$

But both $(-1)^2$ and $(+1)^2$ are just equal to +1, so the squared terms disappear, leaving us with:

$$ABC = +1$$

"Without invoking disreputable elements of reality," Mermin notes, we can also find the product of these three results by a simple quantum-mechanical calculation. If—as noted in the last line of our table—we perform a run on our *gedanken* apparatus where all the detectors measure the *x* components, we find:

$$ABC = -1$$

"So farewell elements of reality!" trumpets Mermin. "And farewell in a hurry. The compelling hypothesis that they exist can be refuted in a single measurement of the three *x* components: The elements of reality require the product of the three outcomes invariably to be +1; but invariably the product of the three outcomes is −1.

"This is an altogether more powerful refutation of the existence of elements of reality than the one provided by Bell's theorem for the two-particle EPR experiment. Bell showed that the elements of reality inferred from one group of measurements are incompatible with the statistics produced by a second group of measurements. Such a refutation cannot be accomplished in a single run, but is built up with increasing confidence as the number of runs increases. Thus in one simple version of the two-particle EPR experiment, the hypothesis of elements of reality requires a class of outcomes to occur at least 55.5% of the time, while quantum mechanics allows them to occur only half the time"—hence the inequality.

"In the GHZ experiment, on the other hand, the elements of reality require a class of outcomes to occur all of the time, while quantum mechanics never allows them to occur."

Entanglement-Based Quantum Cryptographic Key Distribution (Ekert, 1991)

How to distribute a completely secure secret key:

1. Alice and Bob share a long series of entangled photons.

2. They each measure the polarization of their photons, switching *at random* between (ideally) three bases:

    a. vertical or horizontal

    b. diagonal right or left

    c. circular right or left.

3. Alice calls Bob and tells him the order of the measurements she made. (She reads off the bases: "up-down, up-down, diagonal, circular, circular. . . ." They don't discuss the results.)

4. Bob notes each time they used the same base for the same photon pair. Because of entanglement, they know they got the same result for those cases. (If Bob measured "circular, diagonal, diagonal, up-down, diagonal . . . ," the third item—where they both measured their photons on the diagonal—they will keep.)

5. These identical results become the key for the code.

6. To check for eavesdropping, they use pairs measured with *non*identical bases, and analyze them using a Bell inequality.

7. If the inequality tells them that the photons are still entangled, then they know no eavesdropper is present.

8. Alice encodes her message with the key. She can then give it to Bob any way she wants—e.g., by an ad in the newspaper.

NOTES

ABBREVIATIONS USED IN NOTES

" . . . " separates the first identifying words of a quote from the last words of the quote. The absence of quotation marks around a call-out phrase signals a paraphrase rather than a direct quotation.

Letters are credited as "sender-receiver" (i.e., Rutherford-Bohr indicates a letter from Rutherford to Bohr).

All dates in the twentieth century are written in the format *month/day/year*; any twenty-first century dates are written out.

ABBREVIATIONS FOR LONG NAMES

AE	Albert Einstein
AQHP	Archive for the History of Quantum Physics
AZ	Anton Zeilinger
dB	Louis de Broglie
Eh.	Paul Ehrenfest
EPR	Einstein-Podolsky-Rosen
ES	Erwin Schrödinger
FRS	Fellow of the Royal Society
JRO	J. Robert Oppenheimer
QM	Quantum Mechanics
Somm.	Arnold Sommerfeld
vW	Carl Friedrich von Weizsäcker
WH	Werner Heisenberg

(handwritten annotation) → LLG (See p. 384)

A NOTE TO THE READER

xiii *"Science rests on experiments . . . conversations.":* WH, *Physics and Beyond,* xvii.

xv *"Sommerfeld . . . something else":* Bohr 1961 interview in Pais, *Niels Bohr's,* 229.

xv *"I know that you understand . . . quantum theory is":* Bohr-Richardson, 1918 in Pais, *Niels Bohr's,* 192.

xvi *"That is a wall . . . examination":* AE quoted by the astronomer C. Nordmann, 1922 in Clark, 353.

xvii *"I have not found . . .":* Pais, *Niels Bohr's,* 228.

INTRODUCTION: ENTANGLEMENT

3 *A paper is considered famous:* Web of Science, 2006. Stanford's database for high-energy physics preprints, Spires, suggests these guidelines: Known papers: 10–49

total citations. Well-known papers: 50–99. Famous papers: 100–499. Renowned papers: 500 or more (www.slac.stanford.edu/spires/). (S. Redner did a study of *Physical Review* papers that were subsequently cited by other papers in *Physical Review*. He concluded that "nearly 70% of all *PR* articles have been cited fewer than 10 times" and that "the average number of citations is 8.8." June 2005, *Phys. Today.*) The historian of science Olival Freire notes that comparing Spires's guidelines for high-energy physics papers with statistics for papers on the foundations of quantum physics—which is a different field with many fewer physicists— "should be taken with a grain of salt." (Which only highlights how impressive the number of citations for EPR & Bell's theorem are.) Freire, "Philosophy Enters the Optics Laboratory: Bell's Theorem & Its First Experimental Tests (1965–1982)," *Studies in History & Philosophy of Modern Physics*, 37, 577–616.

4 *the most cited paper:* Howard, "Revisiting," 24.

5 Niels Bohr, *a lifelong friend:* the classic 1927 overview of Bohr's ideas is his "Como Lecture" (*The Quantum Postulate & The Recent Development of the Quantum Theory*), Bohr, *Atomic Theory*, 52–91.

5 *"these abstractions . . . space-time view":* Bohr, *Atomic Theory*, 57.

5 *His enthusiastic supporter:* Werner Heisenberg: these views may be found in Heisenberg's earliest philosophy-of-physics papers (dating from the thirties). See *Philosophical Problems of Quantum Physics.* A caveat about reading his (enjoyable) papers on philosophy of physics: as Pauli often complained, Heisenberg had "no philosophy," but believed what suited his physics at the time (which may well have enabled such amazing breakthroughs). All his philosophical works were once public lectures; during the Nazi period, these lectures necessarily had other agendas (e.g., defending the right of theoretical physics to even exist) beyond pure statements of his views. See also his famous 1959 quote that "the elementary particles are not as real [as phenomena in daily life]." (WH, *Physics and Philosophy*, London: Allen & Unwin, 1958; Jammer, *Philosophy of Quantum*, 205.)

5 Wolfgang Pauli: The best summary of Pauli's ideas on the foundations of quantum physics (which are found throughout his *Writings on Physics & Philosophy*) is actually in the final section (on modern physics) of his famous *Kepler* essay. (Pauli, *The Influence of Archetypal Ideas on the Scientific Theories of Kepler*, 1952, reprinted in Pauli, *Writings on Physics & Philosophy*, 258–61.)

6 *Those who are not shocked . . . understood it":* WH, *Physics and Beyond*, 206.

6 Albert Einstein: Don Howard, in a series of wonderful papers—including "*Nicht Sein Kann Was Nicht Sein Darf*, or, the Prehistory of EPR, 1909–1935"— emphasizes that AE's problem with QM was always its lack of separability.

6 *"not . . . mutually independent":* AE-Lorentz, 5/23/09 in Howard, *Nicht*, 75.

6 *"a mutual influence . . . of a quite mysterious nature":* AE, 1925 in Cornell and Weiman, Nobel Prize lecture, 2001, 79.

6 *"spooky action-at-a-distance":* AE-Born, 3/3/47, *B-E Letters*, 157. In 1927, he called it "a very particular mechanism of action at a distance, which would prevent the wave continuously distributed in space from acting out *two* places" (see Howard, *Nicht*, 92).

6 *"a sort of telepathic . . .":* AE-Cassirer, 3/16/37 in Fine, 104n.

6 Erwin Schrödinger: He was a beautiful writer who left behind many great essays on the quantum theory; the best remains the "cat" paper, published in 1935 (Wheeler and Zurek, 152–67).

6 Louis de Broglie: his "pilot-wave theory," or "theory of the double solution," first

appeared in complete form in 1927 in the *Journal de Physique*. Overview in English in dB, 108–25.

6 *"quantum theory without observers"*: Taken from a title of John Bell's, referring to his own extension of David Bohm's hidden variables theory: "Quantum Field Theory Without Observers, or Observables, or Measurements, or Systems, or Wave Function Collapse, or Anything Like That" (Bell, 173).

6 Paul Dirac: Dirac, "Evolution of the Physicist's Picture of Nature," *Sci. American* (May 1963).

6 Max Born: see *B-E Letters*: AE's amused letter 3/18/48 (162–64); Born's 5/9/48 reply to AE's great paper *QM & Reality* (173–75); and Pauli-Born 1954 (221–27).

7 *"Truth . . . complementary"*: Bohr; e.g., Peierls in *NBCV,* 229; cf. Pais, *Niels Bohr's,* 11, 24.

7 *"I'd rather be . . . right"*: Wheeler to Bernstein (Bernstein, *Quantum Profiles,* 137).

CHAPTER ONE: THE SOCKS 1978 AND 1981

8 *when John Bell first met Reinhold Bertlmann:* Bertlmann to LLG, Nov. 10–18, 2000.

8 *plastic shoes:* Renate Bertlmann to LLG, Feb. 28, 2001.

9 *"I'm John Bell"*: Bertlmann to LLG, Nov. 10–18, 2000.

9 *"He did not like . . . 'How do you know?,' "*: Peierls, "Bell's Early Work," *Europhys. News* 22 (1991), 69.

9 *"John always stood . . . reasoning"*: John Perring in Burke and Percival, 6.

9 *"rottenness" . . . "dirtiness" . . . "unprofessional"*: When Davies asked if the difficulties with QM were purely philosophical, Bell replied, "I think there are *professional* problems. That is to say, I'm a professional theoretical physicist & I would like to make a clean theory. & when I look at QM I see that it's a dirty theory" (Davies, *Ghost,* 53–54). "I hesitated to think it might be wrong, but I *knew* that it was rotten" (Bernstein, *Quantum Profiles,* 20).

9 *"Without such an assumption . . . separation"*: AE, *QM & Wirklichkeit,* translated in Howard, "Einstein on Locality," 187. (Note that Howard finds the English translation of this paper in *B-E Letters* flawed in several places and provides his own in this article.)

10 *"Do you really believe . . . looks?"*: As AE asked Pais.

10 *"I'm Reinhold" . . . "working on?"*: Bertlmann to LLG, Nov. 10–18, 2000.

10 *Three years later* (e.g., the location, Ecker's words, and Bertlmann's thoughts): Bertlmann to LLG, Nov. 10–18, 2000. Bell, "Bertlmann's Socks and the Nature of Reality," *Journal de Physique,* Colloque C2, suppl. au #2, Tome 42 (1981), pp. C2 41–61; Bell, 139–58.

13 *"What have you done?" and subsequent phone conversation with Bell:* Bertlmann to LLG, fall 2000.

14 *Bell liked to talk . . . :* Bernstein, *Quantum Profiles,* 63. More on the "Jim Twins" can be found in Lawrence Wright, *Twins: And What They Tell Us About Who We Are,* New York: Wiley, 1997, 43–48.

15 *"to something so simple . . . all"*: Mermin, *Boojums,* xv.

15 *"something between a parable . . . demonstration"*: Ibid., 82.

15 *In the center is a box . . . :* Ibid., 82ff.

16 *"These statistics may . . . floor"*: Ibid., 87–88.

16 *"Given the unconnectedness . . . positions"*: Ibid., 88.

17 *But particles such as these could never produce:* The results only get stranger and more extreme if we add in particles bearing the other configurations of genes. (If

you are so inclined, it is fun to play around with these predictions yourself, comparing them to the "actual" results summarized under "case (1)" and "case (2)" on pages 16–17.)

18 *The number of young . . . otherwise:* d'Espagnat; Bell, "Bertlmann's Socks" (1981); Bell, 157.

19 *"It has been feared . . . light":* Bell, "Atomic-Cascade Photons & QM Nonlocality," 7/10/79, in Bell, 105–6.

19 *Space, Einstein said:* "We entirely shun the vague word 'space,' . . . we replace it by 'motion relative to a practically rigid body of reference' " (AE, *Relativity,* 10).

19 *time is what we measure:* "If, for example, I say that 'the train arrives here at 7 o'clock,' that means, more or less, 'the pointing of the small hand of my watch to 7 and the arrival of the train are simultaneous events.'" (AE, 1905, repr. in Stachel, 125.)

20 *Rigging up something:* A. Aspect, J. Dalibard, and G. Roger, "Experimental test of Bell's inequalities using time-varying analyzers," *Phys. Rev. Letters* 49, 91–94, 1804–7 (1982).

20 *"One of the most . . . appeared":* Feynman-Mermin 3/30/84, in Feynman, *Perfectly Reasonable Deviations from the Beaten Track* (New York: Basic Books, 2006).

20 *At the Massachusetts Institute of Technology:* "Simulating Physics with Computers" in Hey, 133ff.

THE ARGUMENTS 1909–1935

CHAPTER TWO: QUANTIZED LIGHT SEPTEMBER 1909–JUNE 1913

25 *Then Einstein, in 1905:* AE, "On a Heuristic Point of View Concerning the Production & Transformation of Light," *Annalen der Physik* 17: 132–48.

26 *Einstein was embarking on a quest:* Many thanks to Don Howard's beautiful papers, particularly "*Nicht Sein Kann* . . ." for highlighting the theme of separability in Einstein's quantum mechanics papers.

26 *"I am incessantly . . . radiation":* AE-Laub (dated "Monday") 1908 in Clark, 145–46.

26 *"an utterly honest . . . himself":* AE-Zangger, 11/1911 in Einstein and Maric, 98.

26 *"He has, however . . . thought":* AE-Laub, "Monday" 1908 in Clark, 145.

26 *"astonishingly profound":* AE-Laub, "Monday" 1908 in ibid., 145–46.

26 *"I admire this man . . . him":* AE-Laub, 5/19/09 in Pais, *'Subtle,'* 169.

27 *"mutually independent energy quanta":* AE, 1905 in Stachel, 191.

27 *"the ultraviolet catastrophe":* Eh., "Which Features of the Hypothesis of Light Quanta Play an Essential Role in the Theory of Thermal Radiation?" 10/1911 in Klein, *Paul Ehrenfest,* 174, 245–51. Ehrenfest's original wording of the phrase came in a subtitle, "Avoidance of the Rayleigh-Jeans Catastrophe in the Ultraviolet." (The Rayleigh-Jeans law, dealing with radiation—as waves—confined to a box, was derived by the British physicists Lord Rayleigh and James Jeans.)

27 *"This quantum . . . everyone":* AE-Laub, "Monday" 1908 in Clark, 145–46.

27 *"I must have expressed . . . spaces":* AE-Lorentz, 5/23/09 in Howard, *Revisiting,* 6.

27 *Dr. Ruess, who taught . . . :* Folsing, 27.

27 *"It is my opinion . . . theories":* AE, Salzburg, 1909 (Holm, trans.) in Weaver, 295–309.

28 *"I will restrict . . . speaker"* and all Planck's, Stark's, and Einstein's post-speech comments: Weaver, 309–12.

29 *"I have not yet . . . of mine":* AE-Laub, 12/31/09 in Pais, *'Subtle,'* 189.

29 *"The enigma . . . yield"*: AE-Laub, 12/28/10 in ibid.

29 *"I do not ask . . . direction"*: AE-Besso, 5/13/11 in Bernstein, *Quantum Profiles*, 158–59.

29 *"At present . . . gravitation"*: AE-Somm., 10/29/12 in Pais, 'Subtle,' 216.

29 *"one cannot . . . replace it"*: AE-Wien, 5/17/12 in Levenson, 279.

29 *It was a warm June evening*: This gathering of AE, Eh., and Laue is discussed in Klein, *Paul Ehrenfest*, 294–95.

30 *"on your wonderful . . . physics"*: AE-Laue, 6/10/12 in Folsing, 323.

30 *Handsome, thoughtful, and upright*: Nobel biography; Clark, 142, 144, 195.

30 *cigars so awful*: Clark, 144.

30 *"a little masterpiece"*: AE to Alfred Kleiner of the University of Zurich. Ibid., 195, footnotes for this letter missing.

30 *Ehrenfest, on the other*: Klein, *Ehrenfest*, particularly 92–93.

30 "That's *where the frog . . .*" and ". . . *patent claim!*": Casimir, 68.

30 *"You should be careful . . ."*: Laue-Eh., 1/18/12 in Einstein and Maric, xvii.

30 *talking for five days*: Eh.'s diary in Klein, *Ehrenfest*, 294.

30 *"I have understood it!"*: Laue to Seelig, 3/13/52 in Clark, 195.

31 *"The more success . . . looks"*: AE-Zangger, 5/20/12 in Pais, 'Subtle,' 399.

31 *"Those are the madmen . . ."*: AE to P. Frank in Bernstein, *Quantum Profiles*, 204.

CHAPTER THREE: THE QUANTIZED ATOM NOVEMBER 1913

32 *They were hiking:* "It was not long after the publication of Bohr's papers that Stern and von Laue went for a walk up the Uetliberg," as Pais recalled Stern telling him. "On the top they sat down and talked about . . . the new atom model. There and then they made the 'Uetli Schwur:' If that crazy atom model of Bohr turned out to be right, then they would leave physics. It did and they didn't" (Pais, *Inward*, 208). Schiller, in his play *Wilhelm Tell*, famously turned the *Ruetli Schwur* into poetry ("there is a limit to the tyrant's power . . ."). For another example of physicists altering *Wilhelm Tell* for their own purposes, see Born, *My Life*, 87.

32 *UETLIBERG HELL:* as described by Moore, 152.

33 *"you do less damage . . ."*: Frisch, 42–47.

33 *"beautiful days"*: Pais, 'Subtle,' 486.

33 *the inexplicably stable atom*: Bohr, "On the Constitution of Atoms and Molecules," *The London, Edinburgh, and Dublin Philosophical Magazine and Journal of Science*, 26, 1 (July 1913); 476 (September 1913); 857 (November 1913).

34 *"It's just absurd"*: Pais, *Neils Bohr's*, 147, 154. Jeans, 1913: "The only justification put forward for Dr. Bohr's bold postulates is the weighty one of success." Dresden, 24.

34 *"but this is nonsense . . ."* and *"Very curious . . . basic constants"*: Laue and AE's words at this colloquium (where AE first heard of the Bohr atom) were reported to Dresden by an eyewitness in 1964. Dresden, 24.

36 *"At present Kirchhoff . . . spectrum"*: Bunsen in Mary E. Weeks and Henry M. Leicester, "Some Spectroscopic Discoveries," *Discovery of the Elements*, Easton, PA: Journal of Chemical Education, 1968, 598.

36 *"Hitherto unknown elements"*: "Chemical Analysis by Observation of Spectra," Kirchhoff and Bunsen, *Annalen der Physik und der Chemie*, 110, 161–89 (1860); translations on the Internet.

36 *"I have been very fortunate . . . weight"*: Weeks and Leicester, "Some Spectrocospic Discoveries," 599.

36 Handbuch der Spectroscopie: Vol. 1, 800 pgs; Pais, *Niels Bohr's,* 139–41.

36 *"marvelous, but . . . butterfly":* Bohr to Kuhn; Pais, *Niels Bohr's,* 142.

36 *Born in a mansion in Copenhagen:* Ibid., 43–48, etc.

36 *"the ancients . . . firmament":* Bunsen and Kirchhoff, *Chemical News* 2, 281 (Nov. 24, 1860).

37 *"Onward, Christian Soldiers":* Pais, *Inward,* 437.

37 *outdoorsmen:* The 8/26/27 *Sunday Dispatch* describes Rutherford's hearty, smiling, tweedy appearance as "aggressively agricultural" (Eve, 324); Bohr enjoyed hiking and all sorts of manual labor, including chopping down trees.

37 *"is chosen as . . . Einstein":* Bohr-Rutherford, 7/6/12 in Pais, *Niels Bohr's,* 138.

37 *"unless you can . . . barmaid":* as Bohr unnecessarily warned the clear-writing Gamow (GG, *My World Line,* 66–67). Rutherford-Bohr, 1913: "Long papers have a way of frightening readers. It is the custom in England to put things very shortly & tersely in contrast with the Germanic method, where it appears to be a virtue to be as long-winded as possible" (Eve, 220).

38 *"I asked him about . . .":* Hevesy-Bohr, 9/23/13 in Pais, *Inward,* 208.

38 *"faint praise . . . occasions":* Pais, *Niels Bohr's,* 154.

38 *"in exact agreement . . .":* Bohr, *Nature* 92, 231, 1913 in Pais, *Niels Bohr's,* 149.

38 *"The big eyes . . .":* Hevesy-Rutherford, 10/14/13 in Eve, 226.

38 *"He was extremely . . . wright":* Hevesy-Bohr, 9/23/13 in Klein, *Paul Ehrenfest,* 278.

38 *"I felt very happy . . .":* Hevesy-Rutherford, 10/14/13 in Eve, 226.

38 *"Bohr's work . . . physics":* Eh.-Lorentz, 8/1913 in Klein, *Paul Ehrenfest,* 278.

38 *"And now you've pulled . . . soup!":* Casimir, 68.

38 *"completely monstrous":* "Even though I consider it horrible that this success will help the preliminary, but still completely monstrous, Bohr model on to new triumphs, I nevertheless heartily wish physics at Munich further successes along this path!" Eh.-Somm., 5/1916 in Klein, *Paul Ehrenfest,* 286.

38 *"a general point . . . quanta":* Eh., "Adiabatic Invariants and the Theory of Quanta," 1917 in Bolles, 24.

39 *"soft voice . . . considered":* G. P. Thomson, Niels Bohr Memorial Lecture, 1964; French and Kennedy, 285; Bohr "constantly appealed to the understanding of his partner in conversation, with very little hope of success but always with tireless optimism" (vW, "A Reminiscence from 1932" in French and Kennedy, 185).

39 *hypnotizing drawings of electron orbits:* see Kragh, "The Theory of the Periodic System," French and Kennedy, 50–60.

39 *"Ehrenfest writes enthusiastically . . . skeptical fellow":* AE-Borns, 12/30/21; *B-E Letters,* 65.

CHAPTER FOUR: THE UNPICTURABLE QUANTUM WORLD SUMMER 1921

40 *"Do you honestly believe . . . atom?":* WH attributes this to Pauli, but his portrayal of Pauli in their student days as deferring to himself does not ring quite true. At any rate, they discussed the question and agreed on the answer (WH, *Physics and Beyond,* 35–36).

40 *Heisenberg was nineteen:* 12/5/01; Pauli's birthday: 4/25/00.

40 *Laporte was about to turn:* picture and biography in H. R. Crane and D. M. Dennison, *Biographical Memoirs* vol. 50, 268–85 (Washington, D.C.: National Academy Press, 1979).

40 *"fellow sufferers" in W. W.'s physics class:* Enz, 54.

41 *"probably the only . . . world":* WH, *Physics and Beyond,* 29.

41 *"the whole . . . myth"*: Pauli on orbits, ibid., 36.

41 secretive face: ibid., 24.

41 *"like a simple farm boy"*: Born, *My Life*, 212.

41 *In 1920, Pauli had written . . .":* "Pauli's article for the Encyclopedia is apparently finished, & the weight of the paper is said to be 2.5 kilos. This should give some indication of its intellectual weight. The little chap is not only clever but industrious as well": Born-AE, 2/12/21 in *B-E Letters*, 53–54.

41 *Heisenberg liked his directness . . . .*: "When we talked about physics, we were often joined by our friend, Otto Laporte, whose sober, pragmatic approach made him an excellent mediator between Wolfgang and myself. . . . It was probably due to him that the three of us . . . decided to go on a bicycle tour. . . ." WH, *Physics and Beyond*, 28.

41 *"The talks we began . . . effect"*: ibid., 29.

41 *"Bohr has succeeded . . . contradictions"*: Pauli, half a year before this bike trip. Ibid., 26.

42 *Well, said Laporte:* "Philosophy is the systematic misuse of nomenclature specially invented for the purpose. All absolute claims must be rejected *a priori*. We ought only to use such words and concepts as can be directly related to sense perception. . . . It is precisely this return to observable phenomena that is Einstein's great merit." Ibid., 30; see also p. 34.

42 *Ahh, Mach:* "Your suggestion, which sounds so terribly plausible, has already been made by Mach" (Pauli to Laporte; ibid., 34). Pauli's letters to WH in the '20s are full of joking religious metaphors and jargon; the Pauli of these letters is less serious and deferential than the Pauli WH portrays in his memoir. Heisenberg wrote, "Pauli had a very strong influence on me. I mean Pauli was simply a very strong personality. . . . He was extremely critical. I don't know how frequently he told me, 'You are a complete fool,' and so on. That helped a lot" (Cassidy, 109).

42 *"I am baptized antimetaphysical . . . generally valid"*: Pauli-Jung, 3/31/53; Enz, 11 and 13. Pauli's middle name was Ernst, after Mach.

42 *". . . a crude oversimplification"*: Pauli in WH, *Physics and Beyond*, 34.

42 *"Mach did not believe . . . not by chance"*: Pauli to Laporte in ibid.

42 *"Mistakes are no excuse . . . are"*: Laporte to Pauli in ibid.

42 *"atomysticism"*: Somm. "believes in numerical links. . . . That's why many of us have called his science *atomysticism*" (Pauli to WH in ibid., 26).

43 *"atomic music of the spheres"* Somm., *Atombau und Spektrallinien* in Cassidy, 116. Later, Pauli actually got very interested in this aspect of Kepler.

43 *"Success sanctifies the means"*: WH-Pauli on WH's half-quanta, 11/19/21 in Cassidy, 125.

43 *"perhaps it's easier . . . success"*: Pauli to WH, "malicious grin" in WH, *Physics and Beyond*, 26.

43 *How I let myself get dragged:* For WH's recollection of Pauli joking about "living by the principles of St. Jean-Jacques," and of himself making fun of Pauli's late rising, see WH, *Physics and Beyond*, 28; Born describes Pauli's sleeping patterns in *B-E Letters*, 63; compare WH, *Physics and Beyond*, 27.

43 *"He sleeps in a tent"*: WH describes this life beautifully in *Physics and Beyond*, 27–28.

43 *As a child he had been:* Cassidy, 14–17.

44 *Pauli's childhood:* Enz, 7–10, 15, 51, 53. Pauli's childhood *"war ihm immer fad"*— "was to him always stale" ("a uniquely Viennese expression") (Enz, 11).

44 *Born was a self-sacrificing:* Greenspan, 5–8, 155–58.

44 *their "dearest friend":* the final words in *B-E Letters,* 234.

44 *"dark, depressing . . . Einstein":* Born in Greenspan, 75.

45 *"Theoretical physics . . . Germany today":* AE-Born, 3/3/20 in *B-E Letters,* 25.

45 *With a knife-shaped magnet:* On Stern, Gerlach, and their experiment: Pauli-Gerlach: Enz, 78–79; on Gerlach in Frankfurt: Bernstein, *Hans Bethe,* 12; *B-E Letters,* 53; Greenspan, 102–3; on division of labor: Frisch, 24, 44. I. I. Rabi said later, "I found the quantum theory hard to believe—the old quantum theory—until I heard about the Stern-Gerlach experiment. I thought the old quantum theory was stupid. I thought one might be able to invent another model of the atom which had the same properties. But you can't get around the Stern-Gerlach experiment. You are really confronted with something quite new. It goes on in space, and no clever classical mechanism would do, would explain it." Rabi to Bernstein, *The New Yorker,* 10/13/75, p. 75. AE wrote in 1922, "The most interesting thing at the moment is Gerlach's and Stern's experiment. The orientation of atoms without collisions cannot be explained by means of radiation . . . an orientation should, by rights, last more than a hundred years. I made a little calculation about it with Ehrenfest. . . ." (AE-Born, *B-E Letters,* 71).

45 *"the apparent . . . magnetic field":* Bohr, *Atomic Physics,* 37.

46 *"This peculiar mixture . . . students":* WH, *Physics and Beyond,* 35.

46 *"That is absolutely . . . numbers":* Somm. to WH and WH int. in Cassidy, 118–19, 122.

46 *"the whole quantum . . . hands":* Pauli to WH in WH, *Physics and Beyond,* 35.

46 *Heisenberg, on his bike . . . :* "Once we had reached the saddle of the Kesselberg, laboring uphill with our bicycles, we could ride effortlessly along a road boldly cut into the mountain slope past the steep western shore of Walchensee. . . . Across this dark lake, Goethe caught his first glimpse of the snow-covered Alps": Ibid., 29.

47 *But for Heisenberg:* This paragraph owes much to Cassidy's vivid description of WH's teenage years. Particularly interesting is his analysis of the creative and destructive aspects of the Youth Movement (Cassidy, Ch. 2–5).

48 *"A saying occurs . . . my life":* WH-mother, 12/15/30 in Cassidy, 289.

CHAPTER FIVE: ON THE STREETCAR SUMMER 1923

49 *Niels Bohr is riding:* "Sommerfeld was not impractical, not quite impractical; but Einstein was not more practical than I and, when he came to Copenhagen, I naturally fetched him from the railway station. . . . We took the streetcar from the station and talked so animatedly about things that we went much too far past our destination. So we got off and went back. Thereafter we again went too far, I can't remember how many stops, but we rode back and forth in the streetcar because Einstein was really interested at that time; we don't know whether his interest was more or less skeptical—but in any case we went back and forth many times in the streetcar and what people thought of us, that is something else" (Bohr to Aage Bohr and Rosenfeld in 1961 in Pais, *Niels Bohr's,* 229). (The ferry port is less than two miles from Bohr's institute. Bohr seems to be confusing this trip, when AE was coming by ferry from Sweden, with other times when he would—"naturally"— have been coming by train from Berlin.)

49 *suddenly world-famous: New York Times,* 11/18/19 in Pais, 'Subtle,' 309.

49 *"According . . . Japan":* Laue-AE, 9/18/22 in Pais, 'Subtle,' 503.

49 *On June 24, 1922:* Levenson, 270.

49 *"It is no art . . . anyone else"*: Einstein on Rathenau, 1922 in Pais, 'Subtle,' 12.

50 *"miraculous" compass*: AE, "Autobiographical Notes" in Schilpp, ed., *AE: Philosopher Scientist*, 9.

50 *"holy geometry book"*: Ibid., 11.

50 *Einstein walked . . . to Planck's*: Bolles, 57–62.

50 *"Neutralia . . . scientific matters"*: AE-Bohr, 5/2/20 in Pais, *Niels Bohr's*, 228.

50 *"What is so marvellously attractive . . . critical sense"*: AE writing around 1922, *The World as I See It*, 162.

50 *"To me it was . . . to your home"*: Bohr-AE, 5/2/20 in Pais, *Niels Bohr's*, 228.

51 *"I shall not . . . so remarkably suited"*: Bohr, "The Theory of Spectra & Atomic Constitutions" in Dresden, 140.

51 *"Bohr was here . . . trance"*: AE-Eh., 5/4/20 in Pais, 'Subtle,' 416f.

51 *"It is a good omen . . . splendid people"*: AE-Lorentz, 8/4/20 in Pais, *Niels Bohr's*, 228.

51 *"Doesn't he look . . . officer?"* Pauli to WH in WH, *Physics and Beyond*, 24.

51 *If Sommerfeld was hatless*: O'Connor and Robertson, "Sommerfeld," MacTutor Web site, www-groups.dcs.st-and.ac.uk, Oct. 2003.

51 *"He had the rare ability . . . pupils"*: Born, "Arnold Johannes Wilhelm Sommerfeld," Obituary Notices of Fellows of the Royal Society of London 8 (1952), 275–96.

51 *"Such a beautiful . . . unique"*: AE-Somm., 9/29/09 in AE, *Collected Papers*, Vol. 5, 179.

52 *"For me it was the greatest . . . an honor"*: Bohr-AE, 11/11/22 in Pais, *Niels Bohr's*, 229.

52 *"The hypothesis of light-quanta . . . radiation"*: Bohr, Nobel lecture, 1922 in ibid., 233.

52 *"Dear or rather beloved Bohr! . . . Einstein"*: AE-Bohr, 1/11/23 in French and Kennedy, 96; compare Pais, *Niels Bohr's*, 229.

52 *the prize was deferred*: see Pais, 'Subtle', 508–11.

53 *"Number 15"*: Around the time of World War II, Blegdamsvej was renumbered, and the Bohr Institute's address these days is 17.

53 *Bohr's newborn institute*: Pais, *Niels Bohr's*, 170–71.

54 *"Bohr reads the language . . . magazine"*: Hevesy-Rutherford, 5/26/22 in ibid., 385.

54 *"We quite innocently . . . theory"*: "how great a comfort your kind letter was to us all in this terrible muddle about the new element, in which we quite innocently have dropped. We had never dreamt of any competition with the chemists in the hunt for new elements, but wished only to prove the correctness of the theory. . . . Urbain . . . tries however to shift the whole matter, paying no regard to the important scientific discussion of the properties of the element 72, but tries only to claim a priority": Bohr-Rutherford; French and Kennedy, 64.

54 *Dauvillier*: Alexandre Dauvillier was Maurice dB's main assistant throughout the '20s; his results supported the priority of Georges Urbain at the Sorbonne, who had claimed that element 72, like 71, was a rare earth, and not—as Bohr had predicted, and Coster and Hevesy discovered—an analog of titanium (22) and zirconium (40). L. dB and Dauvillier, in a series of papers written together during 1921 through 1925 on this and related subjects, repeatedly found themselves contradicting the ideas and results of Bohr and his institute. See Raman and Forman ("Why Was It Schrödinger . . . ?"), who note that the Broglie-Dauvillier arguments always seemed overly vehement and inadequately scientific to the Bohr Institute;

the whole thing was seen at the time, at least by German observers, to be founded in nationalism (note the time period).

54 *"celtium," "oceanum," "hafnium," "danium," "jargonium"*: Pais, *Niels Bohr's*, 210.

54 *Einstein roared:* "the enormous contrast between [AE's] soft speech and his ringing laughter": Cohen; French, *Einstein*, 40.

54 *"My life from the scientific . . . theory is"*: Bohr-Richardson, 1918; Pais, *Niels Bohr's*, 192.

54 *"That is a wall . . . quanta"*: AE quoted by astronomer Nordmann, 1922; Clark, 353.

54 *"This crazy model"*: using the half-integers which appalled Somm., WH had produced results with an ad hoc model (Cassidy, 120–21).

55 *"very interesting . . . assumptions"*: Bohr on WH's Dec. 1921 model at the "Bohr Festspiele," Göttingen, June 1922 in ibid., 128–30.

55 *"Everything works out . . . philosophy"*: Somm.-AE, 6/11/22 in Cassidy, 123–24; Dresden, 37.

55 *"The question is not . . . help us"*: Bohr, 1921 draft in Pais, *Niels Bohr's*, 193.

55 *"I no longer doubt . . . conviction"*: AE-Besso, 7/29/18 in Dresden, 31.

55 *"The most interesting . . . become invalid"*: Somm.-Bohr, 1/21/23. Saying that he is not certain that Compton could be right, but lectured on the Compton effect wherever he went; Ibid., 160.

55 *"You can understand . . . creed"*: Bohr-Rutherford on Compton, 1/27/24; Ibid., 31.

55 *"The theory of light . . . propagation"*: 1924 Bohr-Kramers-Slater paper; Ibid., 140.

55 *"excludes in principle . . . frequency"*: Bohr, *Zeitschrift für Physik* 13, 117 (1923); Dresden, 140.

55 *"defined by . . . light"*: 1924 BKS paper; Dresden, 140.

55 *"to call . . . new insight"*: Somm.-Bohr, 1/21/23; Ibid., 160.

55 *"even if Einstein . . . radio waves"*: as WH remembered Bohr saying in ibid., 31.

56 *"We have missed our stop"*: see first note for this chapter.

56 *"really able to prove . . . gratings?"*: Bohr interview, 7/12/61 in Pais, *Niels Bohr's*, 232.

56 *"Or conversely . . . photocells?"*: Bohr interview, 7/12/61 in Pais, who notes that Bohr in 1920 was unlikely to have asked this second question; Ibid.

56 *"must I say . . ."*: See Gamow, *Thirty Years*, 215.

56 *"at the present time . . . light and matter"*: Bohr, 2/13/20 Copenhagen lecture; Dresden, 141.

56 *"There are now . . . connection"*: AE in the *Berliner Tageblatt*, 4/20/24 in Pais, '*Subtle,*' 414.

56 *"We must persist . . . the theorist"*: "But as long as the principles capable of serving as starting-points for the deduction remain undiscovered, the individual fact is of no use to the theorist; indeed he cannot even do anything with isolated empirical generalizations of more or less wide application. No, he has to persist in his helpless attitude toward the separate results of empirical research until principles which he can make the basis of deductive reasoning have revealed themselves to him." Einstein's inaugural address to the Prussian Academy of Sciences (1914), AE, *The World As I See It*, 128.

57 *"One thinks of Bohr . . . Newton too"*: Franck interview, 7/10/62 in Pais, *Niels Bohr's*, 4.

57 *"during a stage . . . everything"*: Bohr-Somm., 5/30/22 in Dresden, 43.

57 *"I cannot quite understand"*: What follows is a close paraphrase of Bohr's memory

of what he said, including "famous papers" and "too crazy." Bohr continues: "It was one of the most brilliant strokes of genius, it was almost the decisive point I reported on these discussions [before] . . . a little courteously. But, with Einstein's sense of humor, one could easily say everything possible to him" (1961 interview in Pais, *Niels Bohr's,* 231–32).

57 *"I have full confidence . . . taken":* AE, 1917; Pais, '*Subtle,*' 411.

57 *"I am inclined . . . imaginable":* Bohr-Darwin, 1919, on the interaction between light and matter in Bolles, 47.

58 *"what* possible *calculations":* This is a question physicists were asking Bohr ever since 1913. The terrific results achieved with no apparent calculation is what Einstein is pointing out when he says he admires "the sure instinct" that guides Bohr's work. See, e.g., Pais (*Niels Bohr's,* 205) for Somm., Rutherford, and Franck asking Bohr for the math behind his work on the periodic table.

58 *'mathematical chemistry':* "Bohr . . . had little confidence in the 'mathematical chemistry' that Sommerfeld, Max Born, and other German theorists regarded as an ideal. Instead, he relied on a sort of intuitive understanding": Kragh, "The Theory of the Periodic System"; French and Kennedy, 59.

58 *everything does not always . . . calculations:* Hoyt: "[Bohr] thought that in every *fine point* that came up Sommerfeld was wrong." Franck: "Born was to [Bohr] too much of a mathematician." WH: "Bohr was not a mathematically minded man . . ." Pais, *Niels Bohr's,* 178–79.

58 *"analogy principle" . . . "correspondence principle":* Ibid., 193.

58 *"magic wand":* Somm., *Atombau & Spekrallinien,* 1922 in ibid.

58 *I greatly admire . . . work:* AE-Born about Bohr, 1922 in *B-E Letters,* 71.

58 *"abandonment . . . emergency":* AE, Pais, '*Subtle,*' 420.

58 *"We rode back and forth . . . else":* Bohr interview, 1961 in Pais, *Niels Bohr's,* 229.

59 *"Your work . . . theory":* Somm.-Compton in Moore, 160.

59 *"cares more . . . experiment":* WH-Pauli, 1/15/23 in Dresden, 42.

CHAPTER SIX: LIGHT WAVES AND MATTER WAVES
NOVEMBER 1923–DECEMBER 1924

60 *"You know those difficulties . . . assume":* Slater-mother, 11/8/23 in Dresden, 161.

60 *"decidedly excited":* Slater-parents, 1/2/24 in ibid., 164.

61 *"five to six . . . compute":* Berlingske Tidende, 1/23/24 in Pais, *Niels Bohr's,* 260.

61 *"Prof. Bohr wanted . . . himself":* Slater-parents, 1/2/24 in Dresden, 164.

61 *"The theory of . . . patient":* Kramers, 1923 in ibid., 143.

61 *"Of course they don't . . . have to":* Slater-parents, 1/2/24 in ibid., 164.

61 *"we agree much more than you think":* a famous Bohr saying.

61 *"virtual field":* Dresden, 162.

61 *"virtualization of physics":* WH-Bohr, reporting Pauli's words, 1/8/25 in Cassidy, 190.

61 *the fastest-written paper of his life:* Pais, *Niels Bohr's,* 235 and Dresden, 164.

61 *"I haven't seen it yet":* Slater-parents, 1/13/24 in Pais, *Niels Bohr's,* 235.

61 *"simplicity . . . rational causation":* Slater, *Nature* 116, 278 (1925); Dresden, 165.

62 *"You succeeded . . . perception":* Pauli-Bohr, fall 1924 in Enz, 158.

62 *took him on a trip:* Cassidy, 172–73.

62 *But still better was when Bohr:* WH, *Physics and Beyond,* 46–57.

62 *"Now everything . . . difficulties":* Bohr to Hoyt in Pais, *Niels Bohr's,* 264.

62 *"Bohr's opinion . . . entirely mine"*: AE-Born, 4/29/24 in *B-E Letters*, 82.

62 *newspapers all over Europe*: Dresden notes Danish, German, and Dutch papers, Dresden, 207.

62 *conflict between Einstein and Bohr*: Pais notes that the word "conflict" was used by AE, Pais, 'Subtle,' 420.

62 *"I was in Copenhagen . . . opposition"*: Haber-AE, 1924 in Pais, *Niels Bohr's*, 237.

63 *In April 1924 . . . a month later*: Ibid., 259–60, 262.

63 *known to her friends as*: Dresden, 115ff, 282, 483, 526ff.

63 *"the best beer in the world"*: Gamow, *Thirty Years*, 49, and drawing, "Carlsberg Beer and Its Consequences," 50. Gamow himself arrived penniless at the Institute in 1928 and stayed there on a Carlsberg fellowship. Pais, *Niels Bohr's*, 19, 117, 256ff; Gamow, *My World Line*, 64.

63 *They say that Hans has started a tradition*: On Danish wives of Bohr's students: Pais, *Niels Bohr's*, 168; Weisskopf, 8.

63 *Let me tell you . . . all round since*: Mrs. Kramers's story is as remembered by her children (["insanely excited," "daily no-holds-barred arguments," "exhausted, depressed, and let down"]; Dresden, 289–98, also 479; also Pais, *Niels Bohr's*, 238), and paraphrases of Dresden and Pais.

64 *"I liked Kramers . . . in Copenhagen"*: Slater interview, 10/3/63 in Pais, *Niels Bohr's*, 239. Slater returned to a Copenhagen conference in 1951: "If it had not been for one thing I would have thought that Bohr mellowed completely in his old age and I would have forgotten about my feelings concerning him and Copenhagen more than 25 years earlier. . . . When [Brillouin finished his talk on thermodynamics and information theory], Bohr got up and attacked him with a ferocity that was positively inhuman. I have never heard one grown person castigate another in public, emotionally, and without any reason whatever, as far as I could judge, the way Bohr mistreated Brillouin. After that exhibition, I decided that my distrust of Bohr dating from 1924 was well justified" (Dresden, 169). Compare Gleick, 53–54; Beller, 259. A strange sidenote to the Bohr-Kramers-Slater story is how Slater married Kramers's true love: Dresden, 527–28.

65 *the older de Broglie . . . light-quanta*: Bolles, 188.

65 *"probably a bit astonished . . . ideas"*: dB in Pais, 'Subtle,' 436–38.

65 *"Read it . . . solid"*: AE-Born in Klein's intro for Przibram, xiv.

65 *"He has lifted . . . veil"*: AE-Langevin, no date in Moore, 187.

65 *"not consistent . . . problems"*: Kramers, 1923 in Raman and Forman, 294.

65 *de Broglie had . . . condescended*: Ibid., 295–96.

65 *The minutes . . . nonsense*: Blackett AHQP interview, 1962.

65 *Satyendra Nath Bose . . .*: O'Connor and Robertson, "Bose," *MacTutor History of Mathematics* Web site, www-groups.des.st-and.ac.uk, Oct. 2003. Accessed May 19, 2007.

66 *"If you think . . . für Physik"*: Bose-AE, 6/1924 in Bolles, 205.

66 *"we are all your pupils"*: Ibid.

66 *"I am anxious . . . of it"*: Bose-AE, 6/1924 in Pais, 'Subtle,' 425.

66 *"Bose looked back . . . "*: Chatterjee, Santimay, and Enakshi, *Satyendra Nath Bose* (New Delhi, India: National Book Trust, 1976), 82.

66 *"express indirectly . . . mysterious kind"*: Einstein, Preussische Akademie der Wissenschaften, 1925, in Cornell and Weiman, Nobel lecture, 2001. Eh. and ES had complained about the nonseparability of the particles; this, his second paper on Bose, was AE's reply. ES still wondered; AE (2/28/25) responded, "The quanta . . .

are not treated as *independent of one another.* . . . There is certainly no error in my calculation" (Pais, 'Subtle,' 430; Moore, 183; Howard, *"Nicht,"* 67).

66 *"This mutual influence . . . causes":* Cornell and Weiman, Nobel lecture, 2001.

67 *"From a certain . . . to it?":* AE-Eh., 11/29/24; Pais, 'Subtle,' 432.

67 *"Modern physics . . .* non lignet": Somm., *Atombau;* Dresden, 206. For *non-lignet: The Century Dictionary* (New York: The Century Co., 1913).

67 *"Your frank 'non . . . Slater":* Pauli-Somm., 12/6/24 in Dresden, 206.

67 *"One now . . . quantum world":* Pauli-Somm., 12/6/24 in Cassidy, 194. He had submitted his paper on the exclusion principle four days before.

CHAPTER SEVEN: PAULI AND HEISENBERG
AT THE MOVIES JANUARY 8, 1925

68 January 8, 1925: Cassidy, 190, 582 (note 40). For WH & Pauli as Chaplin fans, see Cassidy, 196.

69 "Schwindlig" "Unsinn": German critics, 1921–22: "Chaplin's incredible nonsense," "silly American clown tricks," "perfect nonsense," "nothingness"—"While critics carped, audiences responded with gales of laughter which by all accounts were without precedent in German cinemas." Saunders, 174–75 (a thorough and fascinating book).

69 *skiing accident:* Cassidy, 193.

69 *"Pope Bohr," "His Eminence Kramers":* WH-Pauli, 6/8/24 is the first reference to Bohr as Pope. Pauli-WH, 2/28/25 is the first reference to Kramers as a Cardinal (7/1925, when WH displaced Kramers at Bohr's side, is the last such reference). Dresden, 268, 272, and compare 137n.

69 *You know . . . met Bohr:* WH said, on meeting Bohr, "My real scientific career only began that afternoon" in WH, *Physics and Beyond,* 38. "A new phase of my scientific life began when I met Niels Bohr": Pauli, "Remarks on the history . . . ," *Science* 103, 213 (1946); Enz, 88. Compare Dresden, 253.

69 *"Bohr is more worried . . . talk of anything else":* WH to Kuhn in Dresden, 262.

70 *"a bit strange to me . . . jokes made":* WH said, "Kramers would not take these difficulties so seriously as Bohr did": WH, Mehra & Rechenberg interview; Ibid., 266.

70 *"the papal blessing":* WH-Pauli, 6/8/24, re: WH's "Zeeman salad" (his semi-quantum, semi-classical explanation of the Zeeman effect), which would be published "with the papal blessing" in ibid., 268.

70 "[komische] *reflection . . . momentum*": Pauli-Landé, 11/10 & 11/24/24 in Enz, 106.

70 *though not by Pauli:* "One also speaks of a 'rotating electron.' However, we consider the notion of a rotating material entity unimportant & it is also not advisable because of the larger-than-light velocities that one then has to take into account. The designation '*Magnetelektron*' shall direct the emphasis to the electromagnetic field of the electron": Pauli, 1928 in ibid., 114.

70 *"Pauli's exclusion principle":* Dirac, "On the theory of Q.M.," *Proc. Roy. Soc. A* 112, 661 (1926); Enz, 128 and Pais, *Inward Bound,* 273.

70 Pauli Verbot: WH, *Z.f.P.* 38 (5/1926) in Enz, 129. "Your *Verbot*": Eh.-Pauli, 1/24/27 ("Dear dreadful Pauli") in Enz, 120.

70 *"Why is a crystal . . . incomprehensible":* Eh.-Pauli, 3/25/31 (partially lost) in Enz, 257–58. See also Casimir, 85–86, for the context—the wonderful story about the "beautiful, new, black, formal suit."

71 *"What I do here . . . to yours":* Pauli-Bohr, 12/12/24 in Enz, 124.

71 *"The physicist . . . truth!":* Pauli-Bohr, 12/12/24 in Cassidy, 192.

71 *Today I have . . . Christmas!!:* WH-Pauli, 12/15/24 in Enz, 124.

71 *"complete* Wahnsinn *. . . so exhaustively"*: Bohr-Pauli, 12/22/24 in Cassidy, 192, and Enz, 124. Cassidy takes Bohr's response straightforwardly as "enthusiastic," but I think that Enz is right: "Usually when Bohr writes something like [that] it means he is rather skeptical" (124).

71 *"If you do not get* schwindlig *. . .":* Bohr in Frisch, 95, Bernstein, *Quantum Profiles,* 20.

72 *"He has 'no philosophy' . . . scientific advances":* Pauli-Bohr, 2/11/24; Dresden, 260.

72 "Quantenmechanik": Born, *Z.f.P.* 26, 379 (1924) in Pais, *Niels Bohr's,* 162: "Born was among the first (perhaps the first) to realize that quantum physics required a new kind of mathematical underpinning." "We [Born and WH] had reason to doubt that Bohr's ingenious but basically incomprehensible combination of quantum rules with classical mechanics was correct. That led us finally to turn our backs on classical mechanics and establish a new quantum mechanics": *B-E Letters,* 78–79.

72 *"Could you perhaps . . . infidel":* Pauli-WH, 2/28/25 in Dresden, 269.

72 *"I feel that we've . . . mechanics":* WH said, "One felt that one had now come a step further in getting into the spirit of the new mechanics. Everybody knew there must be some new kind of mechanics behind it and nobody had a clear idea of it. . . . Almost one had matrix mechanics at this point without knowing it": WH, Kuhn interview in Pais, *Niels Bohr's,* 274.

72 *"the Pauli effect":* See in particular Enz, 149–50 (who quotes Pauli's good friend Fierz: "Pauli himself thoroughly believed in his effect," and Stern: "The number of Pauli effects, the *guaranteed* Pauli effects, is enormously large. [But not in my lab] for he was not allowed to enter").

72 *You know how . . . room:* Gamow said, "It is well known that theoretical physicists cannot handle experimental equipment; it breaks whenever they touch it. Pauli was such a good theoretical physicist that something usually broke in the lab whenever he merely stepped across the threshold": Gamow, *Thirty Years,* 64.

73 *"to keep it in good temper":* An experimentalist friend of Stern's "brought his apparatus a flower every day, in order to keep it in good temper. . . . I had a somewhat higher method. In Frankfurt I had a wooden hammer lying next to my apparatus. And I always threatened the apparatus. One day the wooden hammer had disappeared, whereupon the apparatus did not function, until after three days it was found again . . ." Stern to Jost in Enz, 149.

73 *and we drink wine:* Enz, 147–48.

73 *"When I came to Hamburg . . . champagne":* Pauli to Harry Lehmann, who in turn told Enz. Stern (in the Jost interview) passes the buck to one of the astronomers, Walter Baade, who was to become a good friend of Pauli's, and published a paper on comets with him in 1927. "Baade. He was the main seducer of Pauli concerning wine, alcohol, yes. But when he came to Hamburg Pauli was such that he, in fact, was a strict teetotaler, never had drunk alcohol, in any form, you know, and inveighed horribly against people who [did]" (Enz, 147).

73 *"I have noticed . . . if they are female":* Pauli-Wentzel, 12/5/26 in Enz, 147. It's possible that Pauli had not yet made this discovery by our scene (1/8/25), but he certainly had by Christmas: "I am glad that I have succeeded in drinking wine & rising earlier than you, and that you have not succeeded in converting me [to *spin*]." Pauli-Bohr in Enz, 159.

73 *"At the moment . . . do so":* Pauli-Kronig, 6/21/25 in Enz, 111; Pais, *Niels Bohr's,* 275.

CHAPTER EIGHT: HEISENBERG IN HELGOLAND   JUNE 1925

74 *"Nord und West":* The opening lines of Goethe's Persian poem cycle, *West-östlicher Divan,* Heisenberg's lone companion in Helgoland.

74 *Helgoland:* WH's story is taken from WH, *Physics and Beyond,* 60–62; Pais, *Niels Bohr's,* 275–79.

74 *"Well, you must . . . night":* WH, 1963 interview in Pais, *Niels Bohr's,* 275; see also WH, *Physics and Beyond,* 60.

74 *Heisenberg could barely see:* Ibid.

75 *learning poems . . . by heart:* WH, 1963 interview in Pais, *Niels Bohr's,* 275.

75 *"I always think . . . sea":* Bohr to WH in Sjaelland; WH remembers on Helgoland: WH, *Physics and Beyond,* 52, 60.

75 *"this mathematical . . . relations":* WH in 1967 in Dresden, 247.

75 *"A few days were enough . . . ballast":* WH, *Physics and Beyond,* 61.

75 *"results . . . point of view":* WH-Kronig, 6/5/25 in Cassidy, 201.

75 *"In science you . . . hurt":* WH quoted on the *Playbill* for Frayn's play *Copenhagen.* I have been unable to corroborate this quote.

75 *"I was deeply alarmed . . . too excited to sleep":* WH, *Physics and Beyond,* 61.

76 *"I had been longing . . . sea":* Ibid.

76 *Pauli, "generally my severest critic":* Ibid., 62.

76 *"My own work . . . still unclear":* WH-Pauli, 6/24/25 in Enz, 131.

76 *"In quantum theory . . . in space":* WH 1925 in Beller, 24.

76 *"He had written . . . night":* Born, 1960 int. in Pais, *Niels Bohr's,* 278.

76 *"My entire meager . . . anyway":* WH-Pauli, 7/9/25 in Beller, 54, Cassidy, 197.

76 *"One morning about 10 . . . peculiar formula":* Born, *My Life,* 217.

77 *"the first use . . . science":* Wick, 23. Actually, ES was the first, in a 1922 prefiguring of his great wave-mechanics paper, notes C. N. Yang: "Thus the almost casual introduction in 1922 by Schrödinger of the imaginary unit *i* . . . has flowered into deep concepts that lie at the very foundation of our understanding of the physical world" (Moore, 147).

77 *"radiation-value tables":* WH, "Quantenmechanik," *Naturwissenschaftein* 14, 990 (1926), trans. and quoted in Beller, 27; WH-Pauli: "I am always angry when I hear the theory called nothing but matrix physics. . . . 'Matrix' is certainly one of the most stupid mathematical words in existence" (Wick, 37). WH to Bohr, upon receipt in Copenhagen of Born and Jordan's paper elaborating on his QM: "Here, I got a paper from Born, which I cannot understand at all. It is full of matrices, and I hardly know what they are" (Rosenfeld, 1949 in Greenspan, 127).

77 *"The worst is . . . takes place":* WH-Pauli, 10/23/25; see also *Dreimannerarbeit:* "In the further development of the theory, an important task will lie . . . in the manner in which symbolic quantum geometry goes over into visualizable classical geometry." Born, WH, Jordan, 1926 in Beller, 21n.

77 *"can think far . . . than I":* Born-AE, 7/15/25 in *B-E Letters,* 84.

77 *"kept us breathless for months":* WH, *Physics and Beyond,* 62.

77 *"I am fully . . . complicated":* Born-AE, 7/15/25 in *B-E Letters,* 83–84.

78 *"This was—I remember . . . admirable":* Born, *My Life;* Pais, *Niels Bohr's,* 279.

78 Knabenphysik: Wick, 24; see also Pais, *Inward Bound,* 251.

78 *"the purest soul":* Bohr; ibid., 251.

79 *"the Copenhagen putsch":* Bohr-Franck and Franck-Bohr, 4/21/25 and 4/24/25: Dresden, 210–11; Pauli-Kramers, 7/27/25: Enz, 133; Dresden, 269–70.

79 *"I regard it . . . devoted Pauli":* Pauli-Kramers, 7/27/25 in Pais, *Niels Bohr's,* 238 and Enz, 133 and Dresden, 269–70.

79 *". . . too optimistic":* Kramers to WH, ca. 6/21/25 in Dresden, 276.

79 *"others":* Kramers-Urey, 7/16/25 in ibid., 277.

80 *Spiraling into depression:* See Dresden, pp. 276–85, for the sad story of Kramers' leaving Copenhagen, the close Kramers-Pauli friendship, and the tense, though appreciative, Kramers-WH relations.

80 *"is in a most . . . stage":* Bohr-WH, 6/10/25; Pais, *Niels Bohr's,* 279–80, who notes that Bohr gave a speech in Oslo in late August that contains "no mention of the new quantum mechanics."

80 *"As Kramers perhaps . . . mechanics":* WH-Bohr, 8/31/25 in Pais, *Niels Bohr's,* 280.

80 *"Heisenberg has laid . . . don't)":* AE-Eh. in Woolf, ed., 267.

80 *"excessive Göttingen learnedness":* Enz, 134 and Greenspan, 125.

80 *"Your eternal . . . either!":* WH-Pauli, 10/12/25 in Dresden, 58.

80 *"The most interesting . . . disproof":* AE-Besso, 12/25/25 in Pais, *Niels Bohr's,* 317; "a real witches' calculus": Klein translation; French, *Einstein,* 149.

80 *Heisenberg . . . rejoiced:* "I rejoiced at your new theory of hydrogen and how much I admire that you have made out this theory so fast . . . hearty congratulations!": WH-Pauli, 11/3/25 in Enz, 135.

80 *no longer miserable:* Bohr-Rutherford, 4/18/25, the day after the Bohr-Kramers-Slater paper died (Dresden, 210); Mehra, 467. "Due to . . . [WH], prospects have at a stroke been realized which, although only vaguely grasped, have for a long time been the center of our wishes": Bohr-Rutherford, 1/17/26 in Eve, 314; Pais, *Niels Bohr's,* 280.

CHAPTER NINE: SCHRÖDINGER IN AROSA
CHRISTMAS AND NEW YEAR'S DAY 1925–1926

82 *"actually so* kaput . . . *ideas":* ES-Pauli, 11/8/22 in Moore, 145, who also mentions the chef.

82 *Einstein had brought his younger son:* Pais, *Einstein,* 22.

83 *"On a Remarkable . . . Electron":* ES, *Z.f.P.* 12 (1922) in Moore, 146.

83 *Three years later . . . colloquium:* Moore, 191–92. WH hated how Debye, cigar in teeth, would water his roses when he was supposed to be working. Cassidy, 271.

83 *"Debye casually . . . wave equation":* Bloch, 1976; Moore, 192.

83 *"take seriously . . . world":* ES, "On Einstein's gas theory," *Physikalische Zeitschrift,* 12/15/25 in Moore, 188, and see ES, "Are There Quantum Jumps?," *What Is Life?,* 159.

83 *He was back in Arosa:* Moore, 194–96.

84 *Weyl's wife . . . Scherrer:* Ibid., 175–76.

84 *Growing up in a pink:* Ibid., 10, 12–19.

84 *pearls in his ears:* as Itha Junger remembered he would do; Ibid., 200.

84 *"At the moment . . . assumptions":* ES-Wien, 12/27/25 in ibid., 196.

84 *Weyl, who helped:* Ibid., 196, 200.

85 *"My colleague . . . one!":* Bloch, *Physics Today,* 1976; Ibid., 192.

85 *Very Respected . . . a secret?":* London-ES, 12/7/26 in ibid., 147–48.
85 *"Now, hear that! . . . Perhaps neither is":* Enz, 140.

CHAPTER TEN: WHAT YOU CAN OBSERVE APRIL 28 AND SUMMER 1926

86 *Einstein stood:* This conversation with AE is as in WH, *Encounters,* 112–22; WH
     wrote it slightly differently in WH, *Physics and Beyond,* 62–69.
86 *"You do not mention . . . container":* "He pointed out to me that in my mathemati-
     cal description the notion of 'electron path' did not occur . . . dependent on the
     size of the space": WH, *Encounters,* 113; see also WH, *Physics and Beyond,* 66.
86 *Heisenberg's whole speech . . . now:* "Einstein would . . . be in the audience. . . . I
     wanted . . . to get Einstein interested in the new possibilities": WH, *Encounters,*
     112–13.
86 *"Would you walk . . . further?":* "He invited me to come home with him, so that
     there we might discuss . . .": Ibid., 113. "On the way home, he questioned me about
     my background, my studies . . .": Ibid. "Heisenberg wanted Einstein's opinion on
     whether he should refuse the Leipzig job offer in favor of working with Bohr. Ein-
     stein urged him to work with Bohr." Cassidy, 237.
86 "observing *the electron's path . . . directly observed":* WH, *Encounters,* 113.
86 *"every theory . . . carried out":* Ibid. WH mentions his "astonishment" at this state-
     ment of AE's.
86 *"isn't that precisely . . . relativity?":* WH, *Physics and Beyond,* 63.
86 *"Perhaps I did . . . the same":* Ibid. "When Philipp Franck suggested in 1930 that
     the Bohr-Heisenberg philosophy 'had been invented by you in 1905,' Einstein
     replied, 'A good joke should not be repeated too often' ": Wick, 59.
87 *"The very concept . . . be observed":* WH, *Encounters,* 114.
87 *new doors:* "These considerations were quite new to me, & made a deep impression
     on me at the time; they also played an important part later in my own work . . .":
     Ibid.; see also WH, *Physics and Beyond,* 64.
87 *"So, in your theory":* Heisenberg's memory of Einstein's question: "The electron
     might suddenly and discontinuously leap from one quantum orbit to the other,
     emitting a light quantum as it does so, or it might, like a radio transmitter, beam
     out a wave-motion in continuous fashion. In the first case there is no accounting
     for the interference phenomena that have often enough been observed; in the
     second, we cannot explain the fact of sharp line-frequencies." WH, *Encounters,*
     114.
88 *Heisenberg thought . . . Bohr would say:* "In reply to Einstein's question I fell back
     here on Bohr . . . concepts": Ibid.
88 *"In what quantum . . . taking place":* Ibid.
88 *"a film . . . electron is in":* Ibid., 115.
88 *"You are moving on . . . to the next":* WH, *Physics and Beyond,* 68.
88 *"I admit . . . felt this, too":* Ibid., 69.
88 *"Still, I should . . . laws":* Ibid.
88 *She was a lovely . . . school uniforms:* For the story of Itha and Roswitha, tutored
     and beguiled by ES in the summer of 1926, see Moore, 223–25.
89 *"I believe more . . . Nothing":* as Itha remembered; Ibid., 224.
89 *"a personal God . . . there":* "A personal God cannot be encountered in a world pic-
     ture that becomes accessible only at the price that everything personal is excluded
     from it": ES, *Acta Physica Austriaca* 1 (1948) in Moore, 379.
89 " *'I do not meet . . . Spirit' ":* ES, *Acta Physica Austriaca* 1 (1948) in Moore, 379.

89 *"In summer time . . . brought along":* Alexander Muralt, a student of ES's at the university, 1922–23; Moore, 148–49, and see p. 242.

89 *Schrödinger asked Weyl:* Ibid., 224.

89 *Schrödinger's equation:* ES's interpretation of his equation changed during 1926. *From January to March 1926,* ES interpreted the waves directly as matter, with particles as mere epiphenomena. *From April to June 1926 and on to 1928,* "electrical charge density is given by the square of the wave function"; "what I have called the 'wave equation' up to now is really not the wave equation but rather the equation for the amplitude": ES-Planck, 4/8/26 and 6/11/26; "physical meaning belongs not to the quantity itself but rather to a *quadratic* function of it": ES-Lorentz, 6/6/26; Przibram and Intro. to ES, *Interpretation,* 1–5.

89 *"What do you mean," Itha asked:* ES did tell Itha about his equation, though these two quotations are all that remains of those conversations. "I did not write everything down at once, but kept changing here & there until finally I got the equation. When I got it, I knew I had the Nobel Prize": ES to Itha in ibid. "With reference to the six papers, whose present republication was caused solely by the strong demand for reprints, a young friend of mine [Itha] recently said to the author: 'Hey, you never even thought when you began that so much sensible stuff would come out of it' ": ES, introduction to reprint of wave mechanics papers, Nov. 1926; Ibid., 200.

89 *"waves of matter":* ES, Nobel lecture, 1933 in Weaver, 349.

89 *"and matter is . . . of light:* "the small body itself becomes as it were its own source of light. . . . You are . . . familiar with . . . motes of dust in a light beam falling into a dark room. Fine blades of grass & spider's webs on the crest of a hill with the sun behind it, or the errant locks of hair of a man standing with the sun behind often light up mysteriously by diffracted light": ES, Nobel lecture, 1933 in ibid., 348.

90 *"Just as a mote . . . nucleus of the atom":* ES wrote, "The heavy nucleus of the atom is very much smaller than the atom. . . . The nucleus of the atom must produce a kind of diffraction phenomenon in these [electron] waves, similarly as a minute dust particle does in light waves. . . . We identify the area of interference, the diffraction halo, with the atom; we assert that the atom in reality is merely the diffraction phenomenon of an electron wave captured as it were by the nucleus of the atom": ES, Nobel lecture, 1933; Ibid., 350.

90 *"because of the to me very . . . repelled":* ES, "On the Relation of the Heisenberg-Born-Jordan Quantum Mechanics to Mine," *Annalen der Physik* 79, 734–56 (1926) in Moore, 211.

91 *"The more I think . . . the greatest result":* WH-Pauli, 6/8/26; Ibid., 221: Cassidy brings out the Bohr-echo, Cassidy, 215; see also Dresden, 70.

91 *"mathematically equivalent":* ES, "On the Relation" in Moore, 212.

91 *"cannot and may not . . . this point":* ES, "Quantization as an Eigenvalue Problem, Part IV," *Annalen der Physik* 81, 109–39 (1926) in ibid., 219.

91 *"one gets no answer . . . collision?' ":* Born, "Quantum Mechanics of Collisions," *Zeitschrift für Physik* 37, 863–67 (1926) in Greenspan, 139

91 *"Max Born betrayed us both":* "You have gone over to the other side": WH-Born; Dresden, 75.

92 *"Born overlooks . . . real":* ES-Wien, 8/25/26 in Moore, 225.

92 *"luckily . . . man or beast":* ES, *Mein Leben* in ibid., 81.

92 *"one night . . . barbed wire entanglements":* ES, war diary; Ibid., 81–82. He went on to describe it as "this really enchanting phenomenon."

93 "*a space-time description is impossible ... I reject* a limine. ... *physics does not ... science*": ES-Wien, 8/25/26 in ibid., 225.

93 "*what we cannot ... one of them*": ES, "Quantization . . . , Part II," *Annalen der Physik* 79 (4), 489–527 (1926) in ibid., 208.

CHAPTER ELEVEN: THIS DAMNED QUANTUM JUMPING OCTOBER 1926

94 *Schrödinger lay coughing ... happen*": Heisenberg relates this story in WH, *Physics and Beyond,* 73–76: "While Mrs. Bohr nursed him and brought in tea and cakes, Niels Bohr kept sitting on the edge of the bed talking at Schrödinger: 'But you must surely admit that ...' ": WH, *Physics and Beyond,* 76.

94 *They had been arguing ... Schrödinger:* "Bohr's discussions with ES began at the railway station and were continued daily from early morning until late at night": Ibid., 73.

94 "*Although Bohr was normally ... utterance*": Ibid.; it continues: "All I can hope to do here is produce a very pale copy of conversations in which two men were fighting for their particular interpretation of the new mathematical scheme with all the powers at their command."

All the quotes from Bohr or ES are from WH's memory in *Physics and Beyond,* 73–76, except:

96 "*picturesque pageantry ... parcels*": ES, 1952; ES, *Interpretation,* 29.

96 "*Not to disagree, but merely to understand*": one of Bohr's favorite ways to begin an objection (see, e.g., Gamow, *Thirty Years,* 180, 215–16). Peierls remembers Bohr once saying: "I am not saying this in order to criticize, but your argument is sheer nonsense" (Peierls, "Some Recollections . . . ," French and Kennedy, 229).

97 "*late and looked infinitely ... it was so*": vW, "Reminiscence from 1932," French and Kennedy, 187.

97 "*There will hardly again ... my own work)*": ES-Wien, 10/21/26; Pais, *Niels Bohr's,* 299.

98 "*completely convinced ... defending*": ES-Wien, 10/21/26 in Moore, 228.

98 "*We in Copenhagen ... atomic processes*": WH, *Physics and Beyond,* 76.

98 "*pedagogical ... against the Lords of continuum theory*": WH-Pauli, 11/4/26; Beller, 78.

98 "*We had great pleasure ... theory*": Bohr-Fowler, 10/26/26 in Pais, *Niels Bohr's,* 300.

98 "*How little the words ... correspondence theory*": Bohr-Kramers, 11/11/26 in ibid.

98 *Certain Bose-Einstein condensates:* The 1935 work of London & Tisza "was the first to bring out the idea of Bose-Einstein condensation displaying quantum behavior on a macroscopic size scale, the primary reason for much of its current attraction. Although it was a source of debate for decades, it is now recognized that the remarkable properties of superconductivity and superfluidity in both helium 3 and helium 4 are related to Bose-Einstein condensation": Cornell & Wieman, Nobel Prize lecture, 2001. N.B. It seems to be a coincidence that London was once Schrödinger's assistant (they were never close)—but it is not a coincidence that this epochal work was scoffed at by Lev Landau, one of the greatest physicists of all time, who was Bohr's student.

98 *revive Schrödinger's original ... interpretation:* "So in the situation in which we can have very many particles in exactly the same state, there is possible a new physical interpretation of the wave functions. The charge density and the electric current can be calculated directly from the wave functions and the wave functions take on a physical meaning which extends into classical, macroscopic situations": Feyn-

man's amazing final *Feynman Lecture,* "The Schrödinger Equation in a Classical Context: A Seminar on Superconductivity," *The Feynman Lectures on Physics* (Reading, MA: Addison-Wesley, 1965), p. 21–6.

98 *"It would have been beautiful . . . this world":* Born-ES, 11/6/26, Beller, 36.

98 *"Local Zürich superstitions":* Pauli in Moore, 221.

98 *"Look on the expression . . . even more":* Pauli-ES, 11/22/26 in ibid.

99 *"We are all nice . . . uniformity":* ES-Pauli, 12/15/26 in ibid., 222.

99 *"commander-in-chief . . . army":* JRO "told me jokingly that Born, as if he were the commander-in-chief summoning his army, explained to everybody how false were Schrödinger's ideas": Pascual Jordan interview; Beller, 46.

99 *"put off some . . . way":* JRO interview 1963 in Smith and Weiner, 104.

99 *"They are working . . . successful":* JRO-Ferugusson, 11/14/26 in ibid., 100.

99 *"About me it can . . . time":* Born-AE, 11/30/26 in Pais, 'Subtle,' 443; Pais, *Niels Bohr's,* 288.

99 *"Certainly his favored . . . remarks":* Bohr, "Discussion with Einstein" in Bohr, *Atomic Physics,* 36. Bohr, in writing at least, was certainly not given to "picturesque phrases."

99 Gespensterfeld: Kramers's memories of it (1923) in Dresden, 143. Wigner's memories in Woolf; Pais, *Niels Bohr's,* 287–88.

99 *"Quantum mechanics is . . . at dice":* AE-Born, 12/4/26 in *B-E Letters,* 90.

100 *"came as a hard blow to me":* Born commentary, ibid., 91.

100 *"My heart does not warm . . . primitive":* AE-Eh., 1/19/27 in Fine, 27.

100 *"All attempts . . . failed":* AE & Grommer, meeting report of the *Preussische Akademie,* 1927 in Pais, 'Subtle,' 290.

100 *"At the special . . . your portrait":* Hedi-AE in *B-E Letters,* 94–95.

100 *"What applies to jokes . . . in physics":* AE-Hedi, 1/15/27 in ibid., 95.

CHAPTER TWELVE: UNCERTAINTY WINTER 1926–1927

101 *two men standing in an attic room:* "During the next few months [after ES's visit] the physical interpretation of quantum mechanics was the central theme of all conversations between Bohr and myself. I was then living on the top floor of the Institute, in a cozy little attic flat with slanting walls and windows overlooking the trees at the entrance to Faelled Park. Bohr would often come into my attic late at night, and we constructed all sorts of imaginary experiments": WH, *Physics and Beyond,* 76.

101 *"What the word . . . world":* "When we talk about position or velocity, we always need words that are obviously not defined at all in this discontinuous world. . . . In all of our words that we use for the description of a fact, there are too many c-numbers [*c-number* was Dirac's term for "classical number"]. What the word 'wave' or 'corpuscle' means, one no longer knows": WH-Pauli, 11/23/26 in Beller, 88–89.

101 *"But the mathematics . . . relativity":* "all physical applications of Q.M. that there are now—& as Dirac says, that there ever will be—fall under" Dirac's current work: WH-Jordan, 11/24/26; Cassidy, 236. "One could now do the calculations. That was a definite proof that one had found, at least mathematically, the correct solution": WH, 1963 interview in Pais, *Niels Bohr's,* 302–3. Dirac liked to compare transformation theory (his generalization of Q.M., finished late in 1926) with the completeness of relativity. See Beller, 88; compare Pais, *Inward,* 288–89.

101 *"the nicest . . . deep truth":* "Bohr was trying to allow for the simultaneous existence of both particle and wave concepts, holding that, though the two were mutu-

ally exclusive, both together were needed for a complete description of atomic processes. I disliked this approach. I wanted to start from the fact that Q.M. . . . already imposed a unique physical interpretation" ("Deep truth" was one of Bohr's favorite phrases): WH, *Physics and Beyond,* 76 and 102. Bohr's notes for a speech on 9/16/27 include his insistence on "all information about atoms expressed in classical concepts." (Pais, *Niels Bohr's,* 311; compare Pais, *Niels Bohr's,* 302–3, 309–10. In 1926 this was a cry for help, not yet a firm statement.) "These paradoxes were so in the center of his mind that he just couldn't imagine that anybody could find an answer to the paradoxes, even having the nicest mathematical scheme in the world." "Bohr would say, 'even the mathematical scheme does not help. I first want to understand how nature actually avoids contradictions.' " "He rather felt, 'Well, there's one mathematical tool—that's matrix mechanics. There's another one—that's wave mechanics. . . . But we must first come to the bottom in the philosophical interpretation' ": WH, 1963 interview in Pais, *Niels Bohr's,* 302.

101 *"make any concession to the Schrödinger side . . . probably be wrong":* WH, 1963 interview in Pais, *Niels Bohr's,* 302–3.

102 *"torture . . . mysticism of nature":* Bohr-WH, 4/18/25 in Cassidy, 195.

102 *He wanted to look . . . together:* "Bohr . . . wanted to take the interpretation in some way very serious & play with both schemes": WH, 1963 interview in Pais, *Niels Bohr's,* 303.

102 *"epistemological lessons":* "another of Bohr's favorite terms," ibid., 8, 315.

102 *"I would try . . . angry about it":* WH, 1963 interview in ibid., 303.

102 *a tense and exhausting:* "Both of us became utterly exhausted and rather tense": WH, *Physics and Beyond,* 77.

102 *"So I was alone in Copenhagen":* WH, 1963 interview in Pais, *Niels Bohr's,* 304.

102 *thought-kitchens:* "Ehrenfest tells me many details from Niels Bohr's *Gedanken-küche* [thought-kitchen]; his must be a first-rate mind, extremely critical and far-seeing, which never loses track of the grand design." AE-Plank, 10/23/19 (postcard) in Pais, 'Subtle,' 416.

102 *"Coming nearer . . . exciting thing":* WH, 1963 interview in Pais, *Niels Bohr's,* 302.

102 *"Like a chemist . . . paradox":* WH, 1963 interview in ibid.

102 *In his mind . . . insurmountable:* "I now concentrated all my efforts on the mathematical representation of the electron path in the cloud-chamber. . . . I realized fairly soon that the obstacles before me were quite insurmountable": WH, *Physics and Beyond,* 77; compare Wick, 37. "We have a mathematical scheme which is consistent, . . . if it's right, then anything added to it must be wrong because it is closed in itself": WH, 1963 interview in Pais, *Niels Bohr's,* 303.

103 *Don't you see . . . observed:* "It must have been one evening after midnight when I suddenly remembered my conversation with Einstein and particularly his statement, 'It is theory which first decides what we can observe.' . . . I decided to go on a nocturnal walk through Faelled Park and to think further about the matter": WH, *Physics and Beyond,* 77–78.

103 *"We have always said so glibly . . . experimental difficulties?":* Ibid., 78.

103 *"continually makes the rounds in Copenhagen":* WH-Pauli, 10/28/26; Cassidy, 233; Enz, 144: "Heisenberg answered from Copenhagen 9 days later noting that he, Bohr, Dirac, and Hund scuffled (*rauften sich*) for Pauli's letter."

103 *"The first question . . . schwindlig:* Pauli-WH, 10/19/26 in Pais, *Niels Bohr's,* 304.

104 *Laplace's hypothetical Demon:* See the glossary entry *determinism* on page 338.

104 *Weimar intellectuals . . . Demon:* for a hundred-page analysis of the Weimar scien-

tific climate, see Forman, "Weimar Culture, Causality, and Quantum Theory, 1918–1927: Adaptation by German Physicists and Mathematicians to a Hostile Intellectual Environment," Forman, "Weimar Culture."

104 *"irrationality":* in the '20s, this was one of Bohr's favorite words to describe "the quantum postulate"—he used it three times in the epochal Como lecture (1927) and three times in the introduction to *Atomic Theory* (1929). See Pais, *Niels Bohr's*, 316 and Mermin, *Boojums*, 188.

104 *almost immediately:* "A brief calculation after my return to the Institute showed that one could indeed represent such situations mathematically": WH, *Physics and Beyond*, 78.

105 *"In the strong . . . false":* WH, *Zeitschrift für Physik* 43 (1927) in Beller, 99.

105 *fourteen-page letter:* WH-Pauli, 2/23/27 in Pais, *Niels Bohr's*, 304.

105 *"The solution can now . . . observe it":* WH-Pauli, 2/23/27; Cassidy, 236.

105 *"old snow":* (*alter Schnee*) WH-Pauli, 2/23/27 in Beller, 83.

105 *"relentless criticism" . . . "To make anything clearer":* WH-Pauli, 2/27/27 in ibid., 108–9.

105 *"I believe . . . sent to Pauli":* WH-Bohr, 3/10/27 in Pais, *Niels Bohr's*, 304.

105 *that month . . . Norwegian slopes:* "a month skiing in the Norwegian mountains around Guldbrandsdalen. It was there (as he often told) that the complementarity argument first dawned on him": Ibid., 310. Mrs. Fermi described Bohr's skiing as "elegant": (Pais, *Niels Bohr's*, 497) while von Weizsäcker commented, "In Danish woodlands one cannot learn alpine skiing" (vW, "Reminiscence from 1932," French and Kennedy, 185). These assessments may say as much about their different authors as about Bohr.

105 *"He was very tired . . . exaggerated":* Klein, 1968 int. in Pais, *Niels Bohr's*, 303–4.

106 *Pauli would have said:* Pauli did, in fact, come to Copenhagen later, and according to Kalckar, his visit "played an important role in reconciling WH's with Bohr's point of view" in Pais, *Niels Bohr's*, 310.

106 *there exist particles* and *waves:* WH-Pauli, 5/16/27 in ibid., 309.

106 Why do we imagine that we can observe . . . without disturbing: "Our usual description of physical phenomena is based entirely on the idea that the phenomena concerned may be observed without disturbing them appreciably. . . . The quantum postulate implies that . . . an independent reality can neither be ascribed to the phenomena nor to the agencies of observation": Bohr's Como lecture (1927); Bohr, *Atomic Theory*, 53–54.

106 All our ordinary verbal . . . an irrationality: Bohr (1929), *Atomic Theory*, 19; for "particle, wave; space & time, causality," see, e.g., the Como lecture, Bohr, *Atomic Theory*, 54–57.

106 What if this is a fundamental limitation?: Here follow three quotes from Bohr's Como lecture of 1927 (Bohr, *Atomic Theory*, 53–57). "Quantum theory is characterized by the acknowledgment of a fundamental limitation in the classical physical ideas when applied to atomic phenomena" (53–54). "The very nature of the quantum theory thus forces us to regard the space-time co-ordination & the claim of causality—the union of which characterizes the classical theories—as complementary but exclusive features of the description" (54–55). "[Waves] as well as isolated material particles are abstractions, their properties on the quantum theory being definable & observable only through their interaction with other systems. Nevertheless, these abstractions are . . . indispensable for a description of experience" (56–57).

106 *Bohr felt . . . peace and acceptance:* In a letter to Rutherford (6/3/30), Bohr approvingly summarized these insights—about the "complementarity" of the idea of *wave* and the idea of *particle*—as a "combination of resignation and enthusiasm." The science historian J. L. Heilbron quotes this letter, commenting on Bohr's "exalted pessimism, this eagerness for surrender" (135).

107 *He returned just as Heisenberg:* Heisenberg recalled in 1951: "When I discussed it with Bohr after his return, we were unable to find at once the same language for the interpretation of the theory because in the meantime Bohr had developed the concept of complementarity": Pais, *Niels Bohr's,* 310 (see also pgs. 308–9 and WH, *Physics and Beyond,* 79).

107 *"It ended by my breaking . . . Bohr":* WH, 1963 interview in Pais, *Niels Bohr's,* 308.

107 *"recent investigations by Bohr . . . work":* WH, *Zeitschrift für Physik* 43 (1927); Pais, *Niels Bohr's,* 308–9; compare Wick, 41.

107 *"Anschaulich":* "redefining the word *anschaulich* to refer to the physical or experimentally meaningful, rather than Schrödinger's supposed preference for the merely visualizable or pictoral": Cassidy, 233; compare Pais, *Niels Bohr's,* 304. (The full title is "On the *Anschaulich* Content of the Quantum-Theoretical Kinematics and Mechanics"—*mechanics* is the mathematics of motion and the forces that produce motion; *kinematics* is the branch of mechanics that deals only with motion.)

107 *"Bohr wants to write . . . consistent":* WH-Pauli, 5/16/27 in Pais, *Niels Bohr's,* 309.

107 *"unanschaulicher and more general"* [*than wave mechanics*]: WH-Pauli, 5/31/27; Beller, 70.

107 *"There are essential . . . Anschaulich":* WH-Pauli, 5/16/27 in Pais, *Niels Bohr's,* 309.

107 *This occasion . . . tributes:* AE, *World as I See It,* 146ff; Pais, '*Subtle,*' 15.

107 *"Who would presume . . . given up?":* AE, *World as I See It,* 156.

107 *"significant . . . exceptionally brilliant . . . contribution":* Bohr-AE, re: uncertainty, 4/13/27; Pais, *Niels Bohr's,* 309.

108 *"since the different . . . simultaneously":* Bohr-AE, 4/13/27 in Jammer, *Philosophy of Quantum,* 126.

108 *"only the choice . . . description":* Bohr-AE, 4/13/27 in ibid., 125.

108 *"I think I have refuted . . . causality":* WH in vW, "Reminiscence from 1932," French and Kennedy, 184.

108 *"I handed in a short . . . Kindest regards":* AE-Born in *B-E Letters,* 96.

108 *"Does Schrödinger's . . . Statistics?":* the quote and analysis are from Fine, 27, 99.

108 *"paper in which . . . wish":* WH-AE, 5/19/27 in Pais, '*Subtle,*' 444.

108 *"Perhaps we could . . . experiments":* WH-AE, 6/10/27 in Pais, '*Subtle,*' 467; and compare Jammer, *Philosophy of Quantum,* 125–26.

109 *the experimentalist Walther Bothe:* Nobel Prize biography and Medical Institute of the KWI history (both at www.nobelprize.org). To test the Bohr-Kramers-Slater theory, Bothe and Geiger invented an entirely new technique to measure the "coincidences" between particles after a collision; half a century later this technique would be ready to test Bell's theorem. Dresden, 208.

109 *"the singular good fortune . . . Einstein":* Bothe, Nobel lecture, www.nobelprize.org.

109 *"from a physical standpoint":* AE; Fine, 99.

CHAPTER THIRTEEN: SOLVAY 1927

110 *"Here is Max . . . aggravated looks":* Greenspan, "Surprises in Writing a Biography of Max Born"; *AIP History Newsletter,* Vol. XXXIV, No. 2, Fall 2002. "Ehrenfest making faces" is from her parallel description in *End,* 148.

110 *"The indeterminist . . . follow me at all":* de Broglie, 183.

111 *"I must apologize . . . deeply enough":* AE's first sentence at Solvay in Pais, 'Subtle,' 445.

111 *"Nevertheless, I . . . general remarks":* AE's next sentence at Solvay; Whitaker, *Einstein,* 203.

111 *Imagine we have an electron:* Here I have followed Bohr's description of Einstein's thought-experiment at Solvay (*Atomic Physics,* 41–42; compare Whitaker, *Einstein,* 204 and dB, *New Perspectives,* 150). AE wrote down a more detailed and involved version to be printed in the conference proceedings, but both Bohr and dB remembered how "simple" and brief had been AE's spoken words, which is why I believe Bohr's memory is correct (whereas AE's written version, though conveying the same point, was more formal). (For the same conference proceedings, Bohr sent the 40-page Como lecture, given a month earlier, rather than Kramers's notes of what he had actually said. Pais, *Niels Bohr's,* 318n.) Here is Bohr's memory: "The apparent difficulty, which Einstein felt so acutely, is the fact that, if . . . the electron is recorded at one point A of the plate, then it is out of the question of ever observing an effect of this electron at another point (B), although the laws of ordinary wave propagation offer no room for a correlation between two such events": Bohr, "Discussion with Einstein"; Bohr, *Atomic Physics,* 42. Einstein's written version: "The scattered wave moving towards [the second screen] does not present any preferred direction. If psi-squared was simply considered as the probability that a definite particle is situated at a certain place at a definite instant, it might happen that *one and the same* elementary process would act *at two or more* places of the screen" [i.e., a Schrödinger's cat–type scenario]. "But the [Born] interpretation, according to which psi-squared expresses the probability that *this* particle is situated at a certain place, presupposes a very particular mechanism of action at a distance . . .": AE in Howard, *Nicht,* 92.

111 *"presupposes a very particular . . . screen":* AE, Solvay in Howard, *Nicht,* 92.

111 *"It seems to me . . . relativity":* AE, Solvay in Wick, 54. Compare Howard, *Nicht,* 92.

112 *"The Lord did there . . . earth":* Eh. in Clark, 417.

112 *"I feel myself . . . classical theories":* Bohr, from Kramers's fragmentary notes taken at the time; Whitaker, *Einstein,* 204. Compare Bohr in Pais, *Niels Bohr's,* 318.

112 *"Ominously . . . talking past each other":* Whitaker, *Einstein,* 205.

112 *"once again the awful . . . three a.m.)":* Eh.-Goudsmit et al., 11/3/27 in ibid., 209–10.

112 *"BOHR . . . defeating everybody":* Ibid.

113 *Couldn't you follow the electron:* see Bohr, *Atomic Physics,* 42–47; Fine, 28–29; and Whitaker, *Einstein,* 210.

113 *"Like a game of chess . . . priceless":* Eh.-Goudsmit, 11/3/27 in Whitaker, *Einstein,* 209–10.

113 *"stop telling God how to run the world":* WH, *Physics and Beyond,* 81. Bohr's one (as far as I know) great piece of writing is in "Discussion with Einstein," where he relates the same story, and, far from crowing over his witty comeback, as WH does, Bohr purposely uses his normal convolutions for humor: "In spite of all divergencies of approach and opinion, a most humorous spirit animated the discussions. On his side, Einstein mockingly asked us whether we could really believe that the providential authorities took recourse to dice-playing ('*ob der liebe Gott würfelt*') to which I replied by pointing at the great caution, already called for by ancient

thinkers, in ascribing attributes to Providence in everyday language": Bohr, *Atomic Physics,* 47.

114 *"Einstein, I am ashamed ... theory":* WH, *Physics and Beyond,* 80. "I remember how at the peak of the discussion Ehrenfest, in his affectionate manner of teasing his friends, jokingly hinted at the apparent similarity between Einstein's attitude and that of the opponents of relativity theory; but instantly Ehrenfest added that he would not be able to find relief in his own mind before concord with Einstein was reached": Bohr, *Atomic Physics,* 47. It is characteristic that Bohr reports both quotes while WH only reports the first.

114 *"I must make a choice ... Bohr":* Eh. to Goudsmit, mid-1927 in Pais, 'Subtle,' 443.

CHAPTER FOURTEEN: THE SPINNING WORLD 1927–1929

115 *"the idol of my youth":* de Broglie, 182.

115 *"This exaggerated ... understand them":* AE in ibid., 183–84.

115 *his skepticism:* dB thought that Einstein's comment about a child was "possibly going farther than he might normally have liked to go": Ibid., 184.

116 " 'As an older friend ... believe you' ": Planck to AE, 1913 (AE told Straus, who told Pais); Pais, 'Subtle,' 239.

116 *"carry on! ... right road!":* AE in de Broglie, 184.

116 *"discouraged":* Ibid.

116 *wondering if "position":* ES-AE, 5/30/28, Przibram, 29–30 (summaries of ES-Bohr, 5/13/28 and Bohr-ES, 5/25/28, p. 29).

116 *"What is it that ... communicate":* Bohr to Peterson in Pais, *Niels Bohr's,* 445. "Language is, as it were, a net spread out between people": Bohr, 1933 in WH, *Physics and Beyond,* 138.

116 *"Bohr was not puzzled ... sterile to him":* Peterson, 1963 in Pais, *Niels Bohr's,* 445.

116 *Dear Schrödinger ... A. Einstein:* AE-ES, 5/31/28 in Przibram, 31–32.

117 *"In the depths ... solution":* AE-Weyl, 4/26/27 in Howard, *Nicht,* 87.

117 *"It is very ... do believe it":* AE-Weyl, 4/26/27 in ibid.

117 *"This after all ... heresy":* Pauli-Kramers, 3/8/26 in Dresden, 63.

118 "Geissel Gottes ... proud of it!": Pauli-Kramers, 3/8/26 in Enz, 89.

118 *"peculiar ... two-valuedness":* Pauli, 1925 in ibid., 106–7. The literal word for "two-valuedness" is *Zweiwertigkeit;* Pauli used *Zweideutigkeit,* defined variously as "suggestiveness," "ambiguity," and "double-entendre." (Ein*deutigkeit* is "clarity.")

118 *"a prophet of the electron-magnet gospel":* Bohr-Eh., 12/22/25 in Dresden, 63.

118 *a train trip:* as Bohr told Pais, *Inward,* 278–79 (repeated in *Niels Bohr's,* 242–43).

118 *Heisenberg made a bet ... Kramers:* Dresden, 64.

118 *"crossword puzzles":* Eh.-Dirac, 6/16/27 in Kragh, 46.

118 *"I have trouble ... madness is awful":* AE-Eh., 8/26/26 in Pais, 'Subtle,' 441.

119 *"The saddest chapter ... melancholic":* WH-Pauli, 7/31/28 in Pais, *Inward,* 348.

119 *"a fundamentally new idea":* Pauli-Bohr, 6/16/28 in Cassidy, 282.

119 *Gulliver's Journey to Urania:* Pauli-Klein, 2/18/29 in Enz, 175.

119 *"a difficult ... last-named investigators":* AE-Nobel committee, 9/25/28 in Pais, 'Subtle,' 515.

119 *Dr. Schweitzer:* Born, *My Life,* 240–41.

121 *"Illness has its ... begun to think":* AE, 3/23/29, *Nature* 123, 464–69; Clark, 491.

121 *"Pauli effect!":* GG in Rosenfeld, 3/3/71, "Quantum Theory in 1929: Recollections from the first Copenhagen conference." www.nbi.dk/nbi-history.html#firstconf

122 *"the Pauli effect . . . come singly":* Eh. in Rosenfeld, 3/3/71, "Quantum Theory in 1929."

122 *"Of this inner . . . as ever":* Rosenfeld, 3/3/71, "Quantum Theory in 1929."

122 *"please help me . . . problems are":* Eh.-Kramers, 11/4/28 and 8/24/28 in Dresden, 313.

122 *"a stubby young man . . . contemplation":* Peierls, *Bird of Passage,* 60.

122 *"It was one of the few . . . initiation":* Rosenfeld, 3/3/71, "Quantum Theory in 1929."

122 *"Nothing has done . . . Niels Bohr":* from Wheeler's beautiful memoir, "Physics in Copenhagen in 1934 and 1935," French and Kennedy, 226.

CHAPTER FIFTEEN: SOLVAY 1930

Starting with Bohr, the story of his argument with Einstein at the sixth Solvay conference has gotten very confused over the years, and so I think it is important to start out these notes with the most important quote, from Ehrenfest to Bohr in 1931:

Einstein "said to me that, for a very long time already, he absolutely no longer doubted the uncertainty relations, and that he thus, e.g., had BY NO MEANS invented the 'weighable light-flash box' (let us call it simply L-W-box) '*contra* uncertainty relation,' but for a totally different purpose. . . . It is thus, for Einstein, beyond discussion and beyond doubt, that because of the uncertainty relation, one must naturally choose between the either and the or. But the questioner can choose between them AFTER the projectile is already finally under way": Eh.-Bohr, 7/9/31 in Howard, *Nicht,* 98–99. Don Howard re-examined this letter (which had been mistranslated) and realized that Bohr had simply not understood what Einstein's *gedanken* experiment was for (Howard, *Nicht,* 100).

Two more quotes, one from a friend and follower, and one from an intellectual opponent, might help us to understand how this could have happened. "Bohr had two speeds—not interested or completely interested," said Wheeler. "This applied to everything." (Wheeler to Bernstein, *Quantum Profiles,* 107). And Schrödinger wrote, "Bohr's . . . approach to atomic problems . . . is really remarkable. He is completely convinced that any understanding in the usual sense of the word is impossible. Therefore the conversation is almost immediately driven into philosophical questions, and soon you no longer know whether you really take the position he is attacking, or whether you really must attack the position he is defending" (Schrödinger to Wien, 10/21/26 in Moore, 228).

123 *What if we had a box* (the light-blitz box, part 1): Bohr, "Discussion with Einstein," Bohr, *Atomic Physics,* 53.

123 *no longer burning:* See Rozental, 112; Frisch, 169; and Peat, 185.

124 *"I grant you the consistency":* see Eh-Bohr, 7/9/31; Howard, *Nicht,* 98–99.

124 *"half a light-year away"* (the light-blitz box, part 2): Eh.-Bohr, 7/9/31; Ibid., 99.

124 *"This amounts to . . . whole problem":* Bohr, referring to only the first half of AE's *gedanken* experiment, "Discussion with Einstein," Bohr, *Atomic Physics,* 53.

124 *"When the exigencies . . . phenomena":* "As an objection to the view that a control of the interchange of momentum and energy between the objects and the measuring instruments was excluded—if these instruments should serve their purpose of defining the spacetime frame of the phenomena—AE brought forward the argument that such control should be possible when the exigencies of relativity theory were taken into consideration": Bohr, "Discussion with Einstein," ibid., 52–53.

124 *"superluminal . . . action-at-a-distance"*: "Were that kind of a physical effect from B [the box] on the fleeing light quantum to occur, it would be an action at a distance, that propagates with superluminal velocity": AE-Epstein, 11/5/45 in Howard, *Nicht,* 102.

124 *"Of course it is logically . . . definite color"*: AE-Epstein, 11/5/45 in ibid.

125 *"and the quantum . . . incomplete"*: "Thus I incline to the opinion that the wave function does not (completely) describe what is real, but only a (to us empirically accessible) maximal knowledge regarding that which really exists. . . . This is what I mean when I advance the view that quantum mechanics gives us an *incomplete description* of the real state of affairs": AE-Epstein, 11/5/45; Ibid.

125 *"It was quite a shock . . . excited"*: Rosenfeld; Pais, '*Subtle,*' 446.

125 *"He said to me . . . contra uncertainty-relation,' "*: Eh.-Bohr, 7/9/31; Howard, *Nicht,* 98.

125 *"resignation as regards . . . advance"*: "The resignation as regards visualization and causality . . . might well be regarded as a frustration of the hopes which formed the starting-point of the atomic conceptions. Nevertheless . . . we must consider this very renunciation as an essential advance in our understanding.": Bohr, 1929, *Atomic Theory,* 114–15.

125 *"the demand . . . inconsistencies"*: "In dealing with the task of bringing order into an entirely new field of experience, we could hardly trust in any accustomed principles, however broad, apart from the demand of avoiding logical inconsistencies": Bohr to Einstein in a conversation following his triumph at Solvay in Bohr, "Discussion . . . ," *Atomic Physics,* 56.

125 *"irrationality"*: Forman, "Weimar Culture . . . ," Raman and Forman, 16–19; Pais, *Niels Bohr's,* 316, and Mermin, *Boojums,* 188.

125 *"wholeness"*: "The main point of the lesson given us by the development of atomic physics is . . . the recognition of a feature of wholeness in atomic processes": Bohr, 1957, *Atomic Physics,* 1.

125 *"Any observation . . . description of nature"*: Bohr, 1929, *Atomic Theory,* 115. See also p. 11: "The magnitude of the disturbance caused by a measurement is always unknown."

126 *"the difference between . . . understanding"*: Bohr, "Discussion," *Atomic Physics,* 58.

126 *"Ehrenfest, it can't . . . right"*: "During the entire evening he was extremely unhappy, going from one to the other telling them that it couldn't be true, that it would be the end of physics if Einstein were right": Rosenfeld, 1968; Wick, 56. Pauli and WH "did not pay too much attention: ['ah, well, it will be alright']" (Solvay 1927): Stern, 1961 in Pais, *Niels Bohr's,* 318.

126 *into the club room:* called Fondation Universitaire. Clark, 417.

126 *"The next morning came Bohr's triumph"*: Rosenfeld, 1968; Wick, 57.

126 *Gamow did build a mock version:* "I spent most of the autumn of 1930 in Copenhagen, & we were quite excited about this story. We did not exactly welcome the conquering hero [Bohr] by sounding trumpets & beating drums, but we—I think Gamow, Landau, and that versatile artist Piet Hein—had a nicely designed contraption made in the workshop": Casimir, 315–16; photograph of the contraption: French and Kennedy, 134.

127 *the two of them together:* "At the outcome of the discussion, to which Einstein himself contributed effectively, it became clear, however, that the argument could not be upheld": Bohr, *Atomic Physics,* 53.

INTERLUDE: THINGS FALL APART 1931–1933

128 *"Man only loves . . . anyone else":* Eduard Einstein in Pais, *Einstein*, 24.

128 *"Tetel":* Pais, *Einstein*, 21–25.

128–29 *"save yourself some . . . grown up":* AE-Eduard June 1918 in Michelmore, *Einstein*, 62; www.einsteinwebsite.de/biographies/einsteineduard.html

129 *"coming slowly but . . . youth":* AE-Besso; Levenson, 384.

129 *"slight sense of being absent":* "Edi's enthusiasm for playing has stayed in my memory. All skepticism, insecurity, irony, & the slight sense of being absent that struck me about Edi in school vanished completely while he played": one of Eduard's schoolmates; Pais, *Einstein*, 23.

129 *"frantic":* Brian, 158 and 196; Michelmore, *Einstein*, 59, 123–24.

129 *"rapturous letters":* Eduard in Levenson, 382.

129 *"a really good doctor of the soul":* AE-Eduard 2/5/30 in Brian, 196.

129 *"He has always aimed . . . very hard":* Elsa-Vallentin in Levenson, 383.

129 *"We know from daily . . . dependent":* AE, *The World As . . .* , p. 1; Michelmore, 148.

129 *"The winter here . . . long time":* Elsa-Vallentin in ibid., 149.

129 *"It is well known . . . of a particle":* AE, Tolman, and Podolsky, *Physical Review* 37, 1931, 780–81.

130 *Now Tolman and Podolsky . . . Mechanics":* *Physical Review* 37, 602–15, 1931.

130 *Podolsky had grown up:* Robert Podolsky, personal communication, 2002.

130 *"The purpose of the present . . . mechanics":* AE, Tolman, and Podolsky, *Physical Review* 37, 1931.

130–31 *"In a local pub . . . describes his experiences":* Born-AE, 2/22/31 in *B-E Letters*, 109–10. Born credits the (for him, unusual) political optimism in this letter to his restorative visits with Schweitzer. *B-E Letters*, 112.

131 *"I am always very . . . relativity":* Hedi-AE, 2/22/31 in *B-E Letters*, 109.

131 *"inverse Pauli effect":* Somm. in Enz, 224.

131 *"in a slightly tipsy state":* Pauli-Peierls, 1931 in ibid.

131 *"In spite of the . . . Canadian border).":* Pauli-Peierls, 7/1/31 in ibid., 223–24.

131 *"the cigars Sommerfeld misses very much":* Pauli-Wentzel, 9/7/31 in ibid., 224.

131 " 'Ja, Herr Professor' and 'Nein . . .' ": Pauli to Somm. in ibid., 55.

131 *"Why it was just . . . Herr Bohr":* Pauli-Somm., 12/5/38 (Somm.'s 70th birthday) in ibid., 55–56.

131 *"cheerful student days":* Pauli-Somm., 12/5/38 in ibid., 56.

131 *"If it had been a* bullfighter *. . . chemist!":* Pauli to Franca Pauli (as Franca told Enz in 1971) in ibid., 211.

132 gedanken *experiment . . . to Berlin:* AE was invited by von Laue 11/4/31 to give a colloquium of his choice: "On the Indeterminacy Relations" in Pais, '*Subtle*,' 449.

132 *to Leiden:* winter 1931–32 in Casimir, 316.

132 *back to Brussels:* spring/summer 1933: comment on Rosenfeld in Jammer, *Philosophy of Quantum*, 172–73.

132 *"Without in any way interfering . . . absorption":* AE, unclear source in ibid., 170–71.

132 *Heisenberg set his student . . . 1931:* vW, *Zeitschrift für Physik* 70, 114–30 (1931) in ibid., 178–80.

132 *"exercise in quantum field theory":* vW-Jammer, 11/13/67 in ibid., 179.

132 *"Weigh the box . . . reference frame":* Eh.-Bohr, 7/9/31 in ibid., 171.

132 *"It is interesting . . . (and test).":* Eh.-Bohr, 7/9/31 in Howard, *Nicht*, 99.

132 *"there is* ABSOLUTELY NO NEED FOR AN ANSWER": Eh.-Bohr, 7/9/31 in Jammer, *Philosophy of Quantum,* 172.

132 *Bohr . . . in Bristol:* 10/5/31. Bohr's account of the light-*blitz* box is "essentially the same as in his later recollection" ("Discussion with Einstein"), according to Howard, who points out that if Bohr was manifestly able to misunderstand Eh.'s crystal-clear exposition of AE in 1931, it is much less hard to believe that Bohr had also misunderstood AE's point in 1930. *Nicht,* 100.

132–33 *Nathan Rosen . . . twin:* Asher Peres's obituary for Rosen, 12/24/95.

133 *"Personally, I assess . . . may be wrong":* AE-Nobel committee in Pais, '*Subtle,*' 516.

133 *"after the light quantum . . . distant mirror":* AE, Leiden 1931 in Casimir, 316.

133 *"with the task . . . Einstein had in mind":* Ibid. (compare Pais, '*Subtle,*' 449).

133 *"Today, I made my decision . . . colleagues":* AE, travel diary in Michelmore, 162.

134 *"The five years . . . age of physics":* WH, *Physics and Beyond,* 93.

134 *all of matter . . . seconds:* "gives a mean life for ordinary matter of 10^{-10} seconds": JRO, Feb. 1930 in Pais, *Inward,* 351.

135 *"Nobody took Dirac's theory seriously":* Blackett, as sort-of-interviewed (he refused to be taped), 12/17/62, Archive for the History of Quantum Physics, Berkeley History of Science Department.

135 *"There is only one thing . . . again":* Rutherford to Mott in Weinberg, 109.

135 *The next day . . . positron:* WH, *Physics and Beyond,* 125–29.

135 *"regrettable . . . had been established":* Rutherford in Pais, *Inward,* 363.

135 cardless *poker:* WH, *Physics and Beyond,* 139.

135 *"My suggestion was . . . no reality at all":* Bohr, re: cardless poker in WH, *Physics and Beyond,* 139. Heisenberg remembered the quotation ending with "then it becomes impossible to make credible suggestions."

136 *a telegram to Pauli:* 1956 in Pais, *Inward,* 569.

136 *"two and a half full-time jobs":* one of Cockcroft's students; Hartcup and Allibone, 43.

136 *Walton . . . could repair watches:* Hartcup & Allibone, 39. See also pages 43 and 56.

136 *"like a volcano . . . ash":* Bowdon in Hendry, 17–19.

136 *wrote down different dates:* Hendry, 21.

136 *"Our understanding of the quantum . . . de Broglie":* AE, 9/29/32 in Pais, '*Subtle,*' 516.

137 *"If we recall . . . for this purpose":* Eh., *Zeitschrift für Physik* 78 (1932) in Jammer, *Philosophy of Quantum,* 117–18.

137 *"senseless":* Eh., *Zeitschrift für Physik* 78 (1932) in ibid.

137 *"a source of sheer pleasure":* Pauli, *Zeitschrift für Physik* 80 (1933) in ibid.

137 *"For fear of Bohr . . . drove me to it":* Eh.-Pauli, 10/1932 in Enz, 257.

137 *Itha Junger:* Moore, 223–25; 251–56.

137 *Schrödinger's unpublished notes:* discovered by Ekert's students Matthias Christandl and Lawrence Ioannou of Cambridge University in the Schrödinger archive in Vienna.

138 *"the customary stunt":* Gamow, *Thirty Years,* 167; for a picture of the audience, see p. 156.

138 *"dizzy with success":* a famous phrase of Stalin's in 1930 excusing the excesses of his army.

138 *"fraternize with scientists of the capitalistic countries":* Gamow, *My World Line,* 93.

138 *Inspired by a speech Bohr gave . . . :* The speech, "Light and Life," is included in Bohr, *Atomic Physics,* 3–12. See Daniel J. McKaughan, "The Influence of Niels

Bohr on Max Delbrück: Revisiting the Hopes Inspired by 'Light and Life,' " *Isis* 96, 507–29 (2005).

138 *top hat:* "Max Delbrück, with an imperturbable face and a top hat, was a magnificent master of ceremonies" writes Casimir of other "Copenhagen stunts" ("There was always a good show"). Casimir himself, for a reason he could not remember, missed the *Faust*. Casimir, 119–120.

138 The Blegdamsvej Faust: Reprinted in the back of Gamow, *Thirty Years*. According to vW, it was written "essentially" by Max Delbrück (vW also reports a little of the staging—the lab table and stool, etc., and who played Bohr and Pauli; see French and Kennedy, 188–90) and illustrated, according to Peierls and Rosenfeld, by Gamow himself (French and Kennedy, 228). (As for Gamow, beyond thanking Delbrück for his "kind help in the interpretation of certain parts of the play," he only wrote that "the authors and performers prefer to remain anonymous, except for J. W. von Goethe"—whose rhyme schemes were followed almost exactly. He suggests that, if the author and illustrator did not present themselves, a portion of the royalties from his book be deducted and sent to the Niels Bohr Library.) Barbara Gamow made the English translation. (Gamow, *Thirty Years*, 168–69) See also Segrè, *Faust in Copenhagen*.

139 *"wrapped . . . talking":* vW, in French and Kennedy, 187.

139 *"I say this not to criticize . . .":* Peierls once overheard Bohr's most spectacular version: "I am not saying this in order to criticize, but your argument is sheer nonsense" (French and Kennedy, 229).

141 *"Our laughter about Bohr . . . limit":* vW, "A Reminiscence from 1932" in French and Kennedy, p. 190.

142 *the speed motor boat of Professor Einstein: Vossiche Zeitung,* 6/12/33, quoted on the Einstein Web site, http://www.einstein-website.de/z_biography/tuemmler-e.html (May 19, 2006).

142 *"once again bought by public enemies":* quoted on the Einstein Web site, which describes the whole story, including AE's fruitless postwar search (ibid.).

142 *Einstein, in Pasadena, wrote . . . :* Pais, 'Subtle,' 450.

142 *Back in Berlin, Max Planck . . . :* Heilbron, 153; Einstein–Ludwig Silberstein, 9/20/34; see Cassidy, 307, and *B-E Letters,* 263.

142 *Meanwhile, Heisenberg found himself . . . :* Cassidy, 315; Heilbron, 164.

142 *In April, Pascual Jordan . . . joined the Nazi party:* Cassidy, 316; Enz: "Jordan's unsteady political orientations incited Pauli to issue the remark: 'Alas, good Jordan! He has served all regimes in utmost faithfulness' " (180). Jordan apparently believed that he could better protect his professors Born and Franck as a party member. He joined immediately after Franck resigned from the University of Göttingen in protest over the Nazi policies. See Greenspan, 176.

142 *Things fall apart; the center cannot hold:* Yeats, "The Second Coming."

142–43 *The telephone woke Max Born . . . the Italian border:* Born, *My Life,* 250–54; see also *B-E Letters,* 113–18.

143 *"So we settled . . . lonely life":* Born, *My Life,* 254.

143 *"Hedi said that she longed . . . Alps":* Born, *My Life,* 255–56.

143 *Weyl arrived with Born's . . . join him:* The girls may have come alone with Trixi, Weyl following later. *B-E Letters,* 117; Born, *My Life,* 257.

143 *(she and Erwin . . . were not far away):* Moore, 273.

143 *"Thus . . . the University . . . expeditions":* Born, *My Life,* 258.

143 *"for a little mountain climbing":* Moore, 272–3.

143 *"be affected . . . looks in the fall"*: WH-Born in Cassidy, 308.

144 *precious few of his Jewish friends:* see Cassidy, 483–85. It seems that he did save Guido Beck: Cassidy, 321–22.

144 *"our well-meaning . . . colleagues"*: Born-Eh. in Cassidy, 308.

144 *physics by day and the Zurich nightlife after dark:* see, e.g., a postcard (6/4/28) from Pauli and friends (his assistant Kronig and his senior colleague Scherrer) to Pascual "PQ-QP Jordan," who had just taken over Pauli's position in Hamburg: *"Lieber Herr* Jordan! We are about to study the Zurich night life and try to improve it following a new method of Pauli: by comparison. Many greetings, *Kronig."* "This method, however, may also be used to make things worse!—Greetings, *Pauli."* "I have also heard so many bad things about you that I would like to make your acquaintance. *Scherrer"* (Enz, 196–97).

144 *"slightly disconcerting habit . . . wheel"*: Casimir, 144.

144 *"When I asked him . . . forget it?' "*: Ibid., 145.

144 *"uneasiness"*: Einstein remembered by Rosenfeld quoted in Jammer, *Philosophy of Quantum,* 172–73.

144–45 *"What would you . . . ceased between them?"*: Ibid.

145 *"an illustration . . . phenomena"*: Rosenfeld in ibid., 173.

145 *"What are you doing . . . answers?"*: Besso-AE, 9/18/32 in Brian, 236.

145 *"one of your long journeys"*: Ibid.

145 *"next year"*: AE-Besso, 10/21/32 in ibid.

145 *psychiatry . . . institutionalization:* Elizabeth Einstein, Hans Albert's first wife, as interviewed by Denis Brian: "In early childhood he [Eduard] was a sort of genius who remembered everything he read. He played the piano beautifully. . . . My husband [his brother] thought he was ruined by electroshock treatment" (Brian, 195–96).

145 *In May 1933:* Brian, 247, who also includes the photograph.

145 *"Ehrenfest sent me . . . gone anyway"*: AE-Born, 5/30/33, Oxford, in *B-E Letters,* 113–14.

146 *"I cannot but confess . . . quantum-riddle solved"*: AE, 6/10/33 "Herbert Spencer Lecture" in AE, *World as I See It,* 131ff.

146 *"Many thanks . . . achieving something"*: Born-AE, 6/2/33, Selva-Gardena, in *B-E Letters,* 116.

146 *"I hope you know . . . at a conference"*: Dirac-Bohr, 9/28/33 in Pais, *Niels Bohr's,* 410. Dresden (pg. 313) notes that Eh.'s panic really started to be strong with Dirac's work of 1928. See also Enz, 255–56.

146–47 *"Dirac, what you have . . . force to live"*: Ehrenfest to Dirac in Dirac-Bohr, 9/28/33: Dirac recounts Ehrenfest's every move, and finishes, "The last phrase I remember very well. . . . I was very alarmed about it. . . . I now cannot help blaming myself for not doing anything." Pais, *Niels Bohr's,* 410.

147 *A few weeks later, Ehrenfest . . . :* Ibid.; Segrè, 252. (Pais says the event happened in the waiting room; Segrè, the park nearby. Segrè also notes that Wassik was blinded but not killed.)

147–48 *"My dear friends . . . stay well:* Ehrenfest in Pais, *Niels Bohr's,* 409–10.

148 *"We knew by now . . . very much"*: Born, *My Life,* 264.

148 *"a foreign country . . . future"*: Ibid.

148 *In December, Heisenberg, Schrödinger, and Dirac . . . :* See picture in Moore, 288.

148 *"since he was neither Jewish . . . endangered"*: WH-mother, 9/17/33 in Cassidy, 310.

148 *"I hope . . . my back"*: Schrödinger's toast in Moore, 291.

148 *"anything to do . . . solution":* Dirac's toast in Moore, 290.

148 *"There is no . . . prophet":* Pauli; WH, *Physics and Beyond,* 87.

149 *Heisenberg simply thanked . . . :* Moore, 290.

149 *No prize for peace:* Ibid., 289.

149 *"Concerning the Nobel Prize . . . Born":* WH-Bohr in Cassidy, 325. See Born, *My Life,* 220–21, for Heisenberg's letter to Born, written at the same time as the one to Bohr, from Switzerland: "The fact that I am to receive the Nobel Prize alone, for work done in Göttingen in collaboration—you, Jordan, and I—this fact depresses me and I hardly know what to write to you."

149 *Refugees from Germany . . . :* Pais, *Niels Bohr's,* 543–44; see p. 480 for the story of how Hevesy dissolved von Laue's and Franck's Nobel medals to protect them. See Frisch, p. 95, for one of the many descriptions of the lovableness of Franck.

149 *radioactive cats:* Pais, *Niels Bohr's,* 393.

149 *Before his eyes . . . :* Pais, *Niels Bohr's,* 411–12.

149 *"I had numerous . . . causality in quantum theory":* von Neumann-Klara Dan, 9/18/38 (unpublished letter, collection of Marina von Neumann Whitman). The statuary in Bohr's mansion and greenhouse was by the great Danish neoclassicist, Bertel Thorwaldsen. Many thanks to George Dyson for showing me this wonderful letter.

CHAPTER SIXTEEN: THE QUANTUM-MECHANICAL
DESCRIPTION OF REALITY 1934–1935

I: GRETE HERMANN AND CARL JUNG

150 *In a small basement room:* Cassidy, 271–72; Teller, *Memoirs,* 56, 63; cf. Dresden, 264.

150 *Carl Friedrich von Weizsäcker:* Bernstein, *Hitler's,* 75, 144; Cassidy, 275, 295, 326.

150 *"In that moment . . . this":* vW, "Reminiscence," French and Kennedy, 184.

150 *"did not concern . . . subject":* vW, "Reminiscence," French and Kennedy, 184.

150 *"Physics is an honest . . . about it":* WH to vW; French and Kennedy, 184.

150 *Heisenberg relies completely . . . politics:* Cassidy, 326.

151 *" 'empty of content' ":* WH speaking in 1931 to exact epistemologists in Vienna; in a 1928 speech to philosophers, WH called for the "very difficult task of rolling out the Kantian basic problem of epistemology once again, & . . . starting all over again. . . . But this is your task, not that of the scientist," Cassidy, 256–57.

151 *"Now it is the task . . . situation":* WH, *Berliner Tageblatt,* 1931 in Cassidy, 257.

151 *Her name is Grete Hermann . . . :* WH says she arrived "one or two years" after vW did, and Cassidy corroborates, saying it was 1932, which means she came as soon as she read WH's challenge in the newspaper. But the conversation, as WH remembered it, was closely related to her March 1935 paper, and Jammer (*Philosophy of Quantum,* 207) says she arrived for the 1934 spring semester. WH's chapter on (and ending with) the conversation is dated 1930–34. "The young philosopher Grete Hermann came to Leipzig for the express purpose of challenging the philosophical basis of atomic physics" (WH, *Physics and Beyond,* 117). See Ilse Fischer's biographical essay with photograph, "Von der Philosophie der Physik zur Ethik des Widerstandes: Zum Nachlass Grete Henry-Hermann im Archiv der sozialen Demokratie," found at www.fes.de, and Seevinck, "Grete Henry-Hermann," unpublished manuscript found at www.phys.uu.nl.

152 *"What prevents us . . . mechanics":* Hermann, "*Die Naturphilosophischen Grundlagen der Quantenmechanik,*" *Abhandlungen der Fries'schen Schule,* New Series, Vol. 6, 1935, 99–102, translated for the author by Miriam Yevick, 2004; see also

Harvard Review of Philosophy VII, 37; continuing, "which in combination with the present formal approach might render exact predictions possible again? Everything depends upon the answer to this question."

152 *Von Weizsäcken's face . . . takers:* "although he was a student of physics, [vW] grew unusually animated whenever our talks impinged on philosophical . . .": WH, *Physics and Beyond,* 117.

152 *"Nature in fact . . . them":* WH, *Physics and Beyond,* 120. WH started this quote with "This is nature's way of telling us. . . ."

152 *disturbance "is of such . . . description":* Bohr, *Atomic Physics,* 115.

152 *"is a deterministic . . . possible?":* This question was the title to WH's neverpublished reply to EPR. The next two paragraphs, giving WH's answer to this question, follow Cassidy's summary (261) of the relevant section of this paper.

153 *"a contradiction between the lawlike . . . unavoidable":* WH in a speech in Vienna, 11/27/35. It incorporated the manuscript of his reply to EPR. Cassidy, 261.

153 *"you'll notice that a . . . nonexistence":* "However a necessary step in Neumann's proof is eliminated by [a closer analysis]. On the other hand, if—with Neumann—one does not relinquish this step, it follows that one has implicitly made the unproven assumption that no differentiating characteristics . . . can be attributed. . . . The nonexistence of such characteristics was however the assertion to be proven": Hermann, *Die Naturwissenschaften,* 1935, translated for the author by Miriam Yerick, 2004.

153 *"The possibility of other . . . depends":* "other characteristics upon which the course of the motion depends": Ibid.; *Harvard Review of Philosophy,* 38.

153 $<P+Q> = <P> + <Q>:$ Grete wrote $Erw (R+S) = Erw (R) + Erw (S)$. "*Erw*" is short for *Erwartung,* expectation. Both "*Erw (X)*" and "$<X>$" mean "the expectation value of X." The expectation value of an attribute is a weighted average of its probable values given the circumstances, a basic tool of quantum mechanics.

153 *"With this assumption . . . proof":* Hermann, *Naturwissenschaften;* Seevinck, "Grete Henry-Hermann," unpublished manuscript.

153 *"He's trying to read . . . relations":* "It suggests itself to read off the impossibility of such an increase from the uncertainty relations": Hermann, *Harvard Review of Philosophy,* 37.

153 *"If the position and momentum . . . body?":* Ibid.

154 *"But this argument . . . observation":* Ibid.; *Harvard Review of Philosophy,* 37–38. "This subjective interpretation is incompatible with the derivation of these relations from the dualism of wave and particle pictures": Hermann; *Harvard Review of Philosophy,* 38. When "interpreted merely subjectively" the uncertainty relations "do not seem to say anything about the nature of the physical systems." But this, Grete remarks, is incompatible with WH's derivation of uncertainty from waveparticle dualism: Hermann; *Harvard Review of Philosophy,* 38. "You turn uncertainty into a physical reality, with an objective character": Grete to WH; WH, *Physics and Beyond,* 122.

154 *"If according to this reasoning . . . declare them impossible":* Hermann in *Harvard Review of Philosophy,* 38.

154 *"there can only be a single . . . knows these causes":* Ibid., her italics.

154 *I am here to find them:* The rest of her paper shows the results of this search, in which she believed she had proven, using Bohr's correspondence principle, that "*the characteristics determining the measurement outcome are already provided by quantum mechanics itself*" (Hermann, her italics; *Harvard Review of Philosophy,*

40). This argument, not the one about "Neumann's proof," was the one that impressed WH, and he used it in his reply to EPR. But Bohr found the argument logically inconsistent. Cassidy, 260.

155 *"becoming acquainted . . . new W. Pauli"*: Pauli-Kronig, 10/3/34 in Enz, 240.

155 *"In that interview, . . . archetypical material"*: Jung, discussion after the 5th Tavistock lecture, fall 1935; Enz, 243.

155 *He did not mention . . . source*: Ronald Hayman, *A Life of Jung*, NY: Norton, 2001, p. 327.

155 *"I consulted Herrn Jung . . . treat me"*: Pauli-Rosenbaum, 2/3/32; Enz, 241.

155 *thrilled when Jung . . . "material"*: "I am pleased that you have been able to make so much use of my material. I couldn't help smiling a little when you praised it so much, thinking to myself that it was the first time I had ever heard you address me in such a way. . . . I would like to mention just one point where I had the feeling that your dream interpretation was not entirely accurate. (As you can see, I still won't be 'fobbed off' with just anything.) I am referring to the interpretation of the *seven* of clubs. . . . In my seventh year, my sister was born. *So the 7 is an indication of the birth of the anima*": Pauli-Jung, 2/28/36; *A&A.*

155 *"I feel a certain need . . . outside"*: Pauli-Jung, 10/26/34, including Jordan's paper; *A&A.*

156 *Jung, meanwhile . . . "scientific proof"*: "There are indications that at least a part of the psyche is not subject to the laws of space & time. Scientific proof of that has been provided by the well-known J. B. Rhine experiments": Carl Jung, *Memories, Dreams, Reflections* (New York: Vintage, 1989), p. 304.

156 *"These experiments prove . . . incomplete"*: Ibid., 304–5.

156 *"walking around like . . . possible"*: Kate Goldfinger interviewed by Mehra; Enz, 210.

157 *"the extremely rational . . . personality"*: Franca Pauli in Enz, 287. In 1934, this was Pauli's position on parapsychology: "I certainly do not know of any factual material on it. If I did, God knows whether I would believe any of it": Pauli-Jung, 4/28/34 in Pauli and Jung, 25.

157 *"We had basically agreed . . . causal principle"*: Pauli-Jung, 11/24/50 in ibid.

157 *"Today I do in fact . . . Rhine's experiments"*: Pauli-Jung, 5/27/53 in ibid.

157 *"A man resembling Einstein . . . the archetypes"*: Pauli-Jung, 5/27/53 in ibid.

157 *"the quantum-mechanical . . . knowledge of it"*: Hermann, *Harvard Review of Philosophy*, 41.

2: EINSTEIN, PODOLSKY, AND ROSEN

158 *In 1934, Nathan Rosen*: "Nathan Rosen—the Man & His Life-Work," Israelit, in Mann and Revzen, 5–10; Obituary, Asher Peres, Technion Senate, 12/24/95; Pais, 'Subtle,' 494–95. "I remember Rosen as a gentle soft-spoken man who blended in with the group more than he stood out. By contrast Dirac and Wigner always made their presence felt. If someone said something particularly stupid Dirac would say, 'Now that's interesting.' Then Wigner would say something like, 'How stupid! How can you possibly assert that?' Then Rosen and my father would laugh politely and try to extract from the stupidity some useful grain of truth." Podolsky's son, Bob, describing a 1963 conference in an e-mail to LLG, Feb. 23, 2002.

159 *"Young man . . . together with me?"*: interview with Rosen; Jammer, *Philosophy of Quantum Mechanics*, 181.

159 *Einstein, Podolsky, and Rosen's discussions*: "As Boris told it he was the primary

"LLG" is apparently Louisa [L.?] Gilder — the author herself. (Abbrev. is absent from list on p. 351.)

instigator, the drive behind their collaboration. [Compare with Pais, '*Subtle,*' 494: Rosen told Pais that the main idea of EPR had come from him.] He would come up with new ideas and bounce them off Nathan. . . . One of Boris's great gifts was the ability to take an abstract concept and reduce it to mathematical formula; so it was he who took the idea of correlated particles and described it mathematically. I believe it was Einstein that pointed out the fact that the math implied that a measurement of the state of one particle would instantaneously determine the state of the other particle": Bob Podolsky to LLG by e-mail, Jan. 9 and 25, 2002.

159 *"for reasons of language . . . discussion"*: AE-ES, 6/19/35; Fine, 35.

159 *five hundred English words*: Michelmore, 197.

159 *"when they thought . . . something"*: Bob Podolsky to LLG by e-mail, Jan. 9, 2002.

159 *"We added Einstein's name without asking"*: John Hart to LLG by e-mail, December 7, 2001.

160 "If, without in any way . . . quantity": EPR, *Physical Review* 47, 777.

160 *"One would not arrive . . . theory is possible"*: Ibid., 780.

160 *Podolsky had to leave for California*: "There are no earlier drafts of [EPR] among Einstein's papers and no correspondence or other evidence that I have been able to find which would settle the question as to whether Einstein saw a draft of the paper before it was published. Podolsky left Princeton for California at about the time of submission and it could well be that, authorized by Einstein, he actually composed it on his own." Fine, 35–36.

160 *"EINSTEIN ATTACKS QUANTUM THEORY"*: *New York Times*, 5/4/35 in Jammer, *Philosophy of Quantum*, 189–91.

161 *"Why should we believe in that?"*: Bergmann told Shimony this story in Wick, 286.

161 *"Podolsky always goes . . . problem"*: AE's reference for Podolsky in Bob Podolsky to LLG by e-mail, Jan. 9, 2002.

3: BOHR AND PAULI

161 *"Subtle is the Lord . . . malicious"*: AE in Pais, '*Subtle,*' vi.

162 *"This onslaught came . . . at once"*: Rosenfeld in Wheeler and Zurek, 142.

162 *"the right way to speak about it"*: Ibid.

162 *"No . . . this won't . . . understand it?"*: Ibid.

162 *"Now we have to start . . . work"*: as Bohr reported 11/17/62 in Beller, 145.

162 *"I must sleep on it"*: Rosenfeld in Wheeler and Zurek, 142.

162 *"Einstein has once again . . . sein darf"*: Pauli-WH, 6/15/35 in Enz, 293.

162 The Impossible Fact: Christian Morgenstern, *The Gallows Songs: Christian Morgenstern's* Galgenlieder, Max Knight, trans., University of California Press, 1964.

163 *"I'll grant him . . . YOU to do it"*: Pauli-WH, 6/15/35 in Cassidy, 259, and Rüdiger Schack trans., Fuchs, 549.

163 *"It is probably only . . . popular"*: Pauli-WH, 6/15/35 (translated by Rüdiger Schack) in Fuchs, 550–51.

163 *"Elderly gentlemen . . . content"*: Pauli-WH, 6/15/35 in ibid., 550.

163 *"Quite independently . . . fundamental point"*: Pauli-WH, 6/15/35 in ibid., 297.

163 *finding the all-male . . . stifling*: Moore, 296–98.

163 *"I was very happy . . . in Berlin"*: ES-AE, 6/7/35; Moore, 304; Fine, 66.

164 *was already focusing on . . . entanglement*: See Fine, 67 and Fuchs, 640.

164 *"Podolski! . . . Basiopodolski!"*: Bohr to Rosenfeld in Pais, 430–31.

164 *"The trend of their . . . physics"*: Bohr, *Physical Review* 48, 696.

164 *"criterion of physical reality . . . phenomena"*: Ibid., 696.

164 *"I shall therefore be glad . . . phenomena":* Ibid.

164 *"this new feature . . . relativity":* Ibid., 702. *"Striking analogies which have often been noted"* [between general relativity and complementarity]: Ibid., 701.

165 *"You seem to take a milder . . . morning":* Rosenfeld in Wheeler and Zurek, 142.

165 *"That's a sign . . . problem":* Bohr to Rosenfeld in ibid.

165 *"It all became . . . there": "Let us begin with the simple case . . . diaphragm":* Bohr, *Physical Review* 48, 697.

165 *"My main purpose . . . arrangements and procedures":* Ibid., 699.

165 *"Any comparison . . . irrelevant":* Ibid.

165 *"Indeed we have in each . . . unambiguous way":* Ibid.

165 *"As in the simple . . . concepts":* "These last remarks apply equally well to the special problem treated by E, P, & R. Like the above simple case . . . we are . . . just concerned with a discrimination . . . complementary classical concepts": Ibid.

165 *"We now see that . . . essentially incomplete":* Ibid., 700.

166 *"In* fact, *it is only . . . characterizing":* Ibid.

166 *"necessity of discriminating . . .* under investigation": Ibid., 701.

166 *"It is true that . . . convenience":* "It is true that the place within each measuring procedure where this discrimination is made is in both cases [quantum and classical] largely a matter of convenience": Ibid.

166 *"But it is of fundamental . . . measurements":* "its fundamental importance in quantum theory has its root in the indispensable use of classical concepts in the interpretation of all proper measurements": Ibid.

167 *"It is impossible . . . physics":* "The impossibility of a closer analysis of the reactions between the particle and the measuring instrument is . . . an essential property of any arrangement suited to the study of this kind of phenomena, where we have to do with a feature of *individuality* completely foreign to classical physics": Ibid., 697.

167 *"The very existence . . . physical reality":* Ibid.

167 *"whole argumentation . . . false brilliance":* Rosenfeld in Wheeler and Zurek, 142.

167 *"They do it* smartly . . . right": Bohr to Rosenfeld, in ibid.

167 *"preconceived notions of reality . . . nature herself"* ("as Bohr exhorts us to do"): Rosenfeld in Wheeler and Zurek, 144.

167 *"I want to work out . . . phenomenon":* "As soon as we attempt a more accurate time description of quantum phenomena, we meet with new paradoxes . . .": Bohr, *Physical Review* 48, 700.

4: SCHRÖDINGER AND EINSTEIN

167 *"I consider the renunciation . . . reactionary":* AE-ES, 6/17/35 in Fine, 68.

167 it *"did not come out as well . . . formalism":* AE-ES, 6/19/35 in ibid., 35.

168 *"I don't give a* sausage": AE-ES, 6/19/35 in ibid., 38. AE's words were ". . . ist mir wurst": lit. ". . . is to me a *sausage*"; i.e., "I don't care."

168 *"In front of me . . . lift the covers":* AE-ES, 6/19/35 in Moore, 304.

168 *"Naturally, the second . . . seriously":* AE-ES, 6/19/35 in Fine, 69.

168 *"But the Talmudic . . . expression":* Ibid. and Moore, 304.

168 *"One cannot get at . . . principle":* AE-ES, 6/19/35 in Howard, *Einstein on Locality,* 178.

168 *"The contents . . . incomplete":* AE-ES, 6/19/35 in Moore, 304–5.

169 *as many as three:* On July 2, 1935, Bohr sent WH his response to EPR; WH sent *his* response to EPR to Pauli & AE; ES wrote Pauli about "the Einstein case."

169 *"I'd really like to know . . . too complicated' "*: ES-Pauli, 7/2/35 in Rüdiger Schack trans. in Fuchs, 551–52.

169 *"But I've not yet . . . old Schrödinger"*: ES-Pauli, 7/2/35 in Moore, 306.

169 *"a word that everyone . . . its content"*: Ibid.

169 *"In my opinion . . . Einstein example"*: Pauli-ES, 7/9/35 in Rüdiger Schack trans. in Fuchs, 553.

169 *"an observer . . . new situation"*: Pauli, 1954, in *Writings on Physics*, 33.

169 *It was necessary . . . cause"*: "Like an ultimate fact without any cause, the individual outcome of a measurement is . . . not comprehended by laws": Ibid., 32.

169 *"You have made me extremely . . . very strange"*: ES-AE, 7/13/35 in Fine, 74.

170 *"only with difficulty . . . boot"*: ES-AE, 7/13/35 in ibid., 75.

170 *"prescribed with wise . . . monstrous"*: ES-AE, 7/13/35 in ibid., 76.

170 *"You are the only . . . theory to the facts"*: AE-ES, 8/8/35 in Moore, 305.

170 *"they cannot get out . . . wriggle about inside"*: AE-ES, 8/8/35 in Fine, 59.

170 *He proceeded . . . "the paradox"*: AE-ES, 8/8/35 in Fine, 50 and 47n.

170 *"You, however, see . . . its own two legs"*: AE-ES, 8/8/35 in Moore, 305.

170 *"This point of view . . . macroscopic example"*: AE-ES, 8/8/35 in Fine, 77.

170 *"that, by means of . . . already-exploded systems"*: Ibid., 78.

171 *"Through no art . . . not-exploded"*: Ibid.

171 *"it doesn't work . . . ordinary mechanics"*: ES-AE, 8/19/35 in Fine, 79.

171 *"Discussion . . . Separated Systems"*: ES, *Proceedings of the Cambridge Philosophical Society* 31, p. 555 (1935).

171 *"I would not call . . . entangled"*: Ibid., 555.

172 *"general confession"*: ES, "The Present Situation in QM," *Naturwissenschaften* 23, 807–12; 823–28; 844–49 (first section published 11/29/35) in Wheeler and Zurek, 152–67. The description of the paper as "general confession" appears in ES's footnote to section 12.

172 *"as of twenty-four hours . . . editor"*: ES-AE, 8/19/35 in Fine, 80.

172 *encouraging a young and insecure*: Greenspan, 24 and 67.

172 *"gone to sleep"*: Cäcilie Heidczek–von Laue, 4/6/42 in Greenspan, 243.

172 *"Verschränkung"*: ES, "The Present Situation in QM," *Naturwissenschaften* 23, 827.

173 *"Does one not . . . result?"*: ES, "Present Situation" in Wheeler and Zurek, 155.

173 *"One can even . . . the cat)"*: ES, "Present Situation" in Moore, 308.

173 *"that in the course . . . (pardon the expression)"*: Ibid.

173 *"avoidance of the . . . pleasure trips"*: ES-Bohr, 10/13/35 in Moore, 312–13.

174 *"I must be satisfied . . . beautiful"*: WH-mother, 10/5/35 in Cassidy, 330.

174 *"Recently in London . . . much more peaceful"*: ES-AE, 3/23/36 in Moore, 314.

174 *"It seems hard . . . he does)"*: AE-Lanczos, 3/21/42 in Dukas and Hoffmann, 68.

174 *"God knows I am . . . still is"*: ES-AE, 6/13/46 in Moore, 435.

174 *"You are the only . . . act of observation"*: AE-ES, 12/22/50 in Przibram, 39.

175 *"No reasonable person . . . is vague"*: ES-AE, 11/18/50 in Przibram, 37.

175 *"I am sick of myself"*: Bohr to Pais, 1948 in Pais, *Niels Bohr's*, 12.

175 *"There was Einstein . . . Let him"*: Pais in *Niels Bohr's*, 434.

175 *"I always need . . . system"*: Bohr to Pais, 1948 before dictating "Discussion with Einstein" in Pais, *Niels Bohr's*, 13.

175 *"The discussions . . . Einstein all the time"*: Bohr, "Discussion with Einstein," 1949 in *Atomic Physics*, 66.

176 *"an urchin smile . . . Einstein was up to"*: Pais, *Niels Bohr's*, 13.

176 *"There they were, face . . . speechless":* Ibid.

176 *But you know my doctor . . . tobacco:* Ibid.

176 *"Maxel, you know I love . . . folly?":* ES-Born, 10/10/60 in Moore, 479.

176 *"His private life . . . brain":* Born, *My Life,* 270.

177 *On his blackboard . . . light-filled box:* photographed in French and Kennedy, 304.

THE SEARCH AND THE INDICTMENT 1940–1952

CHAPTER SEVENTEEN: PRINCETON APRIL–JUNE 10, 1949

181 *"Einstein told me . . . sit for a while":* Bohm in Peat, 92.

181 "as a Marxist, he had difficulty believing in quantum mechanics": Gell-Mann, 170.

181 *he is diligently reading Bohr:* see Bohm in Hiley and Peat, 33: "The whole development started in Princeton around 1950, when I had just finished my book *Quantum Theory.* I had in fact written it from what I regarded as Niels Bohr's point of view, based on the principle of complementarity. Indeed, I had taught a course on the quantum theory for three years and written the book primarily in order to try to obtain a better understanding of the whole subject, and especially of Bohr's very deep and subtle treatment of it. However, after the work was finished, I looked back over what I had done and still felt somewhat dissatisfied."

182 *"Oh my God, all is lost":* Lomanitz interview in Peat, 92.

182 *"If you and Oppenheimer will run for President . . .":* Pais, *Einstein,* 95.

182 *eight-year-old girls bribing Einstein with cookies:* Jon Blackwell, the *Trentonian,* 1933, http://capitalcentury.com/1933.html (March 21, 2008).

182 *"I am become Death":* "There floated through my mind a line from the *Bhagavad-Gita* in which Krishna is trying to persuade the Prince that he should do his duty: 'I am become death, the shatterer of worlds' " (JRO in Goodchild, 162).

182 *"there's an FBI man on the committee":* Lomanitz interview in Peat, 92.

182 *Promise me you'll tell the truth:* "I said they should tell the truth" "What did they say?" "They said 'We won't lie' ": JRO interviewed by the HUAC in *In the matter . . .* , 151.

182 *paranoiac:* Rossi interview in Peat, 92.

182 *tossing coins:* Lipkin interview in Peat, 77–8.

182 *"He is totally free . . . precious being":* Gross in Hiley and Peat, 48–49.

183 *"Mr. Bohm . . . League?":* *Hearings Regarding Communist Infiltration of Radiation Laboratory and Atomic Bomb Project at . . . Berkeley,* 5/25/49, 321 in Bohm archive.

183 *"I can't answer . . . as already stated":* Ibid.

183 *"Are you a member" . . . "Democratic Party?":* *Hearings Regarding . . . Berkeley,* 5/25/49, p. 325 in Bohm archive.

183 *"by his Princeton colleagues . . . loyalty":* Peat, 95.

184 *"protect our country":* *Hearings Regarding . . . Berkeley,* 6/10/49, 352 in Bohm archive.

184 *"I believe that in some . . . err at all":* *Hearings Regarding . . . Berkeley,* 6/10/49, 352–53 in Bohm archive.

184 *"He used to make jokes . . . themselves":* Ford to LLG by phone, Dec. 2000.

← see p. 384

CHAPTER EIGHTEEN: BERKELEY 1941–1945

185 *I loved Oppenheimer:* "Bohm's feelings for Oppenheimer extended beyond admiration into what he later described as love. Here was someone who not only understood the passions of Bohm's intellect but who offered him encouragement

and support. It was inevitable that a part of Bohm's nature would look to Oppenheimer, 13 years his senior, as a protective and understanding father." Peat, 43.

185 *J. Robert Oppenheimer was raised:* Goodchild; and Smith and Weiner.

185 *"In addition to a superb . . . unanswered":* Bethe, *Science* 155, 1967.

186 *"He became an almost . . . not follow":* Rabi in Rabi et al., 6–7.

186 *"I didn't start out . . . very rich":* Goodchild, 26.

186 *Oppenheimer's students . . . room:* Ibid., 27–29; Rabi et al., 5, 6, 19; Smith and Weiner, 133.

186 *"many physicisti . . . life it brings":* JRO-brother 1/7/34 in Smith and Weiner, 170.

186 *"Haven't . . . nim-nim-nim boys?":* Pauli in Regis, 133.

186 *"a convinced classicist":* unclear whether these were Bohm's or Weinberg's words. Peat, 50.

186 *"Pythagorean mysticism":* Bohm in Peat, 52.

186 *"Physics has changed . . . formulas":* Bohm interview, *Omni*, 1/87. www.fdavid peat.com/interviews/bohm.htm.

187 *Though he found . . . months:* Peat, 56–58.

187 *"I would say . . . after 1939":* JRO in *In the Matter . . .* , 114.

187 *"We knew . . . a bomb"):* Bohm; interview 6/15/79, Bohm archives, Birkbeck.

187 *In March of 1943 . . . not come: In the Matter . . .* , 119–20.

187 *"we had . . . code":* Ibid.

187 *At the same time:* For a description of the Berkeley investigation, "Scientist X," and Steve Nelson, see *In the Matter . . .* , p. 259; *Hearings of the Committee on Un-American Activities,* pp. v–vi, 1949, Bohm Archive.

188 *"We had very little . . . New York":* Pash in *In the Matter . . .* , 811.

188 *By June of 1943 . . . Berkeley:* "As a result of our study we determined and were sure that Joe was Joseph Weinberg": Pash in *In the Matter . . .* , 811.

188 *"it was known among the physicists . . . very mean job":* All quotes from Lansdale's interview of JRO are from *In the Matter . . .* , pp. 873–83, except "the way he talked . . . war project," p. 121.

189 *"a conspiracy of silence":* The Schecters' reaction to their book's reception in 2002 "Accuracy in Media" conference (www.aim.org).

190 *"truly dangerous":* DeSilva in *In the Matter . . .* , 150.

190 *He named David . . . direct action":* DeSilva in ibid.

190 *"a strange feeling of insecurity": In the Matter . . .* , 149.

190 *"The undersigned answered yes":* Ibid.

190 *to Bohm, they symbolized . . . Marxist state:* See Peat, 66–68; he "made no distinction between his work in physics and his political . . . beliefs," ibid., 135.

CHAPTER NINETEEN: *QUANTUM THEORY* AT PRINCETON 1946–1948

192 *"When I was a boy . . . sense of nature":* Bohm interview, *Omni*, 1/87.

192 *"even as a child . . . movement":* Bohm, *Wholeness,* ix.

192 *"If we think . . . same time":* Bohm, *Quantum Theory,* 146.

192 *"A blurred photograph . . . does not":* Ibid., 145.

193 *"is actually in close . . . picture":* Ibid., 146.

193 *"Many of the ancient . . . be moving":* Ibid., 147.

193 *"once again, quantum theory . . . given direction":* Ibid., 152.

193 *"contrary to general opinion . . . plan":* Ibid., 622.

193 *"Even in the classical . . . abstraction":* Ibid., 167.

193 *bacterium:* "in a few . . . we saw originally?": Ibid., 163.

193 *"In a system whose . . . prediction":* Ibid., 167.

193 *"He was a wonderful . . . being was physics":* Ford to LLG by phone, Dec. 2000.

194 *"We spent a tremendous . . . mentor":* Gross in Hiley and Peat, 46.

194 *"understanding reality . . . that whole:* "In my scientific & philosophical work, my main concern has been with understanding the nature of reality in general & of consciousness in particular as a coherent whole": Bohm, *Wholeness*, p. ix.

194 *"the remarkable point-by-point . . . quantum processes":* Bohm, *Quantum Theory*, 171.

194 *Logic to Thought as . . . mechanics:* Ibid., 169–70.

194 *"too much schmooze":* Wigner; relayed to me by Abner Shimony.

194 *"helpful in giving us . . . quantum theory":* Bohm, *Quantum Theory*, 171.

194 *"If a person tries . . . strong analogy":* Ibid., 169.

194 *"Even if this hypothesis . . . hidden variables":* Ibid., 171.

195 *"(It would not . . . variables exist)":* Ibid., 171.

195 *"every indication . . . hidden variables":* Ibid., 115.

195 *"insofar as the spin . . . at all":* Ibid., 614.

195 *"but not more than one of these":* Ibid.

196 *"Because the wave . . . second atom":* Ibid., 615.

196 *"serious criticism . . . mechanics":* Ibid., 611.

196 *"interwoven potentialities":* Ibid., 159.

196 *"We conclude then . . . quantum theory":* Ibid., 623.

196 *"The term 'quantum mechanics' . . . 'quantum non-mechanics' ":* Ibid., 167n.

CHAPTER TWENTY: PRINCETON JUNE 15–DECEMBER 1949

197 *all over the newspapers:* The Rochester newspaper story is reproduced in *In the Matter . . .* , p. 211. Bohm told Martin Sherwin how disturbed he had been when he heard what JRO had said about Peters. Sherwin interview, Birkbeck College, Bohm Archives.

198 *On a night . . . disloyal":* "The marshal was friendly, and when Bohm asked him for advice, he was told that they could drive to Trenton, the capital, to try to obtain bail. . . . Back en route to Trenton, the marshal talked about science, asking his prisoner about Einstein. He had come from Hungary, he said, and was a loyal American; he hoped that Bohm had not been disloyal." Peat, 98.

198 *"After a brief . . . leave":* Schweber, *In the Shadow . . .* , p. x. Schweber and friends then went to Oppie, "and he graciously offered Bohm a desk at the institute." Peat, Bohm's biographer, tells a different story—that it was Einstein who wanted to bring Bohm to the institute, and Oppie who scotched the plan out of concern for how it would look if he were using the institute to shelter Communists (Peat, 104; see also Peat's long endnote p. 331).

CHAPTER TWENTY-ONE: *QUANTUM THEORY* 1951

199 *He has finished . . . before I wrote the book":* "I . . . sent copies of my book to Einstein, to Bohr, to Pauli, and to a few other physicists. I received no reply from Bohr, but got an enthusiastic response from Pauli. Then I received a telephone call from Einstein. . . .": Bohm in Hiley and Peat, 35. Gell-Mann writes that David "interrupted me excitedly to report that . . . Einstein had already read it and telephoned him to say that David's was the best presentation he had ever seen of the case against him and that they should meet to discuss it. Naturally, when next I saw David I was dying to know how their conversation had gone, and I asked him

about it. He looked rather sheepish and said, 'He talked me out of it. I'm back where I was before I wrote the book' ": Gell-Mann, *The Quark & the Jaguar,* 170.

200 *"He told me that I had explained . . . convinced":* Bohm in Hiley and Peat, 35.

200 *"What came out . . . part is missing":* Ibid.

200 *"What Einstein said . . . theory":* "This was . . . close to my more intuitive sense that the theory was dealing only with statistical arrays": Ibid. "Einstein felt that the statistical predictions of the quantum theory were correct, but that by supplying the missing elements we could in principle get beyond statistics to an—at least in principle—deterministic theory. This encounter with Einstein had a strong effect on the direction of my research, because I then became seriously interested in whether a deterministic extension of the quantum theory could be found."

200 *"It's a theory . . . appearances":* Ibid., 33.

200 *Remember . . . deterministic:* "David told me that as a Marxist, he had had difficulty believing in quantum mechanics. (Marxists tend to prefer their theories to be fully deterministic.)": Gell-Mann, 170.

200 *Dear Professor Blackett . . . Albert Einstein:* David Bohm archive. (The switch between "Dr." and "Mr." is in the original document.)

201 *"I had taught . . . somewhat dissatisfied":* Bohm in Hiley and Peat, 33.

201 *"was that the wave . . . the phenomena":* Ibid.

CHAPTER TWENTY-TWO: HIDDEN VARIABLES AND HIDING OUT 1951–1952

202 *"The usual interpretation . . . of accuracy":* Bohm, *Physical Review* 85, 166, 1952.

203 *He sent the paper . . . 1951:* The paper was received by *Physical Review* on July 5, 1951. Ibid.

203 *"A New Physical . . . Equation":* Ibid., 169.

203 *Never a great . . . particle:* "After this article was completed, the author's attention was called to similar proposals for an alternative interpretation of the quantum theory made by de Broglie in 1926": Ibid., 167.

203 *"a 'quantum-mechanical' potential":* Ibid.

203 *"an objectively real . . . ψ-field":* Ibid., 170.

203 *"The 'quantum-mechanical' forces . . . ψ-field":* Ibid., 186.

203 *"There is . . . not yet been discovered":* Ibid., 170.

204 *"simple explanation" of EPR:* Ibid., 180.

204 *"Einstein . . . has always . . . theory as incomplete":* Ibid., 166.

204 *"Dr. Einstein for several interesting and stimulating discussions":* Ibid., 179.

204 *For the purpose . . . properties of matter:* Ibid., 189.

204 *"leads to precisely . . . interpretation":* Ibid., 166.

204 *"I can't believe . . . see this":* as Miriam Yevick remembered Bohm saying; Peat, 113.

204 *their days at Penn State . . . day-old pies:* Ibid., 28–29.

205 *In college, Bohm . . . strolls):* Ibid., 29–30.

205 *They reminisced . . . kosher food:* Ibid., 23–24.

205 *Oppenheimer told people . . . bury it":* Ibid., 84.

205 *But Oppenheimer had also . . . recommendation:* This fact was later used against JRO during his security hearing. Goodchild, 261.

205 *yellow convertible:* "Weiss had the feeling that his friend was hiding out from the FBI. One evening Bohm asked Weiss to look out the window and see if a yellow convertible was driving back and forth. Weiss saw the car. 'Are you being pursued?' he asked. 'Yes,' Bohm replied, 'they are looking for me'; Peat, 105.

205 *That night Bohm . . . in print!":* Bohm and Weiss "waited until it was dark and

walked to the subway. . . . On the train Weiss noticed a man reading the back page of a newspaper. On the front . . . was a photograph of Bohm, with words to the effect that 'all they ever got from him was his name.' Weiss joked that at last Bohm had gotten his name in the papers"; Peat, 105.

206 *"Your dad always . . . walking at night":* Ibid., 27.

206 *Dave . . . your nose . . . had broken:* Ibid., 30.

206 *Weiss said sincerely . . . never held a grudge:* Ibid.

206 *"It may be that Oppenheimer . . . matter":* AE-Bohm, 12/15/51 in Peat, 116.

206 *The plane will return . . . off the plane:* Ibid., 120.

CHAPTER TWENTY-THREE: BRAZIL 1952

208 *Brazil had seemed awful . . . Avenida Angelica:* Peat, 121–23.

208 VENCERÁS PELA CIÊNCIA: See the U. of São Paulo Web site, which repeatedly emphasizes the "sophistication" of the building.

208 *When he started . . . there:* "It looks as if there will be an opportunity to do much good work": Bohm-Einstein in Peat, 121.

208 the sixth law of thermodynamics: Ibid., 122.

209 *"I like to float . . . furthermost shores":* Bohm-Loewy (Bohm was in Florida at the time, preparing to leave for Brazil) in ibid., 105.

209 *"somewhat shaken . . . my work":* Bohm-Yevick, 11/1951 in ibid., 122.

209 *In December . . . saw Feynman:* Ibid., 125–26.

210 *"there are several good":* Bohm-Einstein in ibid., 121.

210 *"Tell me, Dave . . . meaningful words:* Feynman describes his problems with the Brazilian students learning by rote in "*Surely,*" 211–19.

210 *"I don't believe . . . clearly all by themselves":* Ibid., 165.

210 *"the Princetitute":* Bohm-Yevick in Peat, 131–32.

210 *"But here . . . rather difficult":* "It's a little frightening to realize that the language barrier will for a long time make really close contacts with most people rather difficult. . . . In explaining physics to people who do not understand English very well, the imagination is not stimulated nearly so much. I shall have to guard against . . . stagnation": Bohm-Loewy, postmarked 10/17/51 in Bohm archive.

210 *"School—in the sense . . . Princeton":* "not a school in the sense of education, but in the sense of fish": Feynman, "*Surely*" . . . , 206.

211 *"a toy frying pan . . . little metal stick":* Ibid., 206–8.

211 *Mephistopheles:* Gleick describes Feynman as Mephistopheles in the 1952 Carnaval in *Genius,* 286.

211 *Dick, they took . . . over an hour:* Bohm's passport is taken away under strange pretenses; his roommate watches a circling car: Peat, 124–25.

211 *"I am not really out of the U.S.":* Bohm-Phillips in ibid., 125.

212 *WKB approximation:* The conversation with Einstein started Bohm looking for a deterministic extension of the quantum theory. "I soon thought of the classical Hamilton-Jacobi theory, which relates waves to particles in a fundamental way. . . . When one makes a certain approximation (Wentzel-Kramers-Brillouin), Schrödinger's equation becomes equivalent to the classical Hamilton-Jacobi equation. . . . I asked myself: What would happen, in the demonstration of this equivalence, if we did not make this approximation? I saw immediately that there would be an additional potential, representing a new kind of force, that would be acting on the particle. I called this the quantum potential": Bohm in Hiley and Peat, 35.

212 *Somewhere in there . . . logically consistent:* "When I met Feynman he thought that

the idea was crazy but after enough talk I convinced him that it is logically consistent": Bohm-Yevick, 1/5/52 in Peat, 126.

212 *"terrifically impressed with it"*: Bohm-Yevick, 1/5/52 in ibid.

212 *Feeling the momentum . . . calculator:* Bohm hoped to shake Feynman "out of his depressing trap of doing long & dreary calculations on a theory that is known to be of no use. Instead, maybe he can be gotten interested in speculating about new ideas, as he used to do, before Bethe and the rest of the calculators got hold of him": Bohm-Yevick or Bohm-Loewy (unclear) in Peat, 126.

212 *Feynman looked . . . beats me"*: Feynman describes his calculating games with Bethe in "Surely," 192–95.

212 *As Feynman himself . . . bars?:* "How is it possible that an 'intelligent' guy can be such a goddamn fool when he gets into a bar?": Feynman, "Surely" . . . , 187.

213 *"I like the problems"*: This was Feynman's whole approach to physics. Bohm explained in an interview that he understood the reason that Feynman would not work with hidden variables was that he "could not see a problem in it." Peat, 126.

213 *"I believe that you . . . great"*: Feynman was "convinced that it is a logical possibility and it may lead to something new": Bohm-Loewy, 12/10/51 in Peat, 126.

213 *"practically concedes . . . logical"*: Bohm-Loewy, 12/1951 in ibid., 127.

213 *"For a while . . . mesons"*: Gleick describes this letter exchange with Fermi in *Genius,* 282.

213 *"I can hook up . . . once a week"*: Feynman's story about the blind ham radio operator is in Feynman, "Surely," 211.

213 *"sometimes I think I should get married"*: Feynman's loneliness in Brazil; he proposed by mail, starting a short second marriage: Gleick, 287.

214 *"I now think he is my friend for life"*: Bohm-Yevick, 5/8/52 in Peat, 126: "He was right; even as late as the 1980s, Bohm always visited Feynman during his trips to the U.S."

CHAPTER TWENTY-FOUR: LETTERS FROM THE WORLD 1952

215 *I do not see . . . cannot be cashed"*: Pauli-Bohm, 12/3/51 in Bohm archive.

215 *"instantaneous interactions . . . particles"*: Bohm, *Physical Review* 85, 186; *Physical Review* 108, 1072.

215 *"I want to emphasize . . . important in itself"*: Bohm in Hiley and Peat, 38.

216 *"really read my article . . . article came out"*: Bohm-Yevick in Peat, 128.

216 *"What I am afraid of . . . a big effect"*: Bohm-Yevick, 1/5/52 in ibid., 125.

216 *As January wore . . . stomach troubles worsened:* Ibid., 124.

216 *"passionate desire . . . your heart"*: Bohm-Yevick, 1/9/52 in ibid., 130.

216 *Bernard Peters . . . German citizen again:* Schweber, 129.

216 *"the square root of Bohr times Trotsky"*: Pauli-WH, 5/13/54 in Pais, *Niels Bohr's,* 360.

216 *"the self-anointed . . . que le roi"*: Ibid.

216 *Dear Bohm . . . yours sincerely, L. Rosenfeld:* Bohm archives.

218 *"We are both . . . each other"*: author's interview of M. Yevick, Aug. 2003.

218 *"In his low-key . . . vast panorama"*: Gross in Hiley and Peat, 46.

218 *his wife, Sonia:* M. Yevick interview, Aug. 2003.

218 *Studying in Princeton's . . . give her a start:* Miriam Yevick interview, Aug. 2003.

218 *The men were both more down-to-earth . . . :* "David loved to spin yarns and turn philosophy into physics. He would often be talking about the deep questions and there would be no equations. Eugene was a much more down-to-earth physicist.

He was more skeptical; he had a cynical streak with a sense of humor. George was inclined to think Eugene might be right": Miriam Yevick interview, Aug. 2003.

219 *"He's got to get . . . emphasis:* Ibid.

219 *Gene's right . . . his way around:* "Bohm was very indecisive, & Gross was very decisive and outspoken (like me). It was a big problem that Bohm couldn't get spin in his theory": George Yevick interview, Aug. 2003.

219 *Gross laughed . . . intellectual structures":* "I recall a social evening, where, tongue-in-cheek, he constructed an elaborate and 'convincing' theory of the existence of ghosts and devils": Gross in Hiley and Peat, 47. Preceding sentence: "I marvel at Dave's extraordinary manner of . . . constructing coherent intellectual structures"—the ghost story a "ludicrous" example of this skill.

219 *Gross, suddenly serious . . . secular saint":* "I can only use old-fashioned language to describe his impact on me and others. Dave's essential being was . . . totally engaged in the calm but passionate search into the nature of things. He can only be characterized as a secular saint. He is totally free of guile and competitiveness, and it would be easy to take advantage of him. Indeed, his students and friends, mostly younger than he is, felt a powerful urge to protect such a precious being": Gross in Hiley and Peat, 49.

219 *"One of my best friends . . . tide":* Bohm-Yevick, 3/9/52 in Peat, 131.

219 *"resultlets" . . . "that place":* Bohm-Yevick, 1/1951; 1/1952 and undated, Peat, 131.

220 *"very foolish":* Bohm-Yevick and Bohm-Phillips in Peat, 132.

220 *his initial surprise:* Peat, 129.

220 *When recounting to Miriam . . . bum!)":* "von Neumann thinks the idea consistent, and even 'very elegant' (the unprincipled bum)": Bohm-Yevick, undated; Peat, 132.

220 *"It grants . . . including himself":* Bohm-Yevick, 1/28/52 (unpublished letter, collection of Miriam Lipschutz-Yevick).

220 *"comparable to . . . the translation":* Bohm-Yevick, undated in Peat, 131–32.

220 *"I am convinced . . . right track":* Bohm-Yevick, undated in ibid., 134.

CHAPTER TWENTY-FIVE: STANDING UP TO OPPENHEIMER 1952–1957

221 *Just before the war, a young Max Dresden:* obituary by C. N. Yang, *Physics Today,* June 1998.

221 *There he wrote . . . Kramers:* Dresden, *H. A. Kramers: Between Tradition and Revolution.* New York: Springer-Verlag, 1987.

221 *At first he told them . . . opinion:* Dresden, May 1989 APS speech in Peat, 133.

221 *"We consider . . . waste our time":* JRO to Dresden in ibid., 133.

221 *"juvenile deviationism":* Pais to Dresden in ibid.

221 *"A public nuisance":* As Dresden remembered in ibid.

222 *"If we cannot disprove . . . ignore him":* JRO to Dresden in ibid.

222 *reading von Neumann . . . (in German):* Nasar, *Beautiful . . . ,* 45, 81.

222 *"most physicists . . . nonobservable reality":* Nash-JRO, 1957 in Nasar, *Beautiful Mind,* 220–21.

222 *"it was this attempt . . . destabilizing":* Ibid., 221.

CHAPTER TWENTY-SIX: LETTERS FROM EINSTEIN 1952–1954

223 *"So we old fellows . . . still exist":* Born-AE, 5/4/52; B-E Letters, 190.

223 *Dear Born . . . Yours, Einstein:* AE-Born, 5/12/52; B-E Letters, 192.

223 *Dear friend . . . Your old Hedi:* HB-AE, 5/29/52 in B-E Letters, 193–94.

224 *"Today one hardly . . . de Broglie":* Born in B-E Letters, 193.

224 *12 October, 1953 . . . Yours, A. Einstein:* AE-Born, 10/12/53 in *B-E Letters,* 199.
224 *26 November, 1953 . . . Yours, Max Born:* Born-AE, 11/26/53 in *B-E Letters,* 205–7. "Pauli has come up with an idea which slays Bohm . . .": Whitaker comments, "This appears to have been wishful thinking. Pauli described Bohm's work as 'artificial metaphysics,' but his physical arguments were surprisingly limp. Like Heisenberg, he did not believe that Bohm could modify the quantum formalism without contradicting established experimental results, and he asked for an explanation of the relationship between probability density and wavefunction. Bohm & his associate Vigier were actually able to provide the latter, but even without this reply, one could hardly say that Bohm had been 'slain' ": Whitaker, *Einstein,* 251.
225 *"Dear Bohm, Lilli . . . extended experience":* AE-Bohm, 1/22/54 in Bohm archives.
225 *"Now I want to return . . . new direction":* Bohm-AE, 2/3/54 in Bohm archives.
226 *February 10, 1954 . . . yours, Albert Einstein:* AE-Bohm, 2/1954 in Bohm archives.

EPILOGUE TO THE STORY OF BOHM 1954

227 *I have just heard . . . you think?:* Bohm-Yevick, probably 4/1954, in Peat, 160.
227 *"I would say . . . movement and unfoldment":* Bohm, *Wholeness,* p. ix.
228 *In November 1964 . . . Lectures":* Feynman, *Character of Physical Law,* ch. 6.
228–29 *"our imagination . . . electron is going to go":* Ibid., 127–47.
230 *"Even in its present . . . likely to perform":* Bohm, *Wholeness . . . ,* 109–10.

THE DISCOVERY 1952–1979

CHAPTER TWENTY-SEVEN: THINGS CHANGE 1952

233 *John Bell, a red-haired . . . pieces:* "He lived in Geraldine Road hostel & was one of the . . . young men all with motorcycles which they took to pieces regularly": M. Bell in Bertlmann and Zeilinger, 3.
233 *Bell's new red beard . . . wreck:* Whitaker, *Physics World,* 12/1998, 30.
233 *"must have been just wrong":* Bell in Bernstein, *Quantum Profiles,* 65.
233 *"The kind of professions . . . colonists":* Ibid., 12.
234 *Bell spent much of his childhood . . . finally ended:* Ibid., 12; Whitaker, *Physics World,* 12/1998, 29; and Bertlmann and Zeilinger, 7–9.
234 *"a young man of . . . caliber":* Walkinshaw in Burke and Percival, 5.
234 *"great pleasure . . . with him":* ibid.
234 *"did not mind . . . temperament":* M. Bell in *Europhysics News* memorial edition.
234 *special liking for particle dynamics:* Walkinshaw in Burke and Percival, 5.
234 *"He always liked the theory . . . practical experience":* M. Bell in Bertlmann and Zeilinger, 5.
234 *(Cockcroft was also . . . other colleges):* Peierls describes Cockcroft collecting bricks in *Bird of Passage,* 120.
235 *Among these . . . John Bell:* M. Bell in *Europhys. News* memorial edition.
235 *"the beautiful book . . . more practical things":* Bell, "On the Impossible Pilot Wave" (1982) in *Speakable,* 159–60.
235 *In this he was lucky:* Mandl in Burke and Percival, 10.
235 *"I can remember . . . fierce arguments":* M. Bell in Bertlmann and Zeilinger, 3–4.
235 *"Franz . . . told me . . . unreasonable axiom was":* Bell in Bernstein, *Quantum Profiles,* 65.
236 *accelerators or the foundations of quantum theory:* Whitaker in Bertlmann and Zeilinger, 17.

CHAPTER TWENTY-EIGHT: WHAT IS PROVED
BY IMPOSSIBILITY PROOFS 1963–1964

Bell's conversation with Jauch was a critical one in the history of his thought on hidden variables, for which he subsequently thanked Jauch strongly on several occasions. However, neither man explicitly wrote down more than a vague sense of what was discussed. Jauch's quotes here are all from his book *Are Quanta Real?*, a "dialogue" on the subject of reality in quantum mechanics, set in Geneva in "the fall of 1970"—seven years after the watershed conversation with Bell in Geneva. However, "many of the passages" in this book, "are more or less faithful reproductions of actual conversations" (p. xii). Bell's answers come from his published works, ranging from 1964 to 1986.

"I actually avoided these questions for a number of years," Bell told Davies, "because I saw that people smarter than I had made little progress with them, and I got on with other more practical things. But then in Geneva in 1963 when I was busy with other things I met Professor Jauch at the University. He was concentrating on these issues, and in discussion with him I became determined to do something about them" (Davies, *The Ghost in the Atom,* 56). Bell told Bernstein that Jauch "was actually trying to strengthen von Neumann's infamous theorem. For me, that was like a red light to a bull. So I wanted to show that Jauch was wrong. We had gotten into some quite intense discussions" (*Quantum Profiles,* 67–8). "I am indebted . . . very especially, to Professor J. M. Jauch." (Acknowledgment to "On the problem of hidden variables in QM" [written in 1964], Bell, 11).

238 *comfortable saying "Good morning"*: M. Bell interview, fall 2000.
238 *he had just given a seminar:* Jammer, *Phil. of QM,* 303.
238 *"You see, though I . . . von Neumann's theorem"*: "I saw the impossible done . . . in papers by David Bohm": Bell, "On the Impossible Pilot Wave" (1982), *Speakable,* 160. "I was enormously impressed with [Bohm's paper]. I saw then that von Neumann must have been just wrong": Bell, 1990 in Bernstein, *Quantum Profiles,* 65.
238 *"Bohm's theory, and . . . phenomena"*: de Broglie's theory "is a very ingenious . . . phenomena": Jauch, 74.
239 *"the quest for hidden . . . new evidence"*: Jauch, x. "Yet there are scientific aspects of the [hidden-variable] problem which are extremely interesting and which are worthy of thorough exploration." Jauch, xi.
239 *"The pilot-wave . . . classical ideals"*: Bell, "Six Possible Worlds . . ." (1986), *Speakable,* 194. Bell wrote "almost" before "trivial."
239 *"a closed set . . . mutilated when embarrassing"*: Ibid.
239 *"The subjectivity of . . . eliminated"*: Bell, "On the Impossible Pilot Wave" (1982), *Speakable,* 160.
239 *"The way de Broglie . . . disgraceful"*: "De Broglie . . . was laughed out of court in a way that I now regard as disgraceful. . . . Bohm . . . was rather ignored": Bell (1986) in Davies, *Ghost,* 56.
239 *"arguments were not refuted . . . trampled on"*: Ibid.
239 *"It is possible . . . doing so"*: Jauch, ix. In the original the sentences are in the reverse order.
239 *"Why do people never bring it up? . . . deliberate theoretical choice"*: Bell, "On the Impossible Pilot Wave" (1982), *Speakable,* 160. *Why are you doing it?:* These words stand where Bell footnotes Jauch first among a list of impossibility proofs. Ibid.
240 *"There will almost always be more than one theory"*: In Jauch's book, Simplicio has a dream in which he finds himself in the library of everything that ever will be written in any language and asks for "the theory of elementary particles which

explains all known facts about them." The librarian says, "There are 137 different theories available which . . . agree with all the facts known today." "This is the point," comments Jauch, "where the naked truth is revealed to Simplicio. . . . There are two criteria of truth as Einstein told us, and, as the dream shows, the neglect of the second one leads to absurdity" (see AE, "Autobiographical Notes" in Schlipp, e.g. p. 13). Jauch, 51–53 and 105 (note 15).

240 *'Autobiographical Notes':* They "suggest that the hidden variable program has some interest": Bell, "On the Problem of Hidden Variables . . ." (1966), in *Speakable,* 12 (note 2) (see Schlipp, 81–87).

240 *Undeterred, Jauch . . . observed phenomena":* "The situation presents . . . observed phenomena": Jauch, *Are Quanta Real?,* xi.

240 *"Then as now . . . completely invalid":* Ibid.

240 *"Among the books . . . it becomes 'trivial' ":* Bell (1990); Bernstein, *Quantum Profiles,* 66.

240 *"Even Bohm did not cling":* "It must be admitted that this quantum potential seems rather artificial in form, besides being subject to the criticism that it implies instantaneous interactions between distant particles." Bohm was moving on to "a further new explanation of the quantum theory in terms of a deeper sub-quantum-mechanical level." Bohm & Aharonov, *PR* 108, 1072 (1957).

240 *Why should we not . . . experimenter?:* "nor does it disprove that the phase of the moon, which constellation the sun is in, or the state of my consciousness has nothing to do with the values of these parameters": Jauch, 16.

241 *"If hidden variables . . . causal relationships?":* Ibid., 100, n. 7.

241 *"The door is left wide open":* "It leaves the door wide open for all kinds of theories concerning the origin of your hidden variables": Ibid., 16

241 *"Terrible things . . . the universe":* Bell in Bernstein, *Quantum Profiles,* 72.

241 *"if this is a defect . . . situation":* Ibid. A historical note: Bell told Bernstein that when, later that year, he went on sabbatical to California, "My head was full of the argument of Jauch, & I decided that I would get all that down on paper by writing a review article on the general subject of hidden variables. In the course of writing that I became increasingly convinced that 'locality' was the center of the problem": Ibid., 67–68. This is the most specific reference that I can find to the chronology of Bell's focus on locality, and suggests that the idea may have come to Bell after the conversation with Jauch, rather than during it.

241 *"Of course he did . . . universe":* "All I am trying to do is eliminate as many false routes as possible, so that we are left with fewer possibilities among which we may find the one that will open our understanding for the basic complementarity which pervades all our physical universe." Jauch, 21–22.

241 *"While imagining that I understand . . . Bohr":* Bell, "Bertlmann's Socks . . ." (1981), *Speakable,* 155. He follows with a sentence-by-sentence analysis of Bohr's favorite statement from his reply to EPR. "Indeed I have very little idea what this means." He wondered, "Is Bohr just rejecting the premise—'no action at a distance'—rather than refuting the argument?"

241 *"I have never hidden . . . complementarity":* Jauch, 19.

241 *"Do we not see . . . struggle for understanding":* Ibid., 48.

241 *"rather than being disturbed . . . take satisfaction in it":* Bell, "Six Possible Worlds . . ." (1986), *Speakable,* 189.

241 *"not to resolve these contradictions . . . complementarity' ":* Ibid.

242 *"It is, without... evidence":* "is no doubt the sum and substance... factual evidence": Jauch, 96.

242 *"it seems to me that Bohr... contradictariness":* Bell, "6 Possible Worlds..." (1986), *Speakable,* 190.

242 *"I'm sure you have"... "familiar meaning":* "Bohr seemed to like aphorisms such as: 'the opposite of a deep truth... reverse of its familiar meaning' ": Ibid.

242 *"It emphasizes the bizarre... materialism":* Ibid.

242 *"chance is everywhere... same object":* Jauch, 54.

243 *"seems to have been... applied to":* Bell (1990) in Bernstein, *Quantum Profiles,* 52

243 *"there was no doubt... he has done":* Ibid.

243 *"The principle of complementarity... levels of science":* Jauch, 18.

243 *"I would never deny... immense":* "The justly immense prestige of Bohr": Bell, "6 Possible Worlds..." (1986), *Speakable,* 189.

243 *"strange in Bohr... system occurs":* Bell (1990) in Bernstein, *Quantum Profiles,* 52.

243 *"For me it is the indispensability... mechanics":* Bell, "Six Possible Worlds..." (1986), *Speakable,* 188.

243 *"hidden-variable approach... division":* Bell (1990) in Bernstein, *Quantum Profiles,* 84.

243 *"If you gave definite... little things":* Ibid., 85.

243 *"you are suspected... easier for the physicist":* Jauch, 16.

244 *"Like a hard-pressed... be found":* "hard-pressed prosecutor... culprit must be found": Ibid., 17. The architect of a hidden-variables theory in Jauch's parable is called "Simplicio."

244 *"When we have so many... quest for understanding":* Ibid., 17.

244 *"hideously nonlocal":* Bell (1990) in Bernstein, *Quantum Profiles,* 72.

244 *The Einstein-Podolsky-Rosen... local:* "the Einstein-Podolsky-Rosen setup was the critical one, because it led to distant correlations. They ended their paper by stating that if you somehow completed the quantum mechanical description, the nonlocality would only be apparent. The underlying theory would be local": ibid.

244 *Jauch looked at Bell... admire you:* "How stubborn you are, Simplicio! I cannot help admiring you! Your objections are a challenge to us to think more deeply about the foundations of physics, and so I think we should all be grateful to you": Jauch, 42.

244 *"perhaps, comparable... years ago":* Ibid., xii.

245 *"Many of the passages... opinions are represented":* Ibid., xi–xii.

245 *(quoting Lenin and talking about "wholeness"):* Ibid., 23.

245 *"ideologies and their systems... contemplated by us all":* Ibid., 26.

245 *"In the psychology of C. G. Jung....":* Ibid., 26n (101).

245 *a long dream:* Ibid., 50–51, and notes, 104.

245 *"What is the good?... the two":* Ibid., 104.

246 *"I address these words particularly... shallow"):* Ibid., 97.

246 *Whether this question... will be eclipsed:* Bell, "On the Problem of Hidden Variables in QM" (1964, published 1966), *Speakable,* 1–2.

246 *John F. Kennedy had been shot...:* "It was the worst possible day to come," Bell told Bernstein in *Quantum Profiles,* 67.

247 *"what is proved... imagination":* "long may Louis de Broglie continue to inspire those who suspect that what is proved by impossibility proofs is lack of imagination": Bell, "On the Impossible Pilot Wave" (1982), *Speakable,* 167.

247 *I've just been playing... can't be done:* "So I explicitly set out to see if in some sim-

ple EPR situation I could devise a little model that would complete the quantum-mechanical picture & would leave everything local. I started playing around with the very simple system of two spin ½ particles, not trying to be very serious, but just to get some simple relations between input & output that might give a local account of the quantum correlations. Everything I tried didn't work. I began to feel that it very likely couldn't be done": Bell in Bernstein, *Quantum Profiles,* 72–73.

247 *Josef Jauch is actually trying . . . red light to a bull:* "Not long before I came [to Stanford], I had once again begun considering the foundations of quantum mechanics, stimulated by some discussions with one of my colleagues, Josef Jauch. He, it turned out, was actually trying to strengthen von Neumann's infamous theorem. For me, that was like a red light to a bull. So I wanted to show that Jauch was wrong. We had gotten into some quite intense discussions. I thought that I had located the unreasonable assumption in Jauch's work. Being at Stanford isolated me, & gave me some time to think about quantum mechanics. My head was full of the argument of Jauch, & I decided that I would get all that down on paper by writing a review article on the general subject of hidden variables. In the course of writing that I became increasingly convinced that 'locality' was the center of the problem": Bell in Bernstein, *Quantum Profiles,* 67–68.

247 *He handed it to Mary . . . face:* "When I look through these papers I see her everywhere": Bell's preface to *Speakable,* thanking Mary for her help.

247 *"In his theory . . . would have liked least":* Bell, "On the problem of Hidden Variables in QM" (1966), *Speakable,* 11.

248 *He suggested that Bell . . . measurement:* Jammer, 303.

248 *Bell's edited paper . . . misfiled:* Ibid.

248 *the Bell inequality:* "Probably I got that equation into my head and out on paper within about one weekend. But in the previous weeks I had been thinking intensely all around these questions. And in the previous years it had been at the back of my head continually": Davies, *Ghost* 57; see also Bernstein, *Quantum Profiles,* 72.

249 *"the example considered . . . actually being made":* Bell, "On the EPR Paradox" (1964), *Speakable,* 19.

CHAPTER TWENTY-NINE: A LITTLE IMAGINATION 1969

250 " '*Some consideration . . . perform*' ": Bohm, *Wholeness,* 109–10.

250 *"There are three kinds . . . predict":* Abner Shimony, *Tibaldo and the Hole in the Calendar* (New York: Springer-Verlag, 1998), 1.

251 *Some friends of his . . . of the paper:* "The first draft of the paper was written during a stay at Brandeis University": Bell, "On the EPR Paradox" (1964), *Speakable,* 20.

251 *"just another . . . sound like a kook:* Shimony to LLG, spring 2000.

252 *"He wanted to inoculate . . . the disease":* Ibid.

252 *If you have . . . experiment:* Shimony according to Horne to LLG, June 2005.

252 *He remembered a paper . . . Aharonov:* "Discussion of Experimental Proof for the Paradox of EPR" (1957), *Physical Review* 108, 1070.

252 *letter to the editor . . . Shaknov:* "Angular Correlation of Scattered Annihilation Radiation" (1950), *Physical Review* 77, 136.

253 *"Bohm and Aharonov . . . quantum archaeology!":* Horne, Shimony, and Zeilinger, "Down-Conversion Photon Pairs: A New Chapter in the History of QM Entanglement" (1989), *Quantum Coherence,* 361. Actually, Shimony gives credit for this phrase to Horne; Horne was startled (Optical Society of America meeting in San Jose, CA, Sept. 2007).

253 *When Aharonov visited:* Shimony to LLG, spring 2000.

253 *"The same thing happened . . . believe it":* Clauser to LLG, Oct. 2000.

253 *"I thought, . . . seen in my life":* Ibid.

253 *"I was not yet willing . . . Bell was bluffing":* Clauser, "Early History of Bell's Theorem" (2000) in Bertlmann and Zeilinger, 78.

254 *"My thesis advisor . . . real astronomer":* Clauser to LLG, Oct. 2000.

254 *"We've got a wonderful" . . . "mech anymore":* Shimony to LLG, spring 2000.

254 *"I had been in Abner's . . . Bell's paper":* Horne to LLG, Nov. 2000.

254 *"See if you can design . . . happen":* Shimony to LLG, spring 2000.

254 *"That was the first . . . attention":* Horne to LLG, Nov. 2000.

254 *"Carl, wouldn't" . . . "we did it":* Clauser in Wick, 119.

254 *Kocher's experiment:* Kocher & Commins, "Polarization Correlation of Photons Emitted in an Atomic Cascade" (1967) in *Physical Review Letters* 18, 575.

256 *"Well, why don't . . . ask him":* Clauser to LLG, Oct. 2000.

256 *"crazy American student":* Ibid.; Wick, 120.

256 *"In view of the general . . . shake the world!":* Bell-Clauser, 1969 in Bertlmann and Zeilinger, 80.

256 *"The McCarthy era . . . 'shake the world' ":* Clauser; Ibid.

256 *In Massachusetts . . . the bat cave:* Holt to LLG, Dec. 2001.

257 *"Bread-and-butter physics":* Holt to LLG, Aug. 2004.

257 *"we made pests . . . photons":* Shimony to LLG, spring 2000.

257 *"cute" . . . "haunt me":* Holt to LLG, Dec. 2001.

257 *"I was kind of intrigued . . . real physics' ":* Holt to LLG, Dec. 2001.

257 *"we've been scooped":* Horne, "What Did Abner Do?" at OSA meeting 2007.

257 *"and Wigner said . . . join in":* Shimony to LLG, spring 2000.

257 *"thrilled that somebody was interested":* Horne to LLG, Nov. 2000.

257 *"a hint of shotgun . . . marriage":* Horne at OSA meeting 2007.

258 *"This was not Clauser's . . . astronomy?' ":* Horne to LLG, Nov. 2000.

258 *"It was wonderful . . . little points":* Shimony to LLG, spring 2000.

258 *"heavy machinery":* Horne to LLG, June 2005.

258 *the* Sinergy: Clauser to LLG, Oct. 2000.

258 *"at least we're making waves":* Shimony to LLG, spring 2000.

258 *"The paper was done . . . Berkeley":* Clauser to LLG, Oct. 2000.

258 *"Proposed Experiment to Test Local Hidden-Variable Theories":* Clauser, Horne, Shimony, and Holt (1969), *Physical Review Letters* 23, 880.

259 *"nice demonstration . . . quantum mechanics course":* Commins to LLG, Oct. 2000.

259 *wheeled cart:* Clauser to LLG, Nov. 2000.

259 *"It turned out . . . surprise in the result":* Commins to LLG, Oct. 2000.

259 *"There's nothing . . . waste my time":* Clauser to LLG, Oct. 2000.

259 *"Charlie Townes . . . tolerating the project":* Ibid.

259 *"I was very, very surprised . . . no big deal":* Commins to LLG, Oct. 2000.

259 *"He was very nice . . . end of it":* Ibid.

260 *But Charlie Townes had . . . convince Bohr":* Townes, 69–71.

260 *"Commins told me . . . 'Boy, that's crazy' ":* Freedman to LLG, Oct. 2000.

260 *"I had never really heard . . . wasn't the case":* Ibid.

260 *"How do you tell . . . I can make anything":* Clauser to LLG, Dec. 2001.

261 *"To be an experimental . . . '57 Pontiac":* Ibid.

263 *"Atom beam . . . all there is to it":* Ibid.

264 *"a nice cylindrical little slug":* Ibid.

264 *"I got into a lot . . . lasers now":* Freedman to LLG, Oct. 2000.

264 *2275 Å, 5513 Å, 4227 Å:* Ibid.

264 *"Everything had to be . . . don't exist":* Ibid.

265 *Piles-of-plates . . . not applicable":* Freedman, "Experimental Test of Local Hidden-Variable Theories," Ph.D. thesis, 5/5/72, Berkeley.

265 *who had honed . . . at a time:* Rehder to LLG, Sept. 2005.

265 *"like a big clock . . . engineering problems":* Freedman to LLG, March 2005.

265 *a second trade as a commercial fisherman:* Rehder to LLG, Sept. 2005.

266 *"We had to do . . . phototubes":* Freedman to LLG, Oct. 2000.

266 *"a nice sharp pulse . . . optics nicely":* Clauser to LLG, Dec. 2001.

266 *"O.K. Ideally it wouldn't . . . the damn tubes":* Ibid.

267 *the "red" photon:* "In 'photomultiplier-talk' anything not blue or violet (wavelength longer than 4500 Ang) is [called] 'red.' That is where most photocathodes turn to crap!": Clauser-LLG, Jan. 9, 2002.

267 *"In a Boston winter . . . our experiment":* Clauser to LLG, Dec. 2001.

267 *"People always think . . . equipment you have":* Ibid., March 2002.

267 *"When it was ready . . . hadn't blown up":* Freedman to LLG, Oct. 2000.

268 *"These big mama . . .* Ka-chunk": Clauser to LLG, Dec. 2001.

268 *"We're not getting any" . . . "so I got paid":* Ibid., Oct. 2000.

268 *"battles . . . needed to be done":* Freedman to LLG, Oct. 2000.

268 *"what Clauser devised . . . typical of John":* Ibid., and Shimony to LLG, 2000.

268 *"He's a smart . . . intimidated":* Commins to LLG, Oct. 2000.

268 *"We would go sailing . . . times sailing":* Freedman to LLG, Oct. 2000.

CHAPTER THIRTY: NOTHING SIMPLE ABOUT
EXPERIMENTAL PHYSICS 1971–1975

269 IBM *1620:* Clauser to LLG, March 2002; Freedman-LLG, May 2008.

270 *"It's about Commins . . . permission is denied' ":* Ibid.

270 *If* Gene *is willing:* "Gene Commins is a man of infinite integrity": Freedman to LLG, Oct. 2000.

270 *coincidences:* Actually the green photon always arrived before the purple photon, because it was emitted first. But the spacing of their arrivals could be predicted and a delay built into the electronics on the green-phonton side, so that the arrival of both photons would coincide, simplifying the analysis.

271 exactly what you'd expect: Clauser to LLG, March 2002.

271 *"Wow! . . .* Maybe we've found it!": Ibid.

271 *Just then Commins . . . yet?":* "I got all excited & I think Commins walked in at that moment": Ibid.

271 *"we need more data":* Ibid.

271 *"We're pretty marginal . . . error bars":* Ibid.

271 *"We've obviously got . . . testing":* Ibid.

271 *"Don't worry, John . . . experiment":* Ibid.

272 *"Ooooh, that did not sound good":* Ibid.

272 *"Aw,* shit" . . . "*funny-looking curve":* Ibid.

272 *"if we get . . . quit physics":* Clauser in Freedman to LLG, Oct. 2000.

272 *"I have to get . . . me a Ph.D.":* Holt in Shimony-LLG, Sept. 5, 2003.

272 *"Pipkin was an agnostic . . . smart guy":* Holt to LLG, Aug. 2004.

272 *"I liked tabletop physics . . . part of that":* Ibid.

273 *"a nontrivial glassblowing feat":* Clauser to LLG, March 2002.

273 *"then you have to distill . . . discharge on it":* Ibid.

273 *"Some things I had . . . possible":* Holt to LLG, Aug. 2004.

273 *"one very sad day . . . best specs' ":* Ibid.

273 *"actually brought a sleeping . . . hidden variable":* Ibid.

274 *The red-photon counter has lost it again . . . :* "Occasionally one or the other photo-multiplier developed a greatly increased dark pulse rate. . . . The phenomenon—more often occurring in the C31000E [to catch the pale green 5513Å photon]—was remedied by shutting down the experiment and storing the photo tube at room temperature for a few days; when recooled, the dark rate returns to normal." Freedman, "Experimental Test of Local Hidden-Variable Theories," Ph.D. thesis, 5/5/72, Berkeley note on p. 76.

274 It would be much . . . rushed: "It's much better if you don't have competitors so you don't feel rushed": Clauser to LLG, March 2002.

274 *"Abner called me . . . Italy":* Shimony to LLG, spring 2000; Clauser to LLG, March 2002.

274 *Bell gave a new more general inequality:* "Introduction to the Hidden-Variable Question" (1971), *Speakable,* 29. Compare Jackiw and Shimony, "Depth and Breadth," 90.

274 *Abner was able to show:* Shimony showed that the F-C experiment tested all local realistic theories, not just deterministic ones. Shimony's paper is reproduced in *Foundations of QM, Proceedings of the International School of Physics "Enrico Fermi," Course XLIX,* d'Espagnat, ed. (New York: Academic, 1971), 191; Clauser and Horne (1974), *Physics Review D,* 10, 526.

274 " *'Theoretical physicists . . . such speculation' ":* Bell, "Introduction to the Hidden-Variable Question" (1971), *Speakable,* 29.

275 *"he must be ten feet tall":* Clauser according to Shimony to LLG, spring 2000.

275 *"Anyway, John, I've come up":* Freedman's Inequality can be found in Freedman, "Experimental Test," p. v.

275 *I know that Dick . . . count rate:* Freedman to LLG, Oct. 2000.

275 *"Immediately that evening . . . enthusiasm":* Fry, "Arrogance? Naïveté? Stupidity? An Untenured Assistant Professor Threw Caution to the Wind for a Bell Inequality Experiment" (2007), OSA meeting.

276 *"were a clear window . . . devastated":* Ibid.

276 *"Dick Holt and Pipkin . . . in 1972":* Clauser to LLG, Oct. 2000.

276 *"And it was a very . . . conclusive":* Freedman to LLG, Oct. 2000.

276 *("It was the biggest . . . was big"):* Ibid.

276 *He first found his way . . . reading:* Ibid.

276 *"the linear polarization . . . Clauser et al.":* Holt, "Quantum Mechanics vs. Hidden Variables: Polarization Correlation Measurement on an Atomic Mercury Cascade," 1973 Harvard University preprint, 1.

276 *"This value is in strong . . . prediction":* Ibid.

276 It wasn't a problem . . . two angles: Freedman to LLG, Oct. 2000.

277 *"a sort of Rube Goldberg apparatus":* Holt to LLG, Dec. 2001.

277 *"being young and headstrong . . . simpler":* Ibid.

277 tremendous amount of harm: Freedman to LLG, Oct. 2000.

277 *"debating whether . . . publish":* Holt to LLG, Dec. 31, 2001.

277 *No one has . . . satisfactory suggestions:* Ibid.

277 *"F. M. Pipkin and W. C. Fields":* Ibid.

277 " *'So you're . . . promising young physicist' ":* Pound in Ibid.

278 *"I became pretty good . . . be going on":* Holt to LLG, Dec. 31, 2001.

278 *a team testing . . . in Sicily:* Faraci et al., *Lett. Nuovo Cim.* 9 (1974), 607. Clauser and Shimony: "Their data disagree sharply with the quantum-mechanical predictions. . . . Since their paper is quite condensed, it is difficult to conjecture whether or not a systematic error is responsible for these results." (Clauser and Shimony, "Bell's Theorem," p. 1917.)

278 *"There was no point . . . scheme":* Horne to LLG, June 2005.

278 *Clauser had come . . . glaringly inadequate:* Clauser and Shimony, "Bell's Theorem," p. 1916.

278 *"totally reenergizing!":* Fry's description of the Holt and Pipkin result, "Arrogance? Naïveté? . . ." at OSA meeting, Sept. 2007.

278 *"enough money . . . laser":* Ibid.

278 *"we were standing out . . . confused by it all":* Clauser to LLG, Oct. 2000.

278 *"revealed or* concealed *by QM":* Bell interviewed by a young woman at CERN; this videotape, in the library of the University of Vienna, is dated 11/28/90, which cannot be the date of the interview because Bell had just died that Oct. 1.

278 *"Still am. Still don't understand it":* Clauser to LLG, Oct. 2000.

278 *"we were together . . . a forever paper":* Horne to LLG, June 2005.

279 *"In formulating special . . . 'distance' ":* Clauser to LLG, Oct. 2000.

279 *"Does the chair exist . . . infuriating":* Ibid.

279 *"Well, that's the way it is":* Horne to LLG, June 2005. "The experiments convinced John. But we had different reactions to them. He would say, 'I don't like this,' and I'd say, 'That's the way it is.' "

279 *"It's the whole Bishop . . . weren't there":* Clauser to LLG, Oct. 2000.

280 *Clauser performed a new experiment:* "Experimental Distinction Between the Quantum and Classical Field Theoretic Predictions for the Photo-Electric Effect" (1974), *Physics Review D,* 9, 853.

280 *"of capital importance":* Shimony-LLG, Sept. 5, 2003.

280 *("we need to know . . . Bell's theorem case"):* Clauser to LLG, Oct. 2000.

280 *"It's not a question . . . interrogative 'what' ":* Ibid.

280 *"many valuable . . . Shimony":* Clauser and Horne, "Experimental Consequences of Objective Local Theories" (1974), *Physics Review D,* 10, 535.

280 *experiments of this sort . . . Berkeley:* See ibid., p. 530, for the original discussion; Clauser, "Early History of Bell's Theorem" (2000), for the arguments against these experiments (Bertlmann and Zeilinger, 87–88); and Aspect, "Bell's Theorem: The Naïve View of an Experimentalist" for their defense (Bertlmann and Zeilinger, 141–42).

280 *the "boneyard":* what the attic of LeComte Hall at Berkeley is called.

280 *"on scrounged equipment":* Clauser to LLG, Oct. 2000.

281 *"It was* grim *. . . weeks and weeks":* Ibid., March 2002.

281 *"heaved a sigh of relief":* Ibid., Oct. 2000.

281 his student . . . *Thompson:* Fry and Thompson, "Experimental Test of Local Hidden-Variable Theories" (1976), *Physical Review Letters* 37, 465.

CHAPTER THIRTY-ONE: IN WHICH THE SETTINGS ARE CHANGED 1975–1982

282 coup de foudre: "Gisin uses exactly the same words": Aspect, Optical Society of America meeting in San Jose, CA, Sept. 2007.

282 *"In October 1974 . . . at first sight":* Aspect, (in his abstract), OSA meeting, 2007.

282 *"This was the most exciting . . . dream of":* Aspect, "The Paper That Changed My Life" at OSA meeting, Sept. 2007.

282 *Immediately he decided . . . thesis:* Aspect, "Bell's Theorem: The Naïve View of an Experimentalist" in Bertlmann and Zeilinger, 119.

282 *"I must have applied . . . rejected":* Clauser to LLG, Dec. 2007.

283 *"I don't know" . . . "You can learn":* Ibid.

283 *"Back in the sixties . . . about the theory":* Fry, "Quantum (Un)speakables" conference in Vienna, Nov. 2000.

283 *"If you had sent me . . . bet on his success":* Pipkin in Fry, "Arrogance? Naïveté? . . ."
(2007); OSA meeting.

283 *stigma:* see Clauser's interesting analysis, "Early History of Bell's Theorem" in Bertlmann and Zeilinger, 61ff.

283 *"I had a quite good . . . extremely bad":* Aspect to LLG, Sept. 2007.

283 *The classes . . . stigmas:* "We had just been solving partial differential equations—I knew that I had missed something": Aspect to LLG, Sept. 2007.

284 *"First, it is real . . . washed my brain":* Ibid.

284 *"I was totally convinced by Einstein and Bell":* Aspect, "The Paper That Changed My Life" at OSA meeting, Sept. 2007.

284 *"there was still . . . done":* Ibid.

284 *Conceivably, quantum mechanics . . . crucial":* Bell, "On the EPR Paradox" (1964) in *Speakable,* 20.

284 *and, importantly, frugal:* "I was lucky because water was not expensive": Aspect, OSA meeting, Sept. 2007.

284 *"Each polarizer . . . uncorrelated":* Aspect et al., "Experimental Test of Bell's Inequalities Using Time-Varying Analyzers" (1982), *Physical Review Letters* 49, 1805.

285 *"The switching between . . . nanoseconds":* Ibid.

285 *"since the change is not truly . . . uncorrelated way":* Ibid., 1807.

285 *When Aspect finished . . . position?":* Aspect, "Bell's Theorem: The Naive View of an Experimentalist" in Bertlmann and Zeilinger, 119.

285 *"There will be serious fights . . . danger of becoming crazy":* Bell to Aspect, related by Aspect to LLG, Sept. 2007.

285 *"The Bell experiment . . . nothing interesting":* Freedman to LLG, Oct. 2000.

286 *"I would say . . . asking the wrong question":* Holt to LLG, Dec. 2001.

287 *"showed experimentally . . . equals 4":* Horne, "Quantum Mechanics for Everyone," *Third Stonehill College Distinguished Scholar Lecture,* May 1, 2001.

287 *"Abner and I both . . . years":* Horne to LLG, June 2005.

288 *"only one person . . . in on at the time":* Horne to LLG, June 2005.

288 *"This was my . . . important":* AZ, "Bell's Theorem, Information . . ." in Bertlmann and Zeilinger, 241.

288 *Fascinated, he tried . . . entanglement:* Horne in Aczel, 210.

288 *"I hear you're building . . . fun twelve years":* Horne to LLG, June 2005.

289 *"I'm just sort of a middleman . . . bricks":* Ibid.

289 *"the biggest American car . . . yacht":* Ibid.

289 *"we hit it off wonderfully . . . three of us":* Greenberger to LLG, May 2005.

289 *"a lovely . . . colleagues":* Greenberger, "History of the GHZ . . ." in Bertlmann and Zeilinger, 282.

ENTANGLEMENT COMES OF AGE 1981–2005

CHAPTER THIRTY-TWO: SCHRÖDINGER'S CENTENNIAL 1987

293 *"called the Oracle . . . his office":* Bertlmann, "Magic Moments: A Collaboration with John Bell" in Bertlmann and Zeilinger, 29.

293 *"canteen":* as Bell always called it; Bertlmann to LLG, Nov. 2000.

293 verveine *tea:* "It was like a ritual, at two minutes to four we left the office & went down to the CERN cafeteria. There John ordered in his typical British accent, '*deux infusions verveine, s'il vous plait,*' his favorite tea": Bertlmann, "Magic Moments . . ." in Bertlmann and Zeilinger, 36.

293 *"goodwill in all respects . . . small eruption":* Leinaas, "Thermal Excitations of Accelerated Electrons" in ibid., 402.

294 *"He wrote like that . . . sin to question the orthodox quantum mechanics":* Bertlmann to LLG, Nov. 2000.

294 *"The sun was falling . . . beautiful evening":* Bertlmann to LLG, Nov. 2000.

294 *"John, whether or not . . . mechanics":* Ibid.

294 *"No, I don't . . . benefits mankind":* Ibid.

295 *"I disagree . . . Nobel Prize":* Ibid.

295 *"Who cares about this nonlocality":* Ibid., including Bell's position.

295 *"the radical changes . . . necessitates":* AZ, "On the Interpretation & Philosophical Foundation of QM" in *Vastakohtien Todellisuus, Festschrift for K. V. Laurikainen,* Ketvel, et al., eds., Helsinki University Press, 1996, www.quantum.univie.ac.at/zeilinger/philosop.html (June 2, 2008).

295 *"there is no difference . . . radical as possible":* AZ to LLG, May 2005.

CHAPTER THIRTY-THREE: COUNTING TO THREE 1985–1988

297 *"He knows what . . . can't 'talk' ":* Horne to LLG, June 2005.

298 *"Gee, have you ever" . . . "neutron interferometry":* Ibid.

298 *"We couldn't do neutrons . . . remnant":* Ibid.

299 *"Has anybody ever . . . work on that":* Ibid.

299 *"Good,"* said Zeilinger: "because Mike & I had also once thought about a 3-body problem": AZ to Greenberger, related by Greenberger to LLG, May 2005.

299 *"but M.I.T. had other plans . . . who fit":* Greenberger to LLG, May 2005.

299 Ghosh and Mandel: "Observation of Nonclassical Effects in the Interference of Two Photons" (1987) in *Physical Review Letters* 59, 1903.

299 *"Go get this paper . . . get into":* Horne to LLG, June 2005.

299 *"had never plugged in" . . . "illuminate them?":* Ibid.

299 The same discovery was made: Shih and Alley in *Proceedings of the 2nd Int'l Symposium on Foundations of QM in the Light of New Technology,* Namiki et al., eds. (Tokyo: Physical Society of Japan, 1986).

300 *"and they were* very *. . . than yesterday' ":* Greenberger to LLG, May 2005.

300 *"Did Abner ever . . . with that stuff":* Horne to LLG, June 2005.

300 *"Anton, I am so . . . particles":* Greenberger to LLG, May 2005.

300 *"Well, that can't" . . . "over the head":* Ibid.

301 *"making incomprehensible . . . pair of its own":* Mermin to LLG, Oct. 2005.

301 *"Cornell English department husbands":* Ibid.

301 *"was quite sure . . . paid no attention":* Ibid.

301 With two colleagues, he then wrote: Clifton, Redhead, Butterfield, "Generalization of the Greenberger-Horne-Zeilinger Algebraic Proof of Nonlocality" in *Foundations of Physics* 21, 149–84 (1991).

301 *Greenberger-Horne-Zeilinger result:* Greenberger, Horne, and Zeilinger, "Going Beyond."

301 *"Mathematical rigor . . . view of physics":* Mermin to LLG, Oct. 2005.

301 *"What's Wrong . . . Reality?":* Mermin, *Physics Today,* June 1990, 9.

301 *"new & beautiful twist"*: Ibid.

301 *"In the absence . . . 'disturb' the particle"*: Ibid.

302 *"All six . . . highly heretical"*: Ibid, 9, 11.

302 *"Heresy or not . . . elements of reality"*: Ibid, 11.

302 *"the straightforward . . . single measurement"*: Ibid.

302 *"This fills me with admiration"*: Bell-Mermin, related by Mermin to LLG, Oct. 2005.

CHAPTER THIRTY-FOUR: "AGAINST 'MEASUREMENT' " 1989–1990

303 *"close to being the . . . Cornell"*: Mermin, "Whose Knowledge?" in Bertmann and Zeilinger, 271.

303 *"Against 'Measurement' "*: John Bell, *Physics World,* 8/90, 33–40.

303 *"the article conveys . . . voice"*: Mermin, "Whose Knowledge?" in Bertlmann and Zeilinger, 271.

304 *"second-class . . . first-class difficulties"*: Dirac, "The Evolution of the Physicist's Picture of Nature," *Scientific American,* 5/1963.

309 *"Don't be a sissy!" . . . "elders"*: Mermin, "Whose Knowledge?" in Bertlmann and Zeilinger, 271.

309 *"No prepared . . . wonderful conversations"*: Ibid.

310 *"You old fox! . . . stuff")*: Clauser-Horne 11/25/90, Horne to LLG, June 2005.

310 *"wonderfully barbed . . . superb rhetoric"*: Gottfried, "Is the Statistical Interpretation of Quantum Mechanics Implied by the Correspondence Principle?" in D. Greenberger, W. L. Reiter, and A. Zeilinger, *7th Yearbook Institute Vienna Circle,* 1999.

310 *"Bell claimed . . . theories possessed"*: Mermin-Fuchs, 12/1998 in Fuchs, 321.

310 *"Einstein's equations . . . $\psi / (r_p \ldots t) /^2$ is' "*: Gottfried.

310 *"Kurt never . . . bones"*: Mermin-Fuchs, 12/1998 in Fuchs, 321.

310 *"Bell pressed Gottfried . . . solution"*: Shimony-LLG, Oct. 20, 2005.

310 *"Of all the people . . . view of the world"*: Horne to LLG, June 2005.

310 *"It was a . . . kind of person"*: Greenstein to LLG, Oct. 19, 2005.

310 *Mermin and Gottfried . . . prophet*: Bell "displayed something like the wrath of the Old Testament prophet for those who adhered to positions he judged superficial": Mermin and Gottfried, "John Bell & the Moral Aspect of QM," *Europhysics News* 22 (1991).

310–11 *Shimony "propounded . . . cameras*: Greenstein and Zajonc, *Quantum Challenge,* p. xii.

311 *"agreed that John was . . . topic whatsoever"*: Mermin and Gottfried, "John Bell and the Moral Aspect of Quantum Mechanics," *Europhysics News* 22 (1991).

CHAPTER THIRTY-FIVE: ARE YOU TELLING ME
THIS COULD BE *PRACTICAL?* 1989–1991

312 *In the library . . . :* Ekert to LLG, Sept. 20, 2005.

312 *"At the time . . . time"*: Ibid.

313 *"I just liked public-key . . . fascinated by them"*: Ibid.

313 *"a lovely piece of number theory"*: Ekert et al., "Basic Concepts in Quantum Computation," http://arXiv.org/abs/quant-ph/0011013, 26 (April 22, 2001).

313 *"a big surprise . . . practical application"*: Ekert to LLG, Sept. 20, 2005.

314 *"It just clicked . . . eavesdropping!"*: Ibid.

314 *"Local realism . . . refuted!"*: Ibid.

314 *"If the Clauser-Horne- . . . not touched"*: Ibid.

314 *"I was very happy . . . the whole thing"*: Ibid.

315 *"Are you telling me . . . practical use?"*: Ekert to LLG, May 2005.

315 *"I said, Yes . . . whisk him away"*: Ibid.

CHAPTER THIRTY-SIX: THE TURN OF THE MILLENNIUM 1997–2002

317 *"It was easy . . . did them"*: Horne to LLG, June 2005.

317 *Quantum teleportation:* Bouwmeester, Pan, Mattle, Eibl, Weinfurter, and Zeilinger, "Experimental Quantum Teleportation," *Nature* 390, 575 (1997).

317 *entanglement swapping:* Bennett, Brassard, Crepeau, Jozsa, Peres, Wootters, "Teleporting an Unknown Quantum State via Dual Classical and Einstein-Podolsky-Rosen Channels," *Physical Review Letters* 70, 1895 (1993); Zukowski, Zeilinger, Horne, and Ekert, "Event-Ready-Detectors: Bell Experiment via Entanglement Swapping," *Physical Review Letters* 71, 4287 (1993); Pan, Bouwmeester, Weinfurter, and Zeilinger, "Experimental Entanglement Swapping: Entangling Photons That Never Interacted," *Physical Review Letters* 80, 3891 (1998)

318 *In 2000, Zeilinger . . . background:* Jennewein, Simon, Weihs, Weinfurter, and Zeilinger, "Quantum Cryptography with Entangled Photons," *Physical Review Letters* 84, 4729 (2000).

318 *Meanwhile, among the red . . . wily "Eve":* Naik, Peterson, White, Berglund, and Kwiat, "Entangled State Quantum Cryptography: Eavesdropping on the Ekert Protocol," *Physical Review Letters* 84, 4733 (2000).

318 *"I am a quantum engineer . . . I have principles":* Gisin, "Sundays in a Quantum Engineer's Life," Bertlmann and Zeilinger, 199.

319 *The entangled photons . . . nine and a half miles:* Tittel, Brendel, Gisin, and Zbinden, "Violation of Bell Inequalities More Than 10 km Apart" (1998), *Physical Review Letters* 81, 3563, and "Long Distance Bell-Type Tests Using Energy-Time Entangled Photons" (1999); *Phys. Rev. A*, 59, 4150. entanglement over 31 mi: Marcikic, de Riedmatten, Tittel, Zbinden, Legré, and Gisin, *PRL* 93, 180502 (2004).

319 *"When you do this experiment . . . other data":* Gisin to LLG, May 8, 2002.

320 *Quantum mechanics [has] now existed . . . negative rules":* Ibid.

320 *"in 1991, Artur . . . nonlocality became respectable":* Gisin, "Can Relativity Be Considered Complete?" at the 2005 Quantum Physics of Nature Conference in Vienna; on the Internet at http://arXiv.org/abs/quant-ph/0512168.

320 *"psychological revolution . . . physics in general":* Gisin to LLG, May 9, 2002.

320 *"investigate . . . relativity":* Suarez, www.quantumphil.org/history.htm.

321 *On a really nice Sunday . . . What would John Bell have thought of this?:* Gisin, "Sundays in a Quantum Engineer's Life," Bertlmann and Zeilinger, 202–6.

323 *"Indeed, quantum mechanics . . . mathematical tools":* Gisin to LLG, May 8, 2002.

324 *"The main lesson . . . can be tested":* Gisin, "Sundays . . ." in Bertlmann and Zeilinger, 206.

CHAPTER THIRTY-SEVEN: A MYSTERY, PERHAPS 1981–2006

325 *"Might I say immediately . . . it doesn't look so easy":* Feynman, "Simulating Physics with Computers" in Hey, 136–51; 147–50 cover Bell's Theorem (unnamed).

326 *a breakthrough design for a quantum computer:* Lloyd, "A Potentially Realizable Quantum Computer," *Science* 261, 1569 (1993).

326 *Isaac Chuang, himself . . . 3 and 5.):* Chuang and Gershenfeld, "Bulk Spin-Resonance Quantum Computation," *Science* 275, 350 (1997); Chuang, Vander-

sypen, Zhou, Leung, and Lloyd, *Nature* 393, 143 (1998); Chuang, Gershenfeld, and Kubinec, *PRL* 80, 3408 (1998).

326 *Two men from Bell Labs:* Shor, "Polynomial-Time Algorithms," *SIAM Journal on Computing* 26, 1484 (1997); Grover, *PRL* 79, 325 (1997).

327 *"The universe . . . entangled":* Lloyd, 118–19.

327 *it's a* quantum *one:* Ibid., 3ff.

327 *Uncertainty provides the seeds:* Ibid., 49.

327 *"quantum mechanics . . . out of nothing":* Ibid., 118.

327 *"When we started . . . tax it soon":* Ekert to LLG, Sept. 2005.

328 *"Quantum information . . . information processing":* Nielson, "Rules for a Complex Quantum World," *Scientific American,* Nov. 2002, p. 68.

328 *"After such . . . guts for it":* Smithells-Rutherford, 1932 in Eve, 364.

328 *"All these fifty years . . . mistaken":* AE-Besso 12/12/51 in French, *Einstein,* 138.

328 *"Bohr, my boy . . . ignorance":* Rutherford in Capri, Anton Z., *Quips, Quotes, and Quanta: An Anecdotal History of Physics* (Singapore: World Scientific Pub. Co., 2007), 170; compare Rutherford's statement in 1933: "While the theory of indeterminacy [a.k.a. the uncertainty principle] is of great theoretical interest as showing the limitations of the present wave-theory of matter, its importance in Physics seems to me to have been much exaggerated by many writers. It seems to me unscientific & also dangerous to draw far-flung deduction from a theoretical conception which is incapable of experimental verification, either directly or indirectly" (Rutherford-Samuel in Eve, 378). Another related comment: "Theorists play games with their symbols, but we, in the Cavendish, turn out the real solid facts of Nature" (Eve, 304).

328–29 *"There's the miracle . . . really was":* Rabi in Bernstein, "Physicist."

329–30 *It is easy . . . of its own destruction:* Bell and Nauenberg; "The moral aspect of Quantum Mechanics" (1966), *Speakable,* 26–28.

EPILOGUE: BACK IN VIENNA 2005

331 *"We're like very intelligent" . . . "Being a hidden variable, I see":* all these quotes were heard by LLG at this conference.

333 *"Until quite . . . all about information":* Mermin, "Whose Knowledge?" in Bertlmann and Zeilinger, 273.

333 *"The past decade . . . these pages":* Mermin in Fuchs, p. ii–iii.

333 Quantum mechanics is a law . . . touch: Fuchs, 136, 527.

333 *"It is that zing! . . . that might mean":* Ibid, 336.

334 *"I belong to the crazy . . . way the world is":* Rudolph to LLG, May 2005.

334 *"ontological content" . . . reacts to us:* Fuchs, 45.

334 *"It could still . . . about what?' ":* Rudolph to LLG, Oct. 2005.

334 *"I like the word . . . language":* Fuchs, 333.

334 *"Belief about . . . can't get it":* Ibid., 322.

334 *"It's too anthropocentric . . . glorified monkeys":* Rudolph to LLG, Oct. 2005.

334 *"Almost all the formal . . . know":* Fuchs, 68.

335 *"What we need . . . I'm not around":* Rudolph to LLG, Oct. 2005.

335 *"The distillate that . . . reality' ":* Fuchs, p. v.

335 *"One approach . . . it must be":* Rudolph to LLG, Oct. 2005.

335 *"The task . . . thing about quantum foundations":* Fuchs, *Quantum Foundations in the Light of Quantum Information,* 4 in http://arXiv.org/abs/quant-ph/0106166, 2001.

BIBLIOGRAPHY

Aczel, Amir D. *Entanglement: The Greatest Mystery in Physics.* New York: Four Walls Eight Windows, 2001.

Anandan, Jeeva, ed. *Quantum Coherence: Proceedings of the International Conference on Fundamental Aspects of Quantum Theory to Celebrate 30 Years of the Aharonov-Bohm Effect.* Singapore: World Scientific, 1989, in particular pp. 356–72, where is printed Horne, Michael A., Abner Shimony, and Anton Zeilinger, "Down-Conversion Photon Pairs: A New Chapter in the History of Quantum Mechanical Entanglement."

Aspect, Alain, Phillipe Grangier, and Gérard Roger. "Experimental Tests of Realistic Local Theories via Bell's Theorem," *Physical Review Letters* 47, No 7, 460–63 (1981).

———. "Experimental Realization of Einstein-Podolsky-Rosen-Bohm *Gedankenexperiment:* A New Violation of Bell's Inequalities", *Phys. Rev. Letters* 49, No 2, 91–94 (1982).

Aspect, Alain, Jean Dalibard, and Gérard Roger. "Experimental Test of Bell's Inequalities Using Time-Varying Analyzers," *Phys. Rev. Letters* 49, No 25, 1804–7 (1982).

Bell, John S. *Speakable and Unspeakable in Quantum Mechanics.* Cambridge, England: Cambridge University Press, 1993.

Beller, Mara. *Quantum Dialogue: The Making of a Revolution.* Chicago: University of Chicago Press, 2001.

Bernstein, Jeremy. *Hans Bethe: Prophet of Energy.* New York: Basic Books, 1980.

———. *Hitler's Uranium Club: The Secret Recordings at Farm Hall.* Second Edition. New York: Copernicus Books, Springer-Verlag, 2001.

———. *The Merely Personal: Observations on Science and Scientists.* Chicago: Ivan R. Dee, 2001.

———. "Physicist" [profile of I. I. Rabi]. *The New Yorker,* October 13, 1975, pp. 47ff., and October 20, 1975, pp. 47ff.

———. *Quantum Profiles* [*Bell, Wheeler, and Besso*]. Princeton, NJ: Princeton University Press, 1991.

Bertlmann, R. "Magic Moments with John Bell," lecture at the international conference "Quantum (Un)speakables" in honor of John S. Bell, Vienna, November 10–14, 2000.

Bertlmann, R., and Zeilinger, A., eds. *Quantum (Un)speakables.* Berlin: Springer, 2002.

Bethe, Hans A. "Oppenheimer: 'Where He Was There Was Always Life and Excitement,' " *Science* 155, 1080–84 (March 3, 1967).

Bohm, David. "A Suggested Interpretation of the Quantum Theory in Terms of Hidden Variables" (I and II), *Physical Review* 85, No 2, 166–93 (1952).

————. David Bohm Archives, Birkbeck College, University of London. Bohm Letters: C6, 10–16, 37–41, 42, 44, 46–48, 58 (de Broglie, Einstein, Hanna Loewy, Lomanitz, Pauli, Melba Phillips, Rosenfeld); A116–18: 1979 interview by Martin Sherwin of Lomanitz and Bohm; *Hearings Regarding Communist Infiltration of Radiation Laboratory and Atomic Bomb Project at the University of California, Berkeley, Calif.*: Wednesday, May 25, 1949, Executive Session U.S. House of Representatives Committee on Un-American Activities—Testimony of David Joseph Bohm.; photographs of Bohm and Lomanitz in court; B44: "On the Failure of Communication between Bohr and Einstein." Undated (post-1961) essay by Bohm.

————. *Quantum Theory.* New York: Dover Publications, 1979.

————. *Wholeness and the Implicate Order.* London: Ark Paperbacks, 1983.

Bohm, David, and Yakir Aharonov. "Discussion of Experimental Proof for the Paradox of Einstein, Rosen, and Podolsky," *Physical Review* 108, No 4, 1070–75ff.

Bohr, Niels. *Atomic Theory & the Description of Nature.* Cambridge, England: Cambridge University Press (1934), 1961.

————. *Atomic Physics and Human Knowledge.* New York: John Wiley & Sons, 1958.

Bolles, Edmund Blair. *Einstein Defiant (Bohr Unyielding): Genius versus Genius in the Quantum Revolution.* Washington, D.C.: Joseph Henry Press, 2004.

Born, Max. *The Born-Einstein Letters: The Correspondence Between Albert Einstein and Max and Hedwig Born, 1916–1955.* New York: Walker & Co., 1971.

————. *My Life: Recollections of a Nobel Laureate.* New York: Scribner's Sons, 1978.

Brian, Denis. *Einstein.* New York: John Wiley & Sons, 1996.

Burke, Philip G., and Ian C. Percival. "John Stewart Bell," *Biographical Memoirs of Fellows of the Royal Society, London* 45, 1 (1999).

Casimir, Hendrik B. G. *Haphazard Reality: Half a Century of Science.* New York: Harper & Row, 1983.

Cassidy, David C. *Uncertainty: The Life and Science of Werner Heisenberg.* New York: W. H. Freeman & Co., 1992.

Casti, John L., and Werner DePauli. *Gödel: A Life of Logic.* Cambridge, MA: Perseus Press, 2000.

Clark, Ronald W. *Einstein: The Life and Times.* New York: Avon Books, 1971.

Clauser, John F. "Early History of Bell's Theorem," invited talk, presented at "Quantum (Un)speakables," conference in commemoration of John S. Bell, Vienna, November 10–14, 2000.

Clauser, John F., Michael A. Horne, Abner Shimony, and Richard Holt. "Proposed Experiment to Test Local Hidden-Variable Theories," *Phys. Rev. Letters* 23, No 15, 880–84 (1969).

Clauser, John F., and Michael Horne. "Experimental Consequences of Objective Local Theories," *Physical Review D* 10, No 2, 526–35 (1974).

Clauser, John F., and Abner Shimony. "Bell's Theorem: Experimental Tests and Implications" [review article], *Reports on Progress in Physics* 41, 1881–1927 (1978).

Davies, Paul. *The Ghost in the Atom.* Cambridge, England: Cambridge University Press, 1986.

————, ed. *The New Physics.* Cambridge, England: Cambridge University Press, 1996.

Dawson, John W., Jr. *Logical Dilemmas: The Life and Work of Kurt Gödel.* Wellesley, MA: A. K. Peters, 1997.

de Broglie, Louis. *New Perspectives in Physics: Where Does Physical Theory Stand Today?* A. J. Pomerans, trans. New York: Basic Books, 1962.

d'Espagnat, Bernard. "My Interaction with John Bell." Invited lecture at the international conference "Quantum (Un)speakables" in honor of John S. Bell, Vienna, November 10–14, 2000.

Dirac, P. A. M. "The Evolution of the Physicist's Picture of Nature," *Scientific American,* May 1963.

Dresden, M. *H. A. Kramers: Between Tradition and Revolution.* New York: Springer-Verlag, 1987.

Dukas, Helen, and Banesh Hoffmann. *Albert Einstein, The Human Side.* Princeton, NJ: Princeton University Press, 1989.

Einstein, Albert. *Letters to Solovine 1906–1955.* New York: Citadel Press, 1993.

———. *Out of My Later Years.* New York: Bonanza Books, 1990.

———. *Relativity: The Special & the General Theory.* Robert W. Lawson, trans. New York: Three Rivers Press (1916), 1961.

———. *The World as I See It.* London: John Lane the Bodley Head, 1935.

Einstein, Albert, Boris Podolsky, Nathan Rosen. "Can Quantum-Mechanical Description of Physical Reality Be Considered Complete?" *Physical Review* 47, 777–80 (1935).

Einstein, Albert, and Mileva Maric. *Albert Einstein Mileva Maric: The Love Letters.* Jurgen Renn, ed., Robert Schulmann, ed. & trans., Shawn Smith, trans. Princeton, NJ: Princeton University Press, 1992.

Ellis, John, and Daniele Amati, eds. *Quantum Reflections* (1991 CERN symposium in memory of Bell). Cambridge, England: Cambridge University Press, 2000. In particular, Jackiw's "Remembering John Bell."

Enz, Charles P. *No Time to Be Brief: A Scientific Biography of Wolfgang Pauli.* Oxford: Oxford University Press, 2002.

Eve, Arthur S. *Rutherford: Being the Life and Letters of the Rt. Hon. Lord Rutherford, O.M.* Cambridge, England: Cambridge University Press, 1939.

Feynman, Richard P. *The Character of Physical Law.* Cambridge, MA: M.I.T. Press, 1987.

———. *"Surely You're Joking, Mr. Feynman!"* with Ralph Leighton. New York: Norton & Co., 1985.

Fierz, M., and V. F. Weisskopf, eds. *Theoretical Physics in the Twentieth Century: A Memorial Volume to Wolfgang Pauli.* New York: Interscience Publishers, 1960.

Fine, Arthur. *The Shaky Game: Einstein, Realism, and the Quantum Theory.* Chicago: University of Chicago Press, 1996.

Folsing, Albrecht. *Albert Einstein: A Biography.* Ewald Osers, trans. New York: Penguin, 1997.

Forman, Paul. "Weimar Culture, Causality, and Quantum Theory, 1918–1927: Adaptation by German Physicists & Mathematicians to a Hostile Intellectual Environment." *Historical Studies in the Physical Sciences,* Vol. 3, 1–115 (1971).

Freedman, Stuart Jay. "Experimental Test of Local Hidden-Variable Theories," Ph.D. thesis, University of California, Berkeley, 1972.

Freedman, Stuart J., and John F. Clauser. "Experimental Test of Local Hidden-Variable Theories," *Phys. Rev. Letters* 28, No 14, 938–41 (1972).

French, A. P., ed. *Einstein: A Centenary Volume.* Cambridge, MA: Harvard University Press, 1979.

French, A. P., and P. J. Kennedy, eds., *Niels Bohr: A Centenary Volume.* Cambridge, MA: Harvard University Press, 1985.

Frisch, Otto. *What Little I Remember.* Cambridge, England: Cambridge University Press, 1991.

Fuchs, Chris. *Notes on a Paulian Idea: Foundational, Historical, Anecdotal, and Forward-Looking Thoughts on the Quantum (Selected Correspondence, 1995–2001).* Växjö, Sweden: Växjö University Press, 2003.

Furry, W. H. "Note on the Quantum-Mechanical Theory of Measurement." *Phys. Rev.* 49, 393–99 (1936).

———. "Remarks on Measurements in Quantum Theory" (response to Schrödinger). *Phys. Rev.* 49, 476 (1936).

Gamow, George. *My World Line: An Informal Biography.* New York: Viking Press, 1970.

———. *Thirty Years That Shook Physics: The Story of Quantum Theory.* New York: Dover, 1985.

Gell-Mann, Murray. *The Quark and the Jaguar.* New York: W. H. Freeman & Co., 1994.

Gleick, James. *Genius: The Life and Science of Richard Feynman.* New York: Vintage Books, 1993.

Goodchild, Peter. *J. Robert Oppenheimer: "Shatterer of Worlds."* London: BBC Press, 1980.

Goudsmit, Samuel. "The Discovery of the Electron Spin," 1971 Golden Jubilee of the Dutch Physical Society lecture, www.lorentz.leidenuniv.nl/history/spin/goudsmit.html (June 2, 2008).

Greenberger, D. M., M. A. Horne, and A. Zeilinger. "Going Beyond Bell's Theorem," in *Bell's Theorem, Quantum Theory and Conceptions of the Universe,* M. Kafatos, ed. Dordrecht, Netherlands: Kluwer, 1989, 69–72.

Greenberger, D. M., M. A. Horne, A. Shimony, and A. Zeilinger. "Bell's Theorem Without Inequalities," *American Journal of Physics* 58, 1131–43 (1990).

Greenspan, Nancy Thorndike. *The End of the Certain World: The Life and Science of Max Born.* New York: Basic Books, 2005.

Greenstein, George, and Arthur Zajone. *The Quantum Challenge: Modern Research on the Foundations of Quantum Mechanics.* Sudsbury, MA: Jones and Bartlett Publishers, 1997.

Hartcup, Guy, and T. E. Allibone. *Cockcroft and the Atom.* Bristol, England: Adam Hilger, Ltd., 1984.

Heilbron, J. L. *The Dilemmas of an Upright Man: Max Planck & the Fortunes of German Science.* Cambridge, MA: Harvard University Press, 1996.

Heisenberg, Werner. *Encounters with Einstein & Other Essays on People, Places, & Particles.* Princeton, NJ: Princeton University Press, 1989.

———. *Philosophical Problems of Quantum Physics.* Woodbridge, CT: Oxbow Press, 1979 (originally published 1952 by Pantheon Books as *Philosophical Problems of Nuclear Physics*).

———. *Physics and Beyond: Encounters and Conversations.* New York: Harper and Row, 1971. (a.k.a. *Der Teil und das Ganze* [*The Part and the Whole*]).

Hendry, John, ed. *Cambridge Physics in the Thirties.* Bristol, England: Adam Hilger, Ltd., 1984.

Hermann, Grete. "The Foundations of Quantum Mechanics in the Philosophy of Nature" (originally published in *Die Naturwissenschaften* 41, 721 [1935]; another version in *Abhandlungen der Fries'schen Schule* 6, 2 [1935]), translated from the German by Dirk Lumma, *The Harvard Review of Philosophy* VII, 35–44 (1999).

Hey, Anthony J. G., ed. *Feynman and Computation, with Contributions by Feynman and His Most Notable Successors.* Reading, MA: Perseus Books, 1999.

Hiley, B. J., and F. David Peat, eds. *Quantum Implications: Essays in Honour of David Bohm.* London: Routledge & Kegan Paul, 1987.

Holt, R. A., and F. M. Pipkin. "Quantum Mechanics vs. Hidden Variables: Polarization

Correlation Measurement on an Atomic Mercury Cascade." Ph.D. thesis, Harvard University preprint, 1973.

Horne, Michael Allan. "Experimental Consequences of Local Hidden Variables Theories," Ph.D. dissertation, Boston University, 1970.

Howard, Don. "Einstein on Locality and Separability." *Studies in History and Philosophy Science,* Vol. 16, No. 3, pp. 171–201, 1985.

————. *Revisiting the Einstein-Bohr Dialogue.* 2005 lecture on his Web site, www.nd .edu/~dhoward1/Papers.html

————. "*Nicht Sein Kann Was Nicht Sein Darf,* or, The Prehistory of EPR, 1909–1935: Einstein's Early Worries About the Quantum Mechanics of Composite Systems," in Miller, A. I., ed., *Sixty-Two Years of Uncertainty.* pp. 61–106ff. New York: Plenum Press, 1990.

Jackiw, Roman. "The Chiral Anomaly," *Europhysics News* 22, 76–77 (1991) (Bell Memorial).

Jackiw, Roman, and Abner Shimony. "The Depth and Breadth of John Bell's Physics," *Physics in Perspective* Vol. 4, No. 1, Feb. 2002, 78–116.

Jackiw, Roman, and D. Kleppner. "100 Years of Quantum Physics," *Science* 289, No 5481, 893–98 (Aug. 11, 2000).

Jammer, M. *The Conceptual Development of Quantum Mechanics.* New York: McGraw-Hill, 1966.

————. *The Philosophy of Quantum Mechanics: The Interpretation of Quantum Mechanics in Historical Perspective.* New York: John Wiley & Sons, 1974.

Jauch, J. M. *Are Quanta Real? A Galilean Dialogue.* Bloomington: Indiana University Press, 1973.

Johnson, George. *A Shortcut Through Time: The Path to the Quantum Computer.* New York: Knopf, 2003.

————. *Strange Beauty: Murray Gell-Mann and the Revolution in Twentieth-Century Physics.* New York: Knopf, 1999.

Kafatos, M., ed. *Bell's Theorem, Quantum Theory, and Conceptions of the Universe.* Dordrecht, the Netherlands: Kluwer Academic Publishers, 1989.

Klein, Martin J. *Paul Ehrenfest: Volume 1 The Making of a Theoretical Physicist.* Amsterdam: North-Holland Publishing Co., 1972.

————. "The First Phase of the Bohr-Einstein Dialogue." *Historical Studies in the Physical Sciences,* Vol. 2, 1970.

Kocher, Carl A., and Eugene D. Commins. "Polarization Correlation of Photons Emitted in an Atomic Cascade," *Phys. Rev. Letters* 18, No 15, 575–77 (1967).

Kragh, Helge. *Dirac: A Scientific Biography.* Cambridge, England: Cambridge University Press, 1990.

Lang, Daniel. "A Farewell to String and Sealing Wax" [Profile of Samuel Goudsmit] *The New Yorker,* November 7, 1953, pp. 47ff, and November 14, 1953.

Levenson, Thomas. *Einstein in Berlin.* New York: Bantam, 2004.

Lipschütz-Yevick, Miriam. "Social Influences on Quantum Mechanics?-II" *Mathematical Intelligencer* 23, No. 4 (Fall 2001).

Lloyd, Seth. *Programming the Universe.* New York: Knopf, 2006.

Mann, A., and M. Revzen, eds. *The Dilemma of Einstein, Podolsky, and Rosen—60 Years Later: An International Symposium in Honour of Nathan Rosen.* Bristol, England: Institute of Physics Publishing, 1995.

Mead, Carver. *Collective Electrodynamics: Quantum Foundations of Electromagnetism.* Cambridge, MA: M.I.T. Press, 2000.

Mehra, Jagdish. "Niels Bohr's Discussions with Albert Einstein, Werner Heisenberg, and Erwin Schrödinger: The Origins of the Principles of Uncertainty and Complementarity," *Foundations of Physics* 17, No. 5, 1987.

Mermin, N. David. *Boojums All the Way Through: Communicating Science in a Prosaic Age.* Cambridge, England: Cambridge University Press, 1990.

———. "Is the Moon There When Nobody Looks?" in Richard Boyd, Philip Gasper, and J. D. Trout, eds., *The Philosophy of Science.* Cambridge, MA: M.I.T. Press, 1993.

———. "What's Wrong with These Elements of Reality?," *Physics Today,* June 1990, 9–11.

Michelmore, Peter. *Einstein, Profile of the Man.* New York: Dodd, Mead, 1962.

Miller, A. I., ed. *Sixty-two Years of Uncertainty.* New York: Plenum Press, 1990. An amazing conference proceedings, including John Bell, "Against 'Measurement' "; Michael Horne, Abner Shimony, and Anton Zeilinger, "Introduction to Two-Particle Interferometry"; Don Howard, "*Nicht Sein Kann Was Nicht Sein Darf,* or, The Prehistory of EPR, 1909–1935: Einstein's Early Worries About the Quantum Mechanics of Composite Systems"; Arthur I. Miller, "Imagery, Probability, and the Roots of the Uncertainty Principle."

Moore, Walter. *Schrödinger: Life and Thought.* Cambridge, England: Cambridge University Press, 1990.

Nasar, Sylvia. *A Beautiful Mind.* New York: Touchstone, 1998.

Nolte, David. *Mind at Light Speed.* New York: Free Press, 2001.

In the Matter of J. Robert Oppenheimer: Transcript of Hearing before Personnel Security Board. April 12–May 6, 1954. Cambridge, MA: M.I.T. Press, 1970.

Pais, Abraham. *Einstein Lived Here* (the companion volume to '*Subtle Is the Lord . . .*'). Oxford: Oxford University Press, 1994.

———. *Inward Bound: Of Matter and Forces in the Physical World.* Oxford: Oxford University Press, 1986.

———. *Niels Bohr's Times in Physics, Philosophy, and Polity.* Oxford: Oxford University Press, 1991.

———. '*Subtle Is the Lord . . .*': *The Science and the Life of Albert Einstein.* Oxford: Oxford University Press, 1982.

Pauli, Wolfgang. *Writings on Physics and Philosophy.* Charles P. Enz and Karl von Meyenn, eds. Robert Schlapp, trans. Berlin: Springer-Verlag, 1994.

Pauli, Wolfgang, and C. G. Jung. *Atom and Archetype: The Pauli-Jung Letters 1932–1958.* C. A. Meier, ed. Princeton, NJ: Princeton University Press, 2001.

Peat, F. David. *Infinite Potential: The Life and Times of David Bohm.* Reading, MA: Addison-Wesley, 1997.

Peierls, Rudolf. *Bird of Passage.* Princeton, NJ: Princeton University Press, 1985.

———. *Surprises in Theoretical Physics.* Princeton, NJ: Princeton University Press, 1979.

Penrose, Roger. *The Emperor's New Mind: Concerning Computers, Minds, and the Laws of Physics.* Oxford: Oxford University Press, 2002.

Przibram, K., ed. *Letters on Wave Mechanics: Schrödinger-Planck-Einstein-Lorentz.* New York: Philosophical Library, 1967.

Rabi, I. I., Robert Serber, Victor Weisskopf, Abraham Pais, and Glenn Seaborg. *Oppenheimer.* New York: Charles Scribner's Sons, 1969.

Rajaraman, Ramamurti. "Fractional Charge," invited lecture at the international conference "Quantum (Un)speakables" in honor of John S. Bell, Vienna, November 10–14, 2000.

———. "John Stewart Bell: The Man and His Physics," unpublished paper.

Raman, V. V., and Paul Forman. "Why Was It Schrödinger Who Developed de Broglie's Ideas?" *Historical Studies in the Physical Sciences,* Vol. I, 1969, 294–96.

Regis, Ed. *Who Got Einstein's Office? Eccentricity and Genius at the Institute for Advanced Study.* Reading, MA: Addison-Wesley, 1988.

Rozental, S., ed. *Niels Bohr: His Life and Work as Seen by His Friends and Colleagues.* Amsterdam: North Holland Pub. Co., 1967.

Satinover, Jeffrey. "Jung and Pauli," unpublished paper.

Saunders, Thomas J. *Hollywood in Berlin.* Berkeley: University of California Press, 1994.

Schlipp, P. A., ed. *Albert Einstein: Philosopher-Scientist.* New York: MJF Books, 1970.

Schrödinger, Erwin. "Discussion of Probability Relations between Separated Systems," *Proceedings of the Cambridge Philosophical Society* 31, 555–63 (1935).

———. *The Interpretation of Quantum Mechanics.* Michel Bitbol, ed., Woodbridge, CT: Ox Bow Press, 1995.

———. *What Is Life? And Other Scientific Essays.* New York: Doubleday Anchor Books, 1956.

Schweber, S. S. *In the Shadow of the Bomb.* Princeton, NJ: Princeton University Press, 2000.

Segrè, Gino. *Faust in Copenhagen.* New York: Viking, 2007.

Shimony, Abner. "Metaphysical Problems in the Foundations of Quantum Mechanics," in *The Philosophy of Science.* Richard Boyd, Philip Gasper, and J. D. Trout, eds. Cambridge, MA: M.I.T. Press, 1993.

———. *The Search for a Naturalistic Worldview, Vol. 2.* Cambridge, MA: Cambridge University Press, 1993.

Shimony, Abner, M. A. Horne, and J. F. Clauser, "Comment on the Theory of Local Beables"; Shimony, "Reply to Bell," *Dialectica* 39, No. 2, 97–110 (1985).

Shimony, Abner, Valentine Telegdi, and Martinus Veltman. Obituary: "John S. Bell," *Physics Today* 82 (Aug. 1991).

Smith, Alice Kimball, and Charles Weiner, eds. *Robert Oppenheimer Letters and Recollections.* Stanford, CA: Stanford University Press, 1980.

Stachel, John, ed. *Einstein's Miraculous Year: 5 Papers That Changed the Face of Physics.* Princeton, NJ: Princeton University Press, 1998.

Teller, Edward, *Memoirs: A Twentieth-Century Journey in Science and Politics.* Cambridge, MA: Perseus Publishing, 2001.

Tomonaga, Sin-itiro. *The Story of Spin.* Takeshi Oka, trans. Chicago: University of Chicago Press, 1997.

Townes, Charles. *How the Laser Happened.* Oxford: Oxford University Press, 1999.

Uhlenbeck, George E. "Reminiscences of Professor Paul Ehrenfest," *American Journal of Physics* 24, 431–33 (1956).

Uhlenbeck, George E., and Samuel A. Goudsmit. "Spinning Electrons and the Structure of Spectra," *Nature* 117, 264–65 (1926).

Wang, Hao. *Reflections on Kurt Gödel.* Cambridge, MA: M.I.T. Press, 1987.

Weaver, Jefferson Hane. *The World of Physics: A Small Library of the Literature of Physics from Antiquity to the Present (Vol. II: The Einstein Universe and the Bohr Atom).* New York: Simon & Schuster, 1987.

Weinberg, Steven. *The Discovery of Subatomic Particles.* New York: Scientific American Books, 1983.

Weisskopf, Victor F. *Physics in the Twentieth Century: Selected Essays.* Cambridge, MA: M.I.T. Press, 1972.

Wheeler, J. A., and W. H. Zurek, eds. *Quantum Theory and Measurement.* Princeton, NJ: Princeton University Press, 1983.

Whitaker, Andrew. *Einstein, Bohr, and the Quantum Dilemma.* Cambridge, England: Cambridge University Press, 1996.

―――. "John Bell and the Most Profound Discovery of Science," *Physics World,* December 1998, 29–34.

Wick, David. *The Infamous Boundary: Seven Decades of Heresy in Quantum Physics.* New York: Copernicus, Springer, 1995.

Woolf, Harry, ed. *Some Strangeness in the Proportion: A Centennial Symposium to Celebrate the Achievements of Albert Einstein.* Reading, MA: Addison-Wesley, 1980.

THREE WEB SITES THAT WERE EXTREMELY HELPFUL

The Nobel Prize Web site: www.nobel.se/physics/laureates. This amazing site contains a short biography or autobiography for each laureate, and gives the texts of all the Nobel lectures.

J. J. O'Connor and E. F. Robertson's "MacTutor" Web site: biographies of mathematicians (and many physicists): www-groups.dcs.st-and.ac.uk/~history/. This is a well-researched encyclopedia of well-written short biographies.

The arXiv: http://arXiv.org. The "quant-ph" section of this Web site these days is the first publication site of many of the most important and interesting quantum information theory papers.

ARTICLES ON QUANTUM COMPUTING AND CRYPTOGRAPHY

The Quantum Information issue of *Physics World* 11, No 3, 35–57 (March 1998).

Fitzgerald, Richard. "What Really Gives a Quantum Computer Its Power?," *Physics Today,* 20–22 (January 2000).

Gisin, Nicolas, Grégoire Ribordy, Wolfgang Tittel, and Hugo Zbinden, "Quantum Cryptography," *Reviews of Modern Physics* 74, 145ff. (January 2002).

Grover, Lov. "Quantum Computing," *The Sciences,* 24–30 (July 1999).

Lloyd, Seth. "A Potentially Realizable Quantum Computer," *Science,* 261 (1993).

―――. "Quantum Mechanical Computers," *Scientific American,* 140–45 (October 1995).

Naik, D. S., C. G. Peterson, A. G. White, A. J. Berglund, and P. G. Kwiat. "Entangled State Quantum Cryptography: Eavesdropping on the Ekert Protocol," *Phys. Rev. Letters* 84, 4733 (2000).

Nielsen, Michael A. "Rules for a Complex Quantum World," *Scientific American,* 66–75 (November 2002).

HELPFUL GLOSSARIES/ENCYCLOPEDIAS FOR THE GENERAL READER

Q Is for Quantum: An Encyclopedia of Particle Physics, by John Gribbin. (Mary Gribbin, ed.; Jonathan Gribbin, illus.; Benjamin Gribbin, time lines). New York: The Free Press, 1998.

Oxford Dictionary of Physics. Alan Isaacs, ed. Oxford: Oxford University Press, 2003.

THE MOST ENJOYABLE
INTRODUCTION TO THE QUANTUM THEORY IS

The Einstein Paradox and Other Science Mysteries Solved by SHERLOCK HOLMES, by Colin Bruce. Reading, MA: Helix Books, 1997.

ACKNOWLEDGMENTS

George Gamow's *Thirty Years That Shook Physics* and David Mermin's essay "Quantum Mysteries for Anyone" inspired this book, and it would not have been possible without Abraham Pais's incredible trilogy, *"Subtle Is the Lord," Niels Bohr's Times,* and *Inward Bound.*

There are four people the thought of whom makes me glad I wrote this book, just because it led me to them: to Reinhold and Renate Bertlmann, George Johnson, and Miriam Yevick, I am endlessly grateful.

Many, many thanks to everyone I sought out for memories or explanations. It was a privilege to meet so many wonderful, fascinating people. In particular, thanks to Abner Shimony and Roman Jackiw, who were the first, and crucially guided my thinking and writing.

Thanks to Alain Aspect, Mary Bell, Andy Berglund, John Clauser, Eugene Commins, Artur Ekert, Ed Fry, Ken Ford, Stuart Freedman, John Hart, Dick Holt, Mike Horne, Danny Greenberger, George Greenstein (and his wonderful book with Arthur Zajonc), Nicolas Gisin, Lars Becker Larsen, Seth Lloyd, David Mermin, Robert Podolsky, Dave Rehder, Terry Rudolph, Ramamurti Rajaraman, Jack Steinberger, Jeff Satinover, David Sutherland, and Miriam and George Yevick. Thanks also to Steve Weinstein, who gave me *The Shaky Game;* and especially to Andrew Whitaker, for writing the beautiful *Einstein, Bohr, and the Quantum Dilemma,* and to Chris Fuchs for his wonderful samizdat, *Notes on a Paulian Idea*—these two books sat on my desk rather than on the shelf.

Many thanks to the people who read and commented on part or all of the book: Matt Babineau (whose perceptive comments on the text in 2002 were still guiding me years later), George, Mellie, and Josh Gilder, Donny Fraits, Anne Palmer, Sorina Higgins, Reinhold and Renate Bertlmann, Abner Shimony, Miriam Yevick, Dick Holt, Nicolas Gisin, John Hart, Jeff Satinover, Miles Blencowe, Herschel Snodgrass and his fall 2006 quantum mechanics class at Lewis & Clark, and Patty Karlin, who introduced me to Herschel. Thanks to Henning Makholm for his description of the Copenhagen main railway station and to Barbara Palmer for descriptions of Munich and its movie theaters. Especial thanks to David Mermin, Jacob Barandes, and Herschel Snodgrass for their extremely kind help when I was preparing the text for the paperback.

Thanks to Miles Blencowe, who took on a physics minor in an independent study of entanglement with clarity and good humor. Thanks to Carver Mead for his kindness and his vision of quantum mechanical coherency. Thanks to Sue Godsell and to Birkbeck College for access to and help with the Bohm archives. Thanks to Matyas Koniorczyk, friend and guide during my first conference in Vienna in 2000, and to the late

Fred Balderston, who made Berkeley a welcoming place. Thanks to Alexander Stibor for a wonderful day in Vienna in 2005 and for insisting that I encounter both the Summer Palace and Terry Rudolph; and to Lee Smolin and Herb Bernstein for a beautiful dinner.

Thanks to Louis, America, and Isabella von Harnier, who took me on a magical trip to the Walchensee, where we found Heisenberg's house; and to my cousins Stefan and Helena, Damian, Francesca, and Therese von Gatterberg, for the white asparagus and the two-hour Black Forest motorcycle ride to the church where Born met Schweitzer; and to Constantin and Gundi von Gatterburg, who fed me a magnificent meal and drove me to the airport, all at the last minute.

Thanks to Doug Hagen, who taught me to be a projectionist, which made supporting myself while writing a lot more fun. Thanks to Dakota Reis, who provided moral and feline support; and to my cousin Jeremy Gordinier, who stayed up all night turning my drawings into digital copies on a computer completely unfit for the task. Thanks to Bruce Chapman, who helped me at an hour of need with kindness and financial support, and to Katy Young, for last-minute, long-distance library assistance.

Thanks to my agent and cousin-by-marriage, Nina Ryan, for all her work and sympathetic encouragement.

Thanks to Avelina Truchero, for the homemade fudge! And huge thanks to Roxanne Urrey, without whose kindness and self-sacrificial help in a time of need I would have been lost.

Thanks to Natalia Davis, who gave me space to write for a year and a half in the Oakland hills and several times provided critical moral support. Thanks to Shona Campbell and Brandon Guenther of the Valley Ford Hotel (the best kitchen-saloon-inn in Sonoma County), who provided beautiful work space and terrific locally grown food at various critical moments in the endgame.

Thanks to everyone at Knopf who has worked on those endless stacks of manuscript pages, including Kate Norris for her careful and sensitive copyediting, Meghan Wilson for her calm and indispensable help, and especially thanks to my editor, Jonathan Segal, whose guidance tightened the shaggy, rambling manuscript until it became a united book, and his awesome assistants, Kyle McCarthy and Joey McGarvey.

Finally, thanks to all my friends and family for their support and good humor. Particularly, so many thanks to my mom, Cornelia Brooke Gilder, who wrote two books during the eight years of this book's gestation; for several years we worked side by side in the "Temple" above the garage, and the companionship, accountability, occasional gardening, and lunches on the porch were my best writing experiences; to my sister Mellie for much caffeinated work-companionship, most notably the simultaneous all-nighter on opposite coasts; and to my dad, George Gilder, who was my mainstay as constant critical reader and running companion, and who always saw the big picture.

And thanks to Donny Fraits for accompanying me on the piano as I wrote.

INDEX

Page numbers in *italics* refer to illustrations.
Page numbers beginning with 351 refer to notes.